ALZHEIMER'S DISEASE
Advances in Genetics,
Molecular and Cellular Biology

ALZHEIMER'S DISEASE
Advances in Genetics, Molecular and Cellular Biology

Edited by

Sangram S. Sisodia
The University of Chicago, IL

and

Rudolph E. Tanzi
Massachusetts General Hospital and Harvard Medical School

 Springer

Cover:
Atomic force microscopy images of human chromosomes
Courtesy of Dr. Stefan Thalhammer, GSF-Forschungszentrum for Environment and Health,
Institute of Radiation Protection, AG Nanoanalytics, Ingolstädter Landstraße 1 85764
Oberschleissheim, Germany

Thioflavin S image of plaques and tangles
Courtesy of Dr. Robert D. Terry, Professor Emeritus, Departments of Pathology and
Neurosciences,University of California, San Diego, USA

Library of Congress Control Number: 2006926955

ISBN-10: 0-387-35134-5   e-ISBN-10: 0-387-35135-3
ISBN-13: 978-0-387-35134-6   e-ISBN-13: 978-0-387-35135-3

Printed on acid-free paper.

Printed in the United States of America.

9 8 7 6 5 4 3 2 1

springer.com

Table of Contents

Introduction

Roughly one hundred years ago at a meeting of Bavarian psychiatrists, Dr. Alois Alzhiemer presented the intriguing case of his patient, Auguste D., a 51 year-old female admitted to the local asylum with presenile dementia. He would argue that specific lesions in and around neurons were responsible for dementia. In the ensuing decades, studies of her disorder, which would be named Alzheimer's disease (AD), were largely limited to descriptive neuropathological and psychological assessment of this disease with little understanding of the molecular and cellular mechanisms underlying neurodegeneration and dementia. This would change in the 1980's when the protein components of the major neuropathological hallmarks of the disease, senile plaques (and cerebral blood vessel amyloid) and neurofibrillary tangles were first determined. The identification of the β-amyloid protein (Aβ) and the microtubule-associated tau protein as the main components of plaques and tangles, respectively, would pave the way for the molecular genetic era of AD research. By the late-1980's, the genes encoding the β-amyloid precursor protein (*APP*) and tau (*MAPT*) were identified and would subsequently be shown to harbor autosomal dominant mutations causing early-onset familial AD and frontal temporal dementia (FTD), respectively. Later, in the early 1990's the ε4 variant of the apoliprotein E gene (*APOE*) would be found to be associated with increased risk for late-onset AD. Fundamental differences were soon noted between these two AD genes: *APP* and *APOE*. First, while APP mutations caused AD with virtual certainty, the APOE-ε4 variant increased susceptibility for, but not guarantee onset of AD. Second, while *APP* mutations increased the generation of the neurotoxic peptide, Aβ42, in brain, APOE-ε4 affected aggregation of Aβ into fibrils and its clearance from brain. In 1995, two more familial AD genes, presenilin 1 and 2 (*PSEN1*, *PSEN2*) were identified, and mutations in *MAPT* were linked to frontal temporal dementia. Thus, by 1995, the stage was set for molecular studies of age-related dementias with APP, presenilin 1 and 2, APOE, and tau playing the major roles.

With the turn of the 21st century, the search for novel AD and FTD genes would continue, utilizing high-throughput, chip-based genotyping technologies facilitated by the expansive DNA variant databases and the advent of the HapMap. However, the vast majority of studies addressing the molecular mechanisms underlying dementia would continue to focus on characterizing the five genes already firmly implicated in the etiology and pathogenesis of these dementing disorders. These five genes and the molecules they encode are absolutely critical pieces of the puzzle of age-related dementias, however, exactly how and where they fit into the puzzle remains a subject of intensive investigation. In mathematical terms, these molecules can be considered as the "givens" in the incredibly complex equation underlying the etiology and pathogenesis of AD and FTD. Studies of these molecules have not only elucidated the genetic, molecular, and biochemical basis of AD and FTD, but have already begun to guide the design and development of novel ways to diagnose, treat, and prevent these diseases. The contributors to *Alzheimer's Disease: Advances in Genetics, Molecular and Cellular Biology* cover the remarkable progress that is being made in studies of these and other relevant molecules and the cellular processes in which they participate. The findings described in these chapters have not only enhanced our understanding of the molecular basis of dementia, but have also provided a firm foundation for translational studies that will hopefully serve to take these findings from the bench top to the bedside.

Despite the incredible amount of data that has been garnered by studying the four established AD genes, we also know that they likely account for only 30% of the genetic variance involved in AD. In the Chapter 1, Bertram covers the efforts and challenges in identifying the remaining AD genes. All four of the established AD genes support the amyloid hypothesis of AD, which maintains that the accumulation of Aβ in the brain is the key pathogenic event in AD. The accumulation of Aβ is the net result of production of the peptide versus it's degradation and clearance from brain. The molecular and cell biological studies of the known FAD genes including their effect on Aβ production, are summarized in chapters by Thinakaran and Koo (APP), and Wakabayashi et al (PSEN1 and PSEN2). The role of the β-secretase (BACE) in AD pathogenesis and Aβ production is encapsulated in the chapter by Laird et. al. Huttunen and Kovacs summarize the role of cholesterol and cholesterol pathway enzymes in regulating the Aβ generation. With regard to Aβ clearance, Holtzman and Zlokovic discuss the processes by which Aβ is transported out of the brain, including the role of APOE, while Leissring and Saido summarize the means by which Aβ is degraded in brain. Degradation of Aβ is significantly hampered by its conversion from monomer to oligomeric assemblies and amyloid fibrils. These processes are described in the chapter by Glabe and Bush. With regard to the neurotoxic effects of Aβ, the potentially detrimental effects of the peptide on neuronal signaling are summarized by Cotman and Busciglio, while the *in vivo* effects of Aβ and tau in transgenic mouse models, are presented in the chapter by Ashe. On the other side of the coin, Malinow covers the possible normal physiological roles of Aβ in modulating activity of excitatory synapses. Several excellent chapters on tau, tangles, and the FTDs are also included. Winton et al. provide a summary of the MAPT mutations involved in FTD, while Duff et al. provide an overview of mouse models of FTD and other tauopathies. The Mandelkows discuss the normal and pathogenic roles of the tau protein in axonal transport. Finally, Nagahara and Tuszynski summarize the role of growth factors in AD and prospects for gene therapy using nerve growth factor.

The Editors of *Alzheimer's Disease: Advances in Genetics, Molecular and Cellular Biology* congratulate the authors for their outstanding contributions, and for providing exciting, comprehensive and up-to-date summaries of the most important recent advances in the genetic, molecular, biochemical, and cell biological studies of AD. The last five years have witnessed remarkable progress in all of these areas, and there have certainly been some surprises along the way – Aβ playing a role in modulating synaptic transmission (Malinow); BACE1 contributing to learning and memory processes (Laird et al.); and, neurofibrillary tangles not appearing to cause cognitive deficits in mice (Ashe). Isaac Asimov may have put it best when he said: "The most exciting phrase to hear in science, the one that heralds new discoveries, is not 'Eureka!' (I found it!) but 'That's funny...'". Biotechnology and pharmaceutical companies continue efforts to convert the genetic, molecular, cellular and neurobiological findings described in these chapters into novel therapeutics for treating and preventing AD, FTD, and other dementing disorders of late life. We only hope that the advances described in these pages will help to accelerate this process of rational drug discovery and soon serve to extend and enhance the mental healthspan of our burgeoning elderly population.

Dr. Rudolph E. Tanzi   Dr. Sangram S. Sisodia
Harvard Medical School   University of Chicago

Chapter 1

The Genetics of Alzheimer's Disease

Lars Bertram, MD

Genetics and Aging Research Unit
Department of Neurology
MassGeneral Institute for Neurodegenerative Diseases (MIND)
Massachusetts General Hospital
114 16th Street, Charlestown, MA, 02129
Email: bertram@helix.mgh.harvard.edu

1.   Introduction

Alzheimer's disease (AD), the most common form of age-related dementia, is characterized by progressive and insidious neurodegeneration of the central nervous system that eventually leads to a gradual decline of cognitive function and dementia. The principal neuropathological features of AD are the presence of neurofibrillary tangles and β-amyloid (Aβ) deposited in the form of senile plaques. Although the knowledge of disease pathophysiology still remains fragmentary, it is now widely accepted that inheritance of specific genes plays a critical role in predisposing to onset and/or in modifying disease progression. In fact, familial aggregation had been recognized as a prominent characteristic of AD and several other neurodegenerative disorders. More specifically, the identification of specific, disease-segregating mutations in previously unknown genes has directed attention to specific proteins and pathways that are now considered critical in the pathogenesis of these diseases (e.g. mutant β-amyloid precursor proteins that cause AD; mutant α-synuclein that cause Parkinson's disease; or mutant tau variants that cause frontotemporal dementia (FTD) with parkinsonism; for review see: Bertram and Tanzi, 2005).

Another common feature observed in many neurodegenerative diseases is a dichotomy of familial (rare) vs. seemingly non-familial (common) forms. The latter are also frequently described as "sporadic" or "idiopathic", although there is a growing body of evidence suggesting that a large proportion of these cases are also significantly influenced by genetic factors. These *risk* genes are likely to be numerous, displaying intricate patterns of interaction with each other as well as with non-genetic factors, and – unlike classical Mendelian ("simplex") disorders – exhibit no simple or single mode of inheritance. Hence, the genetics of these diseases has been labeled as "complex". AD is a classic example of a genetically complex disease (Figure 1). Early-onset familial AD (EOFAD), often transmitted as an autosomal dominant trait with onset ages usually below 65 years of age, is caused by rare, but highly penetrant mutations in at least three genes (*APP*, *PSEN1*, *PSEN2*, see following section). However, these cases probably represent not more than 5% of all AD cases. On the other hand, the vast majority of AD occurs after the age of 65 years (late-onset; LOAD), and does not show any overt pattern of familial segregation. Despite intensive efforts to identify the genetic underpinnings of LOAD over the past two decades, there is only a single gene to date (*APOE*) for which genetic variants have been established to significantly increase the risk of developing LOAD across a multitude of independent samples and different ethnicities.

Despite the considerable complexities of AD genetics, tremendous progress has been made over the past two decades towards our understanding of the etiological and pathophysiological mechanisms leading to neurodegeneration. This chapter will outline a brief history of the genetics of AD and discuss the current status and future outlook, with a particular focus on recent findings suggesting the existence of several novel genes that predispose individuals to LOAD.

2.   Early-Onset Familial Alzheimer's disease (EOFAD)

Only 5% (or less) of all AD cases can be explained by early-onset familial AD (EOFAD; Ott et al., 1998; Tanzi, 1999). Despite its rarity, genetic studies of this form of AD are actually facilitated by the availability of large multigenerational pedigrees allowing genetic linkage analysis and subsequent positional cloning. In 1987, EOFAD linkage was reported on the long arm of chromosome 21 that encompassed a region harboring the gene encoding the amyloid precursor protein (APP; gene: *APP*), a compelling candidate gene for AD (Tanzi et al., 1987). In 1991, the first missense mutation in *APP* missense was reported in a family with EOFAD (Goate et al., 1991). Since then, nearly 20 additional AD mutations have been reported in *APP* that account for no more than one tenth of all individuals with early-onset autosomal dominant AD (see Chapter 2, this volume; for an up-to-date overview of AD mutations visit the "AD and FTD Mutation Database"; http://www.molgen.ua.ac.be/ADMutations/). Interestingly, most of the *APP*-variants occur near the putative γ-secretase site between residues 714 and 717, suggesting that alterations in intramembranous,

γ-cleavage of APP (see Chapters 2 and 3, this volume) are critical for the development of AD. A second AD linkage region – on chromosome 14q24 – was reported almost simultaneously by four independent laboratories only one year after the discovery of the first *APP* mutation (Mullan et al., 1992; Schellenberg et al., 1992; St George-Hyslop et al., 1992; Van Broeckhoven et al., 1992). However, it took three more years to clone the responsible gene (*PSEN1*) and identify the first AD-causing mutations (Sherrington et al., 1995). It is now known that *PSEN1* encodes a highly conserved polytopic membrane protein, presenilin 1 (PS1), that plays an essential role in mediating intramembranous, γ-secretase processing of APP to generate Aβ (for review see: Sisodia and St George-Hyslop, 2002), and Chapters 2 and 3, this volume). Even a decade after the original discovery of *PSEN1*, there are several new AD-causing mutations reported annually in this gene, currently totaling more than 140 ("AD & FTD Mutation Database"). Soon after the discovery of *PSEN1* as an AD gene, it became obvious that mutations in this gene, together with *APP*, do not account for all cases of autosomal dominant AD. A database search revealed a second member of the presenilin gene family with significant homology to *PSEN1* (Levy-Lahad et al., 1995; Rogaev et al., 1995), that was subsequently named *PSEN2* (protein: PS2). *PSEN2* maps to the long arm of chromosome 1 and mutations in this gene account for the smallest fraction of all EOFAD cases. On average, individuals in pedigrees with *PSEN2* mutations display a later age of onset and slower disease progression than carriers with *APP* or *PSEN1* mutations.

In conclusion, while all currently known AD causing mutations occur in three different genes located on three different chromosomes, they all share a common biochemical pathway, i.e. the altered production of Aβ leading to a relative overproduction of neurotoxic Aβ42 species, that eventually results in neuronal cell death and dementia (Table 1). Collectively, these discoveries provided the essential connection between the long known familial aggregation of early-onset AD and the increase in Aβ production observed in the brains of autopsied AD patients, findings that gave rise to the "amyloid hypothesis of AD" (for review see: Hardy and Selkoe, 2002 and Tanzi and Bertram, 2005).

Although no additional EOFAD gene has been unequivocally identified since the discovery of *PSEN2* in 1995, several lines of evidence suggest that further genetic factors remain to be identified for this form of AD: I. Numerous early-onset families do not show mutations in *APP*, *PSEN1* or *PSEN2* despite extensive sequencing efforts of open reading frames and adjacent intronic regions (Arango et al., 2001; Lleo et al., 2002; Rademakers et al., 2005; Raux et al., 2005); II. Beyond APP and PS1, there are several additional key proteins involved in γ- and β-secretase cleavage events (BACE1; see Chapter 4, this volume) and in the aggregation (sec Chapter 6, this volume) and deposition of Aβ (e.g. nicastrin, aph-1, pen-2, BACE; see Chapters 3 and 4, this

Table 1. Overview of the established Alzheimer's genes and their functional relevance to pathogenesis.

Gene (Protein)	Chromosomal location	Mode of inheritance	Number of pathogenic mutations (affected families)[a]	Relevance to AD pathogenesis
APP (β-amyloid precursor protein)	21q21.3	autosomal-dominant	20 (60)	increase in Aβ (Aβ_{42}/Aβ_{40}-ratio); mutations close to γ-secretase site
PSEN1 (presenilin 1)	14q24.3	autosomal-dominant	144 (289)	increase in Aβ (Aβ_{42}/Aβ_{40}-ratio); essential for γ-secretase activity
PSEN2 (presenilin 2)	1q31-42	autosomal-dominant	10 (18)	increase in Aβ (Aβ_{42}/Aβ_{40}-ratio); essential for γ-secretase activity (?)
APOE (apolipoprotein E, ε4-allele)	19q13.32	complex (risk increase)	n.a.	increase in Aβ aggregation; decreased Aβ clearance (?); involved in γ-secretase activity (?)

[a](Source: "AD & FTD Mutation Database" [URL: http://www.molgen.ua.ac.be/ADMutations/] current on 10/1/2005).

volume), as well as factors involved in the hyperphosphorylation of tau and the development of neurofibrillary tangles (see Chapters 11 and 12, this volume); III. A recent full genome screen performed by our group has identified at least four early-onset AD linkage regions in addition to the chromosomal location of *PSEN1* on 14q24 (Blacker et al., 2003), and some of these map close to genes encoding components of the γ-secretase complex and β-secretase.

Despite the somewhat unsuccessful quest for novel EOFAD genes to date, recent – and still preliminary – reports have indicated the presence of disease-causing mutations in possibly up to three additional genes, two of which encode proteins that are also strong biochemical candidates for an involvement in AD pathogenesis. First, a linkage study in a large and multigenerational clinically defined multiplex AD family from Belgium indicated the presence of an AD locus near the gene encoding tau on chromosome 17q (*MAPT*). Subsequently, a non-synonymous mutation in exon 13 of *MAPT* (R406W) was reported to co-segregate with AD dementia in this family (Rademakers et al., 2003). While this same mutation was also reported in at least one other family with dementia resembling AD (Ostojic et al., 2004), the majority of cases affected by R406W appear to develop a syndrome fulfilling the criteria of frontotemporal dementia (FTDP-17; Rosso et al., 2003). It therefore remains to be determined whether the clinically assessed Belgian AD family will prove to show neuropathological features at autopsy that confirms a definite diagnosis of AD. Secondly, the same group recently reported evidence of significant linkage with EOFAD to chromosome 7q36 in an extended multiplex AD family from the Netherlands (Rademakers et al., 2005). The same $\sim$10cM haplotype was also found to co-segregate with AD in three additional multiplex families suggesting the presence of a disease-causing mutation in this chromosomal region. A syn-onymous mutation (Ala626) in the gene encoding PAX transcription activation domain interacting protein (*PAXIP1*), located $\sim$400, 000bp downstream of the shared haplotype region, was discovered in AD patients of the index family ("1270"), but absent from 320 control individuals. However, this mutation was absent from the three additional 7q36 haplotype-sharing families, and – according to preliminary analyses – did not show evidence for functional abnormalities in mutation carriers. Thus, the collective evidence supporting *PAXIP1* as a novel EOFAD gene remains relatively weak. Finally, a recent study reported the presence of a D90N mutation in *PEN2*, a gene encoding pen-2 that is a component of the γ-secretase complex (see Chapter 3, this volume; (Sala Frigerio et al., 2005). In addition to being a strong pathophysiological candidate, this gene is also interesting as it maps close to a highly significant linkage region on chromosome 19, approximately 9 Mb proximal of *APOE* (Bertram et al., 2004). However, since the familial transmission of this mutation with AD could not be determined due to a lack of DNA specimens, and since preliminary functional analyses did not reveal an effect of this mutation on

APP metabolism *in vitro*, this finding is probably the least convincing of these putative EOFAD loci.

3. Late-Onset Alzheimer's disease (LOAD)

As outlined above, and in Figure 1, late-onset Alzheimer's (LOAD) is characterized by a considerably more multifaceted and interwoven pattern of genetic and non-genetic factors that are poorly understood. Added to these complexities are methodological difficulties inherent to common diseases, in general, and late-onset diseases like AD, in particular. Family data, for instance, is more often than not only incomplete (e.g. owing to relatives who died before the family-specific age of risk and/or the lack of genotypic information for parents). Another complication is the unknown number of 'phenocopies', i.e. subjects with a non-genetic form of the disease or subjects suffering from other forms of age-related cognitive decline. These and other characteristics largely reduce the power to detect new loci in reasonably sized samples, and continue to hamper the independent replication of proclaimed associations. This is evidenced by the observation that more than a decade after the discovery of *APOE* in AD, no other genetic risk factor has been found to consistently confer

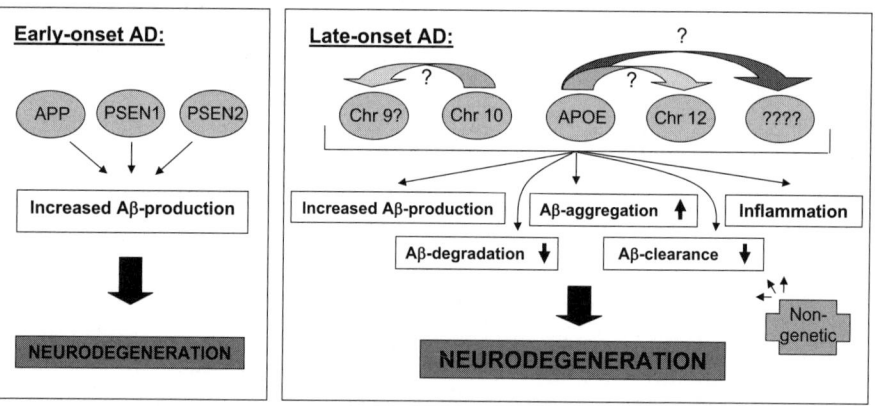

Figure 1. Scheme of contribution and interaction pattern of known and putative AD genes. **Left Panel:** Mutations in the EOFAD genes *APP, PSEN1, PSEN2* all lead to an increase in Aβ-production which almost invariably leads to neurodegeneration and AD. **Right Panel:** Simplified scheme of the interaction pattern of known and proposed late-onset AD genes. Likely, these risk-factor genes each affect one or more of the known or suspected pathogenic mechanisms leading to neurodegeneration in AD. Their effects are further influenced by gene-gene interactions and the contribution of non-genetic risk-factors. Note that the interaction patterns outlined here are for didactic purposes only and have actually not been established. [Figure reprinted with permission from Bertram and Tanzi, 2003 (Bertram and Tanzi, 2003)].

susceptibility to AD, despite intensive efforts in many laboratories worldwide (for reviews see: Finckh, 2003; Rocchi et al., 2003; Bertram and Tanzi, 2004). Despite the fact that although almost none of the well over 300 genes that have been tested for association with AD in nearly 1,000 independent publications over the past 20 years have yielded consistent results, several lines of evidence suggest that further gene hunting in LOAD is indeed worthwhile. First, there are regions on at least 10 chromosomes showing evidence for genetic linkage or association with AD in at least two studies (Bertram and Tanzi, 2004). This number does not include chromosome 19, for which several groups have also suggested the presence of one or more risk genes in addition to *APOE* (Poduslo and Yin, 2001; Bertram et al., 2004; Wijsman et al., 2004; Adighibe et al., 2005). Secondly, a recent simulation study on a well characterized sample of AD families found evidence for the existence of four to seven additional AD susceptibility loci influencing the age of onset for AD besides *APOE* (Daw et al., 2000) across a variety of different inheritance models. One of these loci was even predicted to show effect sizes similar to that of *APOE-ε4*. Finally, systematic meta-analyses on all published AD genetic association studies show significant summary odds ratios for a few genes when all available genotype data is summed across studies ("AlzGene", see below; Bertram et al., 2005). However, in nearly all cases, the associated effects – a measure of the risk increase conferred by a specific polymorphism – are only modest and lack a solid functional/biochemical foundation.

Apolipoprotein E (APOE). The first and to date only proof of principle for applying the "positional candidate gene strategy" (i.e. testing biologically plausible candidate genes in promising linkage regions) in AD was provided by the identification of *APOE* as a risk gene. The *positional* evidence was delivered almost simultaneously with the identification of the first *APP* mutations, when Pericak-Vance and colleagues reported marginal linkage of a locus on chromosome 19q to cases of predominantly LOAD (Pericak-Vance et al., 1991). Two years later, after Aβ was found to bind apoE (Strittmatter et al., 1993), and suggesting a functional involvement of apoE in AD, a common polymorphism in *APOE* that maps near the 19q linkage region, was tested and shown to be associated with increased risk for AD (Saunders et al., 1993; Strittmatter et al., 1993). In contrast to all other reports of genetic association in AD, this result has been overwhelmingly replicated in a large number of studies across many ethnic groups worldwide (for meta-analysis see: Farrer et al., 1997). Three major alleles occur at the *APOE* locus – $\varepsilon2$, $\varepsilon3$ and $\varepsilon4$ – that translate into combinations of two amino acid changes at residues 112 and 158 of the apoE-protein ($\varepsilon2$: Cys/Cys; $\varepsilon3$: Cys/Arg; $\varepsilon4$: Arg/Arg, respectively). While the $\varepsilon4$ allele has been demonstrated to significantly increase the risk for AD, its minor allele – $\varepsilon2$ – has been (less consistently) associated with a decreased risk for AD (Corder et al., 1994).

Although the association between $\varepsilon4$ and AD is robust, it is not specific. In contrast to the genetic variants in EOFAD, the presence of $\varepsilon4$ is neither necessary, nor sufficient, to actually *cause* the disease. Rather, it appears to be a genetic risk-modifier, that predominantly acts through decreasing the age of onset in a dose-dependent manner, in the sense that, homozygous carriers show a younger onset age than carriers of just a single copy (Blacker et al., 1997; Meyer et al., 1998). Expressed in terms of disease risk, this effect translates into an increased relative risk (as measured by the odds ratio) of roughly 3-fold for heterozygous, and almost 15-fold for homozygous $\varepsilon4$-carriers as compared to the $\varepsilon3/3$ genotype (Farrer et al., 1997; Bertram et al., 2005). Similar trends can be seen across different ethnic groups, ranging from odds ratios of around 2-fold in Hispanic to more than 33-fold in Japanese $\varepsilon4/4$ individuals. However, due to the non-specific nature of the $\varepsilon4$-effect, there is widespread consensus within the research, as well as the clinical community, that it is presently inadvisable to use *APOE* genotyping as a sole diagnostic or prognostic test for AD (Burke et al., 2001).

Despite the established genetic role of *APOE* in AD, only little is known about the potential pathophysiological effects and mechanisms that are affected by the encoded protein, apoE. The most straightforward hypothesis assumes that the different polymorphic variants directly influence $A\beta$-accumulation (Strittmatter et al., 1993), a notion supported by several lines of evidence: the number of $A\beta$-plaques in the brain of $\varepsilon4$-allele carriers vs. non-carriers is elevated; the presence or absence of human *APOE* markedly effects the $A\beta$ deposition in transgenic mice over-expressing *APP* (for a recent review on apoE function see: Poirier, 2000; see Chapter 10, this volume); and *APOE*-$\varepsilon4$ decreases the onset age in EOFAD caused by *PSEN1* mutations (Nacmias et al., 1995). Unfortunately, no consensus has yet been reached regarding the predominant *mechanism(s)* underlying these latter observations. Current hypotheses include the involvement of apoE in the aggregation of $A\beta$ (Bales et al., 1999), the clearance of $A\beta$ via low-density lipoprotein receptor related-protein (LRP; Beffert et al., 1999; see Chapter 10, this volume), or via a systemic dysfunction in lipid transport based on the observation that high plasma cholesterol levels are associated with increased β-amyloid in the brain (Poirier, 2000). This latter hypothesis has been countered by several recent epidemiological studies on large cohorts that did not find a link between increased serum cholesterol and risk for AD (Reitz et al., 2004; Li et al., 2005). Finally, it has been proposed that apoE is involved in γ-secretase cleavage of APP (Irizarry et al., 2004), although this effect was similar in extent for all three protein isoforms and, hence, is unlikely to explain the increase in AD risk observed for the $\varepsilon4$-allele.

Other Putative LOAD Loci. As outlined in the introduction to this section, no less than 300 genes have been tested as positional, or functional candidate AD genes (or both) in nearly 1,000 independent peer-reviewed publications over the past two decades. Currently, ten AD genetic association studies are published on average *each month* (Bertram and Tanzi, 2004). For the public, and the AD research community alike, this wealth of information is becoming increasingly more difficult to follow, evaluate and – most importantly – to interpret, in particular as virtually none of the initially positive association findings have been unequivocally replicated in independent follow-up analyses. However, successful replication – just as the primary detection of disease association – is dependent on a number of interrelated factors, including: locus/allelic heterogeneity (i.e. a different set of genes/alleles contributes to AD risk across different samples); small effect size and underpowered sample sizes (i.e. for moderate odds ratios of ∼1.5, sample sizes of more than 500 cases and 500 controls are needed in most situations); linkage disequilibrium with unknown variants (the initially observed association may have been caused by another and as yet unknown variant nearby); and, population stratification and poor case-control matching (both of which can lead to biased results). In the case of multiple conflicting reports, meta-analysis estimating summary odds ratios across all published studies on any specific polymorphism can help distinguish relevant disease associations from false positive ones and those playing only a negligible role.

Over the past two years, we have begun to create a centralized and publicly available internet database ("AlzGene"; Bertram et al., 2005) that is maintained on a daily basis and systematically identifies and samples the most crucial information of every peer-reviewed genetic association study in AD (i.e. study design, sample size, results, and polymorphism details and allele frequencies). In addition, we perform systematic meta-analyses on all polymorphisms in implicated genes that have published data available for at least three independent samples (Figure 2). The resource is available at http://www.alzgene.org and, at the time of this writing (November 2005), contains detailed information of over 700 studies investigating nearly 300 genes.

Data collection for each gene is updated as soon as new studies are published and, therefore, the results of meta-analyses can change over time. Based on the current dataset (October 1st, 2005), "AlzGene" has uncovered a number of overall significant associations, while other genes, some of which have been discussed as putative AD risk factors for years, do not appear to make a significant contribution to disease risk in the general population. In the remainder of this chapter, I will discuss some of the positive and negative highlights that our systematic meta-analyses of the published data to date. The reader is encouraged to visit the AlzGene website (http://www.alzgene.org) to check whether any of these results have changed over time.

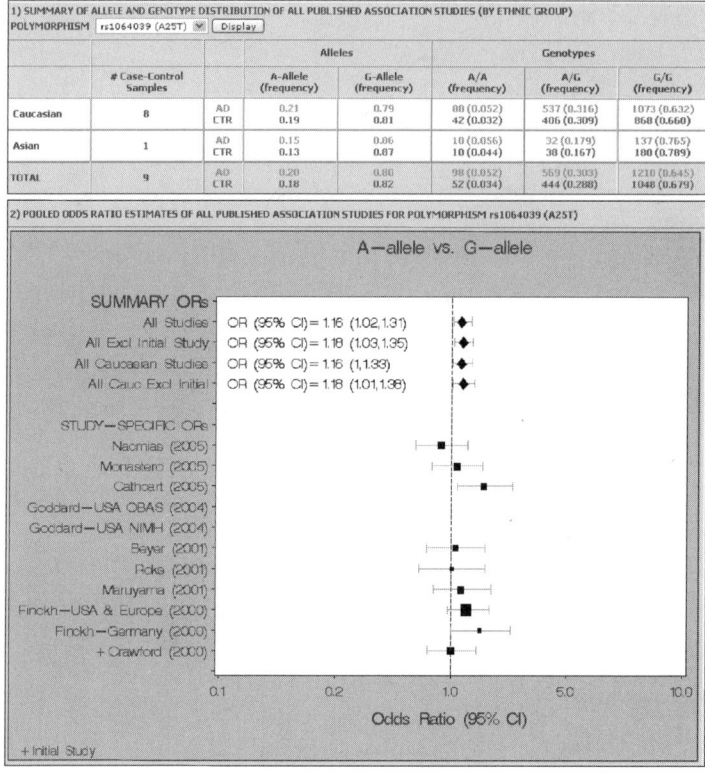

Figure 2. Screenshot of "AlzGene" meta-analysis using cystatin C (*CST3*) as example. **Upper Panel:** Pooled genotype and allele frequencies in published cases-control samples for SNP rs1064039 (Ala25Thr) in *CST3*, divided by ethnic group. As can be seen, the A-allele (encoding threonine) is slightly overrepresented in cases as compared to controls, across both ethnic groups that have been studied to date. In total, this table and the meta-analyses (see below) are based on the published genotypes or allele frequencies of 1,877 AD cases and 1,544 control subjects. **Lower Panel:** Study-specific and summary allelic odds ratios on published genotype- or allele-frequencies. For these analyses, study specific crude odds ratios are calculated from the published data, and then summed using a random-effects model following four paradigms: "All Studies" includes all available data; "All Excl Initial Study" excludes the initial (and usually positive) study from the summary odds ratio; "All Caucasian Studies" and "All Cauc Excl Initial" repeats the above analyses limited to case-control samples of Caucasian ethnicity, which is usually the ethnic group studied most frequently. Diamonds and squares represent summary and study-specific odds ratio estimates, respectively, and horizontal bars represent 95% confidence intervals. Note that while only two studies (Finck-Germany [2000] and Cathcart [2005]) show significant study specific odds ratios, the summary estimate is significant because most studies actually observed a (non-significant) overrepresentation of the A-allele in cases vs. controls (with the exception of Nacmias [2005]). The numbers displayed were current on October 1st, 2005; for an up-to-date version of these analyses please visit the "AlzGene" database at www.alzgene.org (Bertram et al., 2005). [Figure reprinted with permission of the Alzheimer Research Forum].

Genes currently showing significant association in "AlzGene". One of the first polymorphisms that was analyzed in "AlzGene" was the ε2/3/4 variant in *APOE*. Since the ε4 variants was already an established AD risk factor, we elected to limit our analyses to the 42 papers originally analyzed in the meta-analysis by Farrer and colleagues (Farrer et al., 1997). This allowed us to determine how our overall approach of data inclusion and analysis would fare compared to the earlier comprehensive, but conventional analysis. Surprisingly, our calculated odds ratios were almost identical to those reported previously, including the finding of an ∼40% decrease in AD risk in carriers of the ε2 allele (see "AlzGene" for details). In addition to these results, we also detected significant, but weaker, effects for polymorphisms in the *APOE* promoter, as well as in the *APOC1* gene that maps 5 kb 3' of *APOE*. Association with both of these loci has only been inconclusively observed previously, and attributed to varying degrees of linkage disequilibrium (LD) between these variants and ε2/3/4. While the amount of LD cannot be estimated from the data available in "AlzGene", it clearly shows that even very modest (i.e. odds ratios around 1.5), and LD-related effects can be successfully identified by applying our inclusion criteria, data collection and meta-analysis methodology.

Other genes which currently show evidence of significant association in "AlzGene" include a very common insertion/deletion polymorphism in intron 16 of the angiotensin converting enzyme (*ACE*) on chromosome 17. While the association in "AlzGene" is statistically significant across the 34 studies (that are comprised of ∼6, 000 cases and more than 10,000 controls), it is weak in the sense that the odds to develop AD in carriers of at least one insertion allele are only increased by ∼10%. Interestingly, this small effect size is quite comparable to this variant's effect on myocardial infarction in a meta-analysis of comparable size (Keavney et al., 2000). Also significantly associated in "AlzGene" is a non-coding SNP in intron 8 of *PSEN1* (Wragg et al., 1996), suggesting that variants in this gene – in addition to causing EOFAD – could also be responsible for contributing to the risk of LOAD. Note, however, that the *PSEN1*-related effect is even less pronounced than that for *ACE*, and may be difficult to detect in sample sizes of less than 10,000 cases and 10,000 controls. While the results obtained with *ACE* and *PSEN1* can be considered quite robust based on the large number of samples studied, other promising findings in "AlzGene" should be considered more provisional as they are usually based on 10 or less studies. These genes include *MTHFR* (5,10-methylenetetrahydrofolate reductase)on chromosome 1p36, *TF* (transferrin) on 3q2, *ESR1* (estrogen receptor 1) on 6q25, and *PRNP* (prion protein) and *CST3* (cystatin C; Figure 2) on chromosome 20p13 and 20p11, respectively. Interestingly, the gene currently showing the strongest and most significant effects (i.e. transferrin, allelic summary odds ratio ∼1.3), does not map to any of the major linkage regions to emerge from full genome linkage screens

(Bertram and Tanzi, 2004), reinforcing the notion that in genetically complex diseases, association designs are inherently more powerful to detect genes of small effect than linkage studies (Risch and Merikangas, 1996).

Genes with negative meta-analysis results in "AlzGene". Despite these few positive findings, the vast majority of genes analyzed in "AlzGene" do not show evidence for significant association in any of the analyses performed. Some of the more prominent candidates in this category currently include the genes of the interleukin-1 cluster (*IL1A* and *IL1B*) on chromosome 2q12, *BCHE* (butyrylcholinesterase) on 3q26, *IL6* (interleukin 6) on 7p15, *VLDLR* (very low-density lipoprotein receptor) on chromosome 9p24, *PLAU* (urokinase type plasminogen activator) on 10q22, *CTSD* (cathepsin D) and *BDNF* (brain-derived neurotrophic factor) chromosome 11p15 and 11p14, *LRP* (low density lipoprotein receptor related protein) and *A2M* (alpha-2 macroglobulin) on 12p13 and 12q13, *ACT* (alpha-1 antitrypsin) on 14q32, and *MAPT/STH* (tau/saitohin) on 17q21. For all of these genes, the "AlzGene" meta-analyses included the data of at least 2,000 cases vs. 2,000 controls (for several genes these numbers are much higher), and thus it is unlikely that these genes harbor common variants that make a relevant contribution to AD risk in the general population.

Regardless of their negative outcome in "AlzGene", two of these genes show some interesting aspects beyond the default analyses. *A2M*, for instance, shows consistently negative meta-analyses with the two most frequently studied variants (i.e. a 5 bp insertion/deletion 5' of exon 18, and an alanine to valine substitution at codon 1000) in the case-control samples, but the situation is quite different when only family-based analyses are considered, i.e. a study design in which affected and unaffected individuals stem from the same families and which are therefore less prone to bias due to population heterogeneity. It is noteworthy that all of these studies actually report a significant, or at least suggestive, result for either or both of these polymorphisms in *A2M*, with the exception of two small samples from Duke and Canada (see "AlzGene" for details). Thus, *A2M* could actually represent a risk factor for AD in families multiply affected by the disease, an effect which would be nearly impossible to detect in a conventional sample of unrelated cases and controls.

The second finding worth mentioning is that of the tau locus (*MAPT* and *STH*). While earlier studies usually investigated a varying number of polymorphisms throughout the entire gene in order to reconstruct the underlying ancestral haplotype structure (usually denoted as "H1" and "H2"), more recent studies have also tested a novel non-synonymous SNP (Q7R) in a novel gene of unknown function, that is located in intron 9 of the tau gene (*STH*, encoding saitohin; Conrad et al., 2002). In most populations, the glutamine (Q) allele is part of the H1-haplotype, while arginine (R) is invariably associated with

H2 (Conrad et al., 2004). Meta-analyses of the available association data in "AlzGene" reveal a marginally protective effect of the H2-allele in studies specifically investigating variants in *MAPT*, whereas meta-analyses limited to *STH* actually failed to detect such an effect. Accordingly, when both datasets are combined, the suggestive results of H2 disappear altogether. A recent study on this subject suggested that the discrepancies may be due to recombination events on both the H1 and H2 backgrounds and suggest the existence of a sub-haplotype on the H1-background (denoted "H1c"), presumably harboring the actual disease-causing variant(s) (Myers et al., 2005). On its own, however, the association data published on H1c is relatively weak and – just as any other proclaimed association finding –needs to be independently replicated and meta-analyzed before any further conclusions can be drawn.

4.   Conclusion

Despite the great progress in the field of AD genetics that has led to the discovery and confirmation of three autosomal-dominant early-onset genes (*APP, PSEN1, PSEN1*) and one late-onset risk-factor (*APOE*), strong evidence exists suggesting the presence of additional AD genes for both forms of the disease. The hunt for these genes is aggravated by several factors that generally complicate the identification of complex disease genes: locus and/or allelic heterogeneity; small effect sizes of the underlying variants; unknown and dif-ficult to model interaction patterns; population differences; insufficient sample sizes/sampling strategies; and, linkage disequilibrium among polymorphisms other than those initially associated with the disease. The emergence of more powerful and efficient genotyping technologies (e.g. whole genome association screening) as well as analysis tools (e.g. systematic and continuously updated meta-analyses) should enable us to disentangle the genetics of AD and other complex diseases. Eventually, the insights gained from such studies will lead to a better understanding of the pathophysiological mechanisms leading to neurodegeneration. This knowledge will lay the foundation to developing new treatment strategies that will ultimately allow to cure, delay or even prevent this devastating disease.

References

Adighibe O, Arepalli S, et al. (2005) Genetic variability at the LXR gene (NR1H2) may contribute to the risk of Alzheimer's disease. Neurobiol Aging.

Arango D, Cruts M, et al. (2001) Systematic genetic study of Alzheimer disease in Latin America: mutation frequencies of the amyloid beta precursor protein and presenilin genes in Colombia. Am J Med Genet 103(2):138–143.

Bales KR, Verina T, et al. (1999) Apolipoprotein E is essential for amyloid deposition in the APP(V717F) transgenic mouse model of Alzheimer's disease. Proc Natl Acad Sci USA 96(26):15233–15238.

Beffert U, Cohn JS, et al. (1999) Apolipoprotein E and beta-amyloid levels in the hippocampus and frontal cortex of Alzheimer's disease subjects are disease-related and apolipoprotein E genotype dependent. Brain Res 843(1–2):87–94.

Bertram L, McQueen MB, et al. (2005) The AlzGene Database, Alzheimer Research Forum. Avalaible at: http://www.alzgene.org; last accessed 10/1/2005.

Bertram L, Menon R, et al. (2004) PEN2 is not a genetic risk factor for Alzheimer's disease in a large family sample. Neurology 62(2):304–306.

Bertram L, Tanzi R, (2003) Genetics of Alzheimer's disease. Neurodegeneration: The Molecular Pathology of Dementia and Movement Disorders. D. Dickson. Basel, ISN Neuropath Press.

Bertram L, Tanzi RE, (2004) Alzheimer's disease: one disorder, too many genes? Hum Mol Genet 13 Spec No 1:R135–141.

Bertram L, Tanzi RE, (2005) The genetic epidemiology of neurodegenerative disease. J Clin Invest 115(6):1449–1457.

Blacker D, Bertram L, et al. (2003) Results of a high-resolution genome screen of 437 Alzheimer's Disease families. Hum Mol Genet 12(1):23–32.

Blacker D, Haines JL, et al. (1997) ApoE-4 and age at onset of Alzheimer's disease: the NIMH genetics initiative. Neurology 48(1):139–147.

Burke W, Pinsky LE, et al. (2001) Categorizing genetic tests to identify their ethical, legal, and social implications. Am J Med Genet 106(3):233–240.

Conrad C, Vianna C, et al. (2002) A polymorphic gene nested within an intron of the tau gene: implications for Alzheimer's disease. Proc Natl Acad Sci USA 99(11):7751–7756.

Conrad C, Vianna C, et al. (2004) Molecular evolution and genetics of the Saitohin gene and tau haplotype in Alzheimer's disease and argyrophilic grain disease. J Neurochem 89(1):179–188.

Corder EH, Saunders AM, et al. (1994) Protective effect of apolipoprotein E type 2 allele for late onset Alzheimer disease. Nat Genet 7(2):180–184.

Daw EW, Payami H, et al. (2000) The number of trait loci in late-onset Alzheimer disease. Am J Hum Genet 66(1):196–204.

Farrer LA, Cupples LA, et al. (1997) Effects of age, sex, and ethnicity on the association between apolipoprotein E genotype and Alzheimer disease. A meta-analysis. APOE and Alzheimer Disease Meta Analysis Consortium. Jama 278(16):1349–1356.

Finckh U, (2003) The future of genetic association studies in Alzheimer disease. J Neural Transm 110(3):253–266.

Goate A, Chartier-Harlin MC, et al. (1991) Segregation of a missense mutation in the amyloid precursor protein gene with familial Alzheimer's disease. Nature 349(6311):704–706.

Hardy J, Selkoe DJ, (2002) The amyloid hypothesis of Alzheimer's disease:progress and problems on the road to therapeutics. Science 297(5580):353–356.

Irizarry MC, Deng A, et al. (2004) Apolipoprotein E modulates gamma-secretase cleavage of the amyloid precursor protein. J Neurochem 90(5):1132–1143.

Keavney B, McKenzie C, et al. (2000) Large-scale test of hypothesised associations between the angiotensin-converting-enzyme insertion/deletion polymorphism and myocardial infarction in about 5000 cases and 6000 controls. International Studies of Infarct Survival (ISIS) Collaborators. Lancet 355(9202):434–442.

Levy-Lahad E, Wasco W, et al. (1995) Candidate gene for the chromosome 1 familial Alzheimer's disease locus. Science 269(5226):973–977.

Li G, Shofer JB, et al. (2005) Serum cholesterol and risk of Alzheimer disease:A community-based cohort study. Neurology 65:1045–1050.

Lleo A, Blesa R, et al. (2002) Frequency of mutations in the presenilin and amyloid precursor protein genes in early-onset Alzheimer disease in Spain. Arch Neurol 59(11):1759–1763.

Meyer MR, Tschanz JT, et al. (1998) APOE genotype predicts when–not whether–one is predisposed to develop Alzheimer disease. Nat Genet 19(4):321–322.

Mullan M, Houlden H, et al. (1992) A locus for familial early-onset Alzheimer's disease on the long arm of chromosome 14, proximal to the alpha 1-antichymotrypsin gene. Nat Genet 2(4):340–342.

Myers AJ, Kaleem M, et al. (2005) The H1c haplotype at the MAPT locus is associated with Alzheimer's disease. Hum Mol Genet 14(16):2399–2404.

Nacmias B, Latorraca S, et al. (1995) ApoE genotype and familial Alzheimer's disease:a possible influence on age of onset in APP717 Val–>Ile mutated families. Neurosci Lett 183(1–2):1–3.

Ostojic J, Elfgren C, et al. (2004) The tau R406W mutation causes progressive presenile dementia with bitemporal atrophy. Dement Geriatr Cogn Disord 17(5):298–301.

Ott A, Breteler MM, et al. (1998) Incidence and risk of dementia. The Rotterdam Study. Am J Epidemiol 147(6):574–580.

Pericak-Vance MA, Bebout JL, et al. (1991) Linkage studies in familial Alzheimer disease:evidence for chromosome 19 linkage. Am J Hum Genet 48(6):1034–1050.

Poduslo SE Yin X (2001) A new locus on chromosome 19 linked with late-onset Alzheimer's disease. Neuroreport 12(17):3759–3761.

Poirier J (2000) Apolipoprotein E and Alzheimer's disease. A role in amyloid catabolism. Ann NY Acad Sci 924:81–90.

Rademakers R, Cruts M, et al. (2005) Linkage and association studies identify a novel locus for Alzheimer disease at 7q36 in a dutch population-based sample. Am J Hum Genet 77(4):643–652.

Rademakers R, Dermaut B, et al. (2003) Tau (MAPT) mutation Arg406Trp presenting clinically with Alzheimer disease does not share a common founder in Western Europe. Hum Mutat 22(5):409–411.

Raux G, Guyant-Marechal L, et al. (2005) Molecular diagnosis of autosomal dominant early onset Alzheimer's disease:an update. J Med Genet 42(10):793–795.

Reitz C, Tang MX, et al. (2004) Relation of plasma lipids to Alzheimer disease and vascular dementia. Arch Neurol 61(5):705–714.

Risch N, Merikangas K (1996) The future of genetic studies of complex human diseases. Science 273(5281):1516–1517.

Rocchi A, Pellegrini S, et al. (2003) Causative and susceptibility genes for Alzheimer's disease:a review. Brain Res Bull 61(1):1–24.

Rogaev EI, Sherrington R, et al. (1995) Familial Alzheimer's disease in kindreds with missense mutations in a gene on chromosome 1 related to the Alzheimer's disease type 3 gene. Nature 376(6543):775–778.

Rosso SM, Donker Kaat L, et al. (2003) Frontotemporal dementia in The Netherlands:patient characteristics and prevalence estimates from a population-based study. Brain 126(Pt 9):2016–2022.

Sala Frigerio C, Piscopo P, et al. (2005) PEN-2 gene mutation in a familial Alzheimer's disease case. J Neurol 252(9):1033–1036.

Saunders AM, Strittmatter WJ, et al. (1993) Association of apolipoprotein E allele epsilon 4 with late-onset familial and sporadic Alzheimer's disease. Neurology 43(8):1467–1472.

Schellenberg GD, Bird TD, et al. (1992) Genetic linkage evidence for a familial Alzheimer's disease locus on chromosome 14. Science 258(5082):668–671.

Sherrington R, Rogaev EI, et al. (1995) Cloning of a gene bearing missense mutations in early-onset familial Alzheimer's disease. Nature 375(6534):754–760.

Sisodia SS, St George-Hyslop PH (2002) gamma-Secretase, Notch, Abeta and Alzheimer's disease:where do the presenilins fit in? Nat Rev Neurosci 3(4):281–290.

St George-Hyslop P, Haines J, et al. (1992) Genetic evidence for a novel familial Alzheimer's disease locus on chromosome 14. Nat Genet 2(4):330–334.

Strittmatter WJ, Saunders AM, et al. (1993) Apolipoprotein E: high-avidity binding to beta-amyloid and increased frequency of type 4 allele in late-onset familial Alzheimer disease. Proc Natl Acad Sci USA 90(5):1977–1981.

Tanzi RE (1999) A genetic dichotomy model for the inheritance of Alzheimer's disease and common age-related disorders. J Clin Invest 104(9):1175–1179.

Tanzi RE, Bertram L (2005) Twenty years of the Alzheimer's disease amyloid hypothesis:a genetic perspective. Cell 120(4):545–555.

Tanzi RE, Gusella JF, et al. (1987) Amyloid beta protein gene: cDNA, mRNA distribution, and genetic linkage near the Alzheimer locus. Science 235(4791):880–884.

Van Broeckhoven C, Backhovens H, et al. (1992) Mapping of a gene predisposing to early-onset Alzheimer's disease to chromosome 14q24.3. Nat Genet 2(4):335–339.

Wijsman EM, Daw EW, et al. (2004) Evidence for a novel late-onset Alzheimer disease locus on chromosome 19p13.2. Am J Hum Genet 75(3):398–409.

Wragg M, Hutton M, et al. (1996) Genetic association between intronic polymorphism in presenilin-1 gene and late-onset Alzheimer's disease. Alzheimer's Disease Collaborative Group. Lancet 347(9000):509–512.

Chapter 2

APP Biology, Processing and Function

Gopal Thinakaran[1] and Edward H. Koo[2]

[1]*Department of Neurobiology Pharmacology and Physiology*
The University of Chicago
Chicago, Illinois 60637
Email: gopal@uchicago.edu

[2]*Department of Neurosciences*
University of California
San Diego, La Jolla, California 92093
Email: edkoo@ucsd.edu

The amyloid precursor protein (APP) plays a central role in Alzheimer's disease (AD) pathogenesis and in AD research. In large part, this is because APP is the precursor to the amyloid-β-protein (Aβ), the 40-42 amino acid residue peptide that is at the heart of the amyloid cascade hypothesis of AD. Consequently, intracellular trafficking and proteolytic processing of APP have been the focus of numerous investigations over the past two decades. Tremendous progress has been made since the initial identification of Aβ as the principal component of brain senile plaques of individuals with AD and the subsequent cloning of *APP* cDNA. Specifically, molecular characterization of the secretases involved in Aβ production has facilitated cell biological investigations on APP processing, and advanced efforts to model AD pathogenesis in animal models. In this chapter, we will review the recent developments in APP trafficking, discuss salient features of amyloidogenic processing of APP in organelles and membrane microdomains, and examine the putative biological functions of APP. The latter focus is essential because APP clearly plays physiological roles in the nervous system, some of which may contribute to neurodegeneration. Details concerning the pathways mediating production, aggregation, and degradation of Aβ will be covered extensively in Chapters 6, 8, 9 and 10, and will be mentioned here only in passing for the purposes of clarity.

1.   APP gene family

The human *APP* gene was first identified in 1987 by several laboratories independently using partial protein sequence information obtained by the Glenner and Beyreuther/Masters laboratories several years earlier. The two *APP* homologues, *APLP1* and *APLP2*, were discovered several years later. The identification of APP led to several early surprises. First, APP is a type I membrane protein whereby two predicted cleavages, one in the extracellular domain (β-secretase cleavage) and the other in the transmembrane region (γ-secretase cleavage) are necessary to release Aβ from the precursor molecule. Second, APP is located on chromosome 21 (21q21.2-3). This provided an immediate connection to the almost invariant development of AD pathology in trisomy 21 (Down's syndrome) individuals (see below). Finally, although no typical functional motifs were seen, it was speculated that APP might function as a cell surface receptor. Almost twenty years later, this prediction has yet to be definitively fulfilled. As an historical aside, it should be pointed out that the first mutations that were found to be causative in inherited forms of familial AD and a related inherited condition, hereditary cerebral haemorrhage with amyloid angiopathy, Dutch type, were found in the *APP* gene (Levy et al., 1990; Hardy, 1997). Although mutations in *APP* are rare in comparison to mutations in *PSEN* genes (which encode presenilin-1 and -2) (see Chapter 1), they are nevertheless important because they provided early and seminal evidence that APP plays a central role in AD pathogenesis.

APP is now known to be one of three members of a larger gene family. These include *APLP1* and *APLP2* in humans, *APPL* (fly), and *APL-1* (worm) (Coulson et al., 2000). All genes encode type I membrane proteins with a large extracellular domain and a short cytoplasmic region that undergo similar processing (see below). Importantly, only *APP*, but not any of the other *APP*-related genes, contains sequence encoding the Aβ domain. Therefore, *APLP1* and *APLP2* are not the precursors to Aβ and if these two genes contribute to AD pathogenesis, then their roles must be indirect. *APP* and *APLP2* are ubiquitously expressed although alternative splicing generates isoforms that may be expressed in a cell type specific manner; for example, APP695 (the 695 amino acid isoform) is neuron-specific. On the other hand, *APLP1* is expressed selectively in the nervous system.

2.   APP processing

2.1   APP secretases

Full-length APP undergoes sequential proteolytic processing as outlined in Figure 1. APP is first cleaved by α-secretase (non-amyloidogenic pathway) or β-secretase (amyloidogenic pathway) within the lumenal domain, resulting in the shedding of nearly the entire ectodomain and generation of membrane-tethered α- or β-C-terminal fragments (CTFs). The major neuronal β-secretase is a transmembrane aspartyl protease, termed BACE1 (β-site APP cleaving enzyme; also called Asp-2 and memapsin-2). BACE1 cleaves APP within the ectodomain, generating the N-terminus of Aβ (Vassar, 2004). However, the principal BACE (β') cleavage site in native APP is between Glu +11 and Val +12 of the Aβ peptide. Several zinc metallopreoteinases such as TACE/ADAM17, ADAM9, ADAM10 and MDC-9, and an aspartyl protease, BACE2, can cleave APP at, or near, the α-secretase site (Allinson et al., 2003), located within the Aβ domain (between residues Lys16 and Leu17 of the Aβ peptide), essentially precluding the generation of intact Aβ.

The second proteolytic event in APP processing involves intramembranous cleavage of α- and β-CTFs by γ-secretase, that liberates p3 (3 kDa) and Aβ (4 kDa) peptides, respectively, into the extracellular milieu. The minimal components of γ-secretase include presenilin-1 or -2 (PS1 or PS2), nicastrin, APH-1, and PEN-2 (Edbauer et al., 2003; Iwatsubo, 2004) (see Chapter 3). Protein subunits of the γ-secretase assemble early during biogenesis and co-operatively mature as they leave the endoplasmic reticulum. Biochemical and pharmacological evidence are consistent with PS1 (or PS2) as the catalytic subunit of the γ-secretase. A pair of conserved aspartate residues within the predicted transmembrane domains 6 and 7 of PS1 and PS2 is crucial for γ-secretase activity. APH-1 and PEN2 are thought to stabilize the γ-secretase complex and nicastrin to mediate the recruitment of APP CTF to the catalytic site of the γ-secretase. The major sites of γ-secretase cleavage correspond to positions 40 and 42 of Aβ. Greater than 90% of secreted Aβ ends in residue 40, and Aβ42 accounts for less than 10% of total Aβ. In addition, γ-secretase cleavage at a distal site generates a cytoplasmic polypeptide, termed APP intracellular domain (AICD). Familial AD-linked mutations in APP near the γ-secretase cleavage site affect cleavage specificity at Aβ40/42 sites, favouring cleavage at position 42. Intriguingly, familial AD-linked mutations in PS1 and PS2 influence γ-secretase cleavage by an elusive mechanism that also modulates the proteolysis of APP to selectively enhance the generation of Aβ42 peptides.

Amyloidogenic processing is the favoured pathway of APP metabolism in neurons largely due to the greater abundance of BACE1, and non-amyloidogenic pathway is predominant in all other cell types. Commitment of APP to these pathways can be differentially modulated by the activation of cell-surface receptors such as serotonin 5-hydroxytryptamine (5-HT$_4$) receptor, metabotropic glutamate receptors, muscarinic acetylcholine receptors, and platelet-derived growth factor receptor. Signalling downstream of these receptors regulate APPsα and Aβ secretion by engaging intermediates including

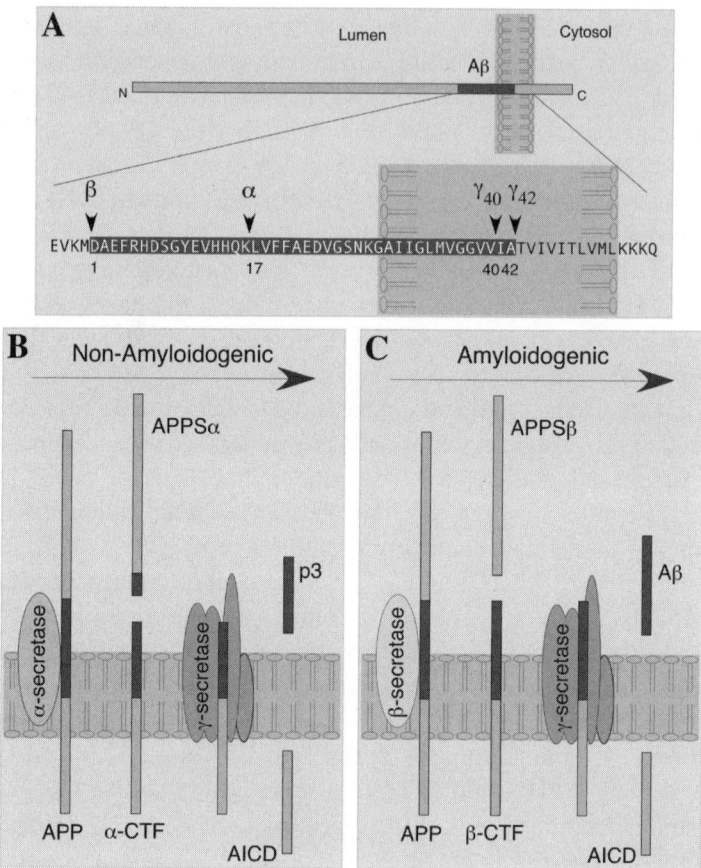

Figure 1. Proteolytic processing of APP. **A)** Schematic structure of APP is shown with Aβ domain shaded in red and enlarged. The major sites of cleavage by α-, β-, and γ-secretases are indicated along with Aβ numbering from the N-terminus of Aβ (Asp1). **B)** Non-amyloidogenic processing of APP refers to sequential processing of APP by membrane-bound α- and γ-secretases. α-secretase cleaves within the Aβ domain, thus precluding generation of intact Aβ peptide. The fates of N-terminally truncated Aβ (p3) and APP intracellular domain (AICD) are not fully resolved. **C)** Amyloidogenic processing of APP is carried out by sequential action of membrane-bound β- and γ-secretases.

PKC, PKA, phosphatidylinositol 3 kinase, mitogen-activated protein kinase kinase, extracellular signal-regulated kinase, Src tyrosine kinase, small GTPase Rac, inositol 1,4,5-trisphosphate, cAMP, and calcium. However, whether APP is a direct substrate of these intermediates has not been established. Whereas secreted APPSα has been reported to have neurotrophic properties, Aβ peptides have adverse effects on neuronal survival.

It appears that none of the aforementioned secretases have unique substrate specificity towards APP. Besides APP, several transmembrane proteins such as pro-TNFα and pro-TGFα undergo ectodomain shedding by enzymes with α-secretase activity. The relatively low affinity of BACE1 toward APP led to the suggestion that APP is not its sole physiological substrate. In support of this idea, α2,6-sialyltransferase and low density lipoprotein receptor-related protein (LRP) have been identified as additional substrates that are processed by BACE1. Similarly, PS1 and PS2 play a crucial role in intramembranous γ-secretase cleavage of several type I membrane proteins other than APP, including the Notch1 receptor and its ligands, Delta and Jagged2, cell-surface adhesion protein CD44, the receptor tyrosine kinase ErbB4, netrin receptor DCC, LRP, lipoprotein receptor ApoER2, cell adhesion molecules N- and E-cadherins, synaptic adhesion protein nectin-1α, cell surface heparin sulphate proteoglycan syndecan-3, p75 neurotrophin receptor etc (Koo and Kopan, 2004). Like APP, a signature of γ-secretase cleavage of these additional substrates is the requirement of an ectodomain shedding event.

Ever increasing number of transmembrane substrates and intracellular domains released by the proteolytic cleavage of these substrates indicate that in addition to being a modulator of many cell signaling paradigms via cleavage of proteins such as Notch, γ-secretase could simply be a proteasome or secretosome that catabolises membrane-bound protein "stubs" of type I membrane proteins (Kopan and Ilagan, 2004). Outcome of γ-secretase cleavage of substrates can either be activation of signaling as is the case in Notch receptor cleavage and the release of Notch intracellular domain, or termination of signaling as described for intramembranous cleavage of DCC (Parent et al., 2005). Apart from its essential role in the proteolytic function of the γ-secretase, PS1 and PS2 have been shown to participate in fundamental physiological functions including calcium homeostasis, neuronal signaling, protein trafficking, protein degradation, fine-tuning of immune system, neurite outgrowth, apoptosis, memory and synaptic plasticity (Sisodia et al., 1999; Koo and Kopan, 2004; Thinakaran and Parent, 2004). Hence, although APP secretases and factors regulating their activity in amyloidogenic pathway have long been considered as therapeutic targets for the treatment of AD, it is unclear whether secretase inhibitors will be free from serious side effects.

2.2   Intracellular itinerary and processing of APP

During its transit from the ER to the plasma membrane through the constitutive secretory pathway (Figure 2), nascent APP undergoes post-translational modification by *N*- and *O*-glycosylation, ectodomain and cytoplasmic phosphorylation, and tyrosine sulphation. In cultured cells, it is estimated that only about 10% of nascent APP molecules are successfully delivered to the plasma membrane, based primarily on overexpression systems. APP can be proteolytically processed at the cell surface mainly by α-secretases, resulting in the shedding of APPsα ectodomain (Sisodia, 1992). Activation of protein kinase C increases APPsα secretion by mechanisms involving the formation and release of secretory vesicles from the trans-Golgi network, thus enhancing APP (and possibly α-secretase) trafficking to the cell surface.

Unlike many cell surface receptors, full-length APP does not reside for considerable length of time at the cell surface. Approximately 70% of surface-bound APP is internalized within minutes of arriving at the plasma membrane. A "YENPTY" internalization motif located near the C-terminus of APP (residues 682-687 of APP695 isoform) is responsible for this efficient internalization. Following endocytosis, APP is delivered to late endosomes and a fraction of endocytosed molecules is recycled to the cell surface. Measurable amounts of internalized APP also undergo degradation in the lysosome.

At steady-state, the majority of BACE1 localizes to late Golgi/TGN and endosomes, consistent with amyloidogenic cleavage of wild-type APP during endocytic/recycling steps (Koo and Squazzo, 1994) (Figure 2). BACE1 activity is optimal at acidic pH *in vitro*, supporting the notion that BACE1 likely cleaves APP during transit in acidic endocytic compartments. Available data indicate the presence of γ-secretase complex and enzyme activity in multiple compartments including the ER, late-Golgi/TGN, endosomes and the plasma membrane (Cook et al., 1997; Xu et al., 1997; Greenfield et al., 1999; Kaether et al., 2002; Takahashi et al., 2002; Chyung et al., 2004; Vetrivel et al., 2004). Recent estimates suggest only a minor presence of γ-secretase activity at the cell-surface, whereas the majority of the mature components of γ-secretase complex are found, and shown to be enzymatically active, in intracellular organelles such as ERGIC, Golgi apparatus, the TGN, and late endosomes.

As discussed below, in neurons APP is trafficked anterogradely along peripheral and central axons, and proteolytically processed during transit (Koo et al., 1990; Buxbaum et al., 1998). Reduced Aβ deposition in BACE1 transgenic mice illustrates how subcellular site of amyloidogenic processing in neurons can greatly influence Aβ production and deposition *in vivo* (Lee et al., 2005). Nevertheless, the intracellular organelles/transport vesicles where Aβ is generated in neurons are not fully characterized.

Studies conducted in non-neuronal and neuroblastoma cell lines show that Aβ is mainly generated in TGN as APP is trafficked through the secretory and recycling pathways (Figure 2). Attempts to address the role of endocytic APP trafficking by expression of dominant-negative mutant of dynamin, an important component of the endocytic machinery, resulted in discrepant findings. This is not surprising, since overexpression of mutant dynamin causes pleiotropic effects on endocytic trafficking of numerous proteins including APP secretases. Nevertheless, mutations within the APP cytosolic YENPTY

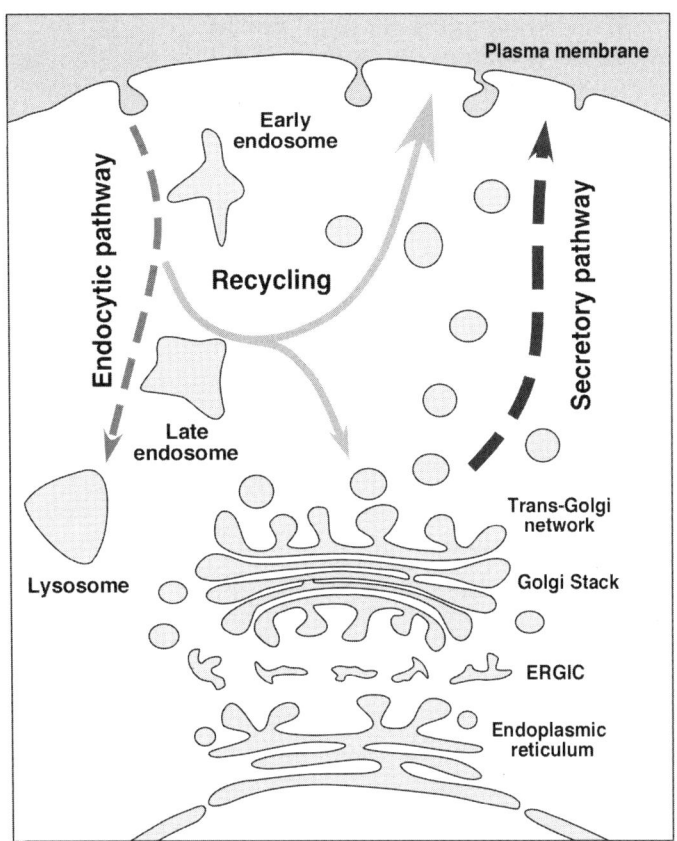

Figure 2. Intracellular trafficking of APP. Nascent APP molecules mature through the constitutive secretory pathway. Once APP reaches the cell surface, it is rapidly internalized and subsequently trafficked through endocytic and recycling compartments back to the cell surface or degraded in the lysosome. Non-amyloidogenic processing mainly occurs at the cell surface where α-secretases are present. Amyloidogenic processing involves transit through the endocytic organelles where APP encounters β- and γ-secretases.

motif selectively inhibit internalization of APP and decrease Aβ generation (Perez et al., 1999). Several cytosolic adaptors with phosphotyrosine-binding domains, including Fe65, Fe65L1, Fe65L2, Mint 1 (also called X11α), Mint 2, Mint3, and Dab1 bind to the APP cytoplasmic tail at or near the YENPTY motif, and regulate APP trafficking and processing (King and Turner, 2004). Mint proteins (so named for their ability to interact with Munc18) can directly bind ADP-ribosylation factors, raising the intriguing possibility that vesicular trafficking of APP may be regulated by Mints serving as coat proteins (Hill et al., 2003). Interestingly, Fe65 acts as a functional linker between APP and LRP (another type I membrane protein containing two NPXY endocytosis motifs) in modulating endocytic APP trafficking and amyloidogenic processing (Pietrzik et al., 2004). A conformational change introduced by phosphorylation at Thr-668 (14 amino acids proximal to the YENPTY motif) interferes with Fe65 binding to APP, and facilitates BACE1 and γ-secretase cleavage of APP (Ando et al., 2001; Lee et al., 2003). In addition, Fe65 stabilizes the highly labile AICD, which may serve as a regulatory step in modulating the physiological function of AICD (see below). Despite the elaborate regulatory mechanisms that modulate cell surface transport and endocytic trafficking of APP, transit through these compartments are not essential for generation of Aβ as shown by amyloidogenic processing of APP in cells expressing syntaxin 1A mutants defective in exocytosis (Khvotchev and Sudhof, 2004). Still, the overexpression of Mint 1, Mint 2 or Fe65 causes reduction in Aβ generation and deposition in the brains of transgenic mice, suggesting a physiological role for these adaptors in regulating amyloidogenic processing of APP in the nervous system.

2.3   Axonal transport

Neurons are unique in their morphology with a long axonal compartment and a rich dendritic arbor that have to be sustained bioenergetically almost entirely from the perikaryon. Protein processing and trafficking are therefore often modified in neurons, just as can happen in polarized epithelial cells. Indeed, the axonal compartment has been compared to the apical compartment while the somatodendritic compartment may be functionally analogous to the basolateral compartment in epithelial cells. Accordingly, APP trafficking in neurons and epithelial cells take on an extra layer of complexity (Haass et al., 1994). Further, neurons are believed to be the major source of Aβ in brain, an idea supported by the observation that APP expression is highest in neurons. Therefore, if amyloid deposits are deposited at sites removed from the neuronal cell body, then APP or Aβ must be axonally transported form the perikaryon to distal processes. Indeed, APP is transported in axons via the fast anterograde transport machinery such that at least one documented source of amyloid deposits originate from synaptically released Aβ pool (Koo et al., 1990; Lazarov et al., 2002). Because anterograde transport of APP requires conventional kinesin, it is not surprising that APP has been found in complexes with kinesin light chain (KLC) subunit, a component of the kinesin-1 transport machinery (Kamal et al., 2000). Indeed, it has been shown that overexpression of the *Drosophila* APP homolog, APPL, in *Drosophila* neurons disrupts axonal transport, a phenotype similar to that seen in flies lacking components of the kinesin motor (Torroja et al., 1999). Taken together, these findings led to the hypothesis that APP may represent a kinesin cargo receptor, linking kinesin-1 to a unique subset of transport vesicles because different motor proteins are known to carry different membranous cargos. This model is consistent with the observation that the microtubules that carry APP anterogradely in axons are different from the transport carrier of synaptophysin (Kaether et al., 2000). However, enrichment of APP in Rab 5-positive vesicles from synaptosomal preparations, but not in synaptic vesicles, likely reflects APP sorting after internalization from the axonal plasmalemma and is probably not indicative of anterograde transport (Marquez-Sterling et al., 1997). APP was reported to interact directly with KLC but recent evidence is more consistent with the view that the interaction is mediated indirectly through adaptor proteins, of which JIP-1, a member of the JNK-interacting protein family (JIP), is a likely candidate as it is known to interact with both KLC and APP (King and Turner, 2004). Taken together, the data suggest that while still embedded within the membrane of a cargo vesicle, APP interacts with KLC either directly or more likely indirectly, to facilitate the anterograde movement of the membranous cargo along the axon. Unresolved by this model is how APP is initially sorted into a particular class of vesicles. The potential importance of the initial sorting of APP is underscored by the report that BACE1 and presenilins are contained within the same kinesin-1 dependent APP transport vesicles (Kamal et al., 2001). This finding led to the suggestion that not only is APP required for the delivery of the enzymatic machinery necessary for $A\beta$ production, but $A\beta$ generation also occurs enroute from the cell body to the nerve terminals within the transport cargo that is carried by APP. However, the report that APP is a kinesin-1 receptor and a common vesicular compartment carried all the processing machinery necessary for $A\beta$ generation has not been confirmed by others (Lazarov et al., 2005). Nevertheless, KLC deficient animals, when crossed with APP transgenic mice, showed axonal pathology manifested by axonal swellings and increased amyloid levels and deposits in brain (Stokin et al., 2005). The latter argue that perturbations of axonal transport during aging may predispose to the development of AD pathology. This suggestion is in line with observations that disruption of slow axonal transport is associated with neuronal death in animal models of motor neuron disease (LaMonte et al., 2002).

2.4 Raft association of secretases and amyloidogenic processing

Growing evidence indicates a functional relationship between cellular cholesterol level and efficiency of amyloidogenic processing. Cholesterol depletion of cultured cells by lovastatin treatment and methyl-β-cyclodextrin extraction inhibits APP processing by BACE1 and lowers $A\beta$ production (Simons et al., 1998; Ehehalt et al., 2003). Furthermore, the above treatments stimulate non-amyloidogenic processing of APP by α-secretase ADAM10 (Kojro et al., 2001), raising the intriguing possibility that cholesterol levels may determine the balance between amyloidogenic and non-amyloidogenic processing of APP. However, moderate, but not complete, reduction of cholesterol leads to increased amyloidogenesis in neuronal cells (Abad-Rodriguez et al., 2004). Thus, cholesterol regulation of APP secretases may be more complex than previously understood.

Evidence from a variety of *in vitro* and *in vivo* studies indicates that specialized membrane microdomains termed lipid rafts, which are rich in cholesterol and sphingolipids, might be the critical link between cholesterol levels and amyloidogenic processing of APP. Lipid rafts function in the trafficking of proteins in the secretory and endocytic pathways in epithelial cells and neurons, and participate in a number of important biological functions (Simons and Toomre, 2000). Disruption of rafts by depletion of cellular sphingolipids increases secretion of APPsα and Aβ42, but not Aβ40 (Sawamura et al., 2004), suggesting that at least certain aspects of amyloidogenic processing of APP can be modulated by raft microdomains. High *versus* low cholesterol or sphingolipid depletion may cause selective alterations in the association of APP and secretases with cholesterol-rich membrane domains (discussed below), thus causing the apparent discrepancy in the outcome of APP processing.

Lipid rafts are biochemically defined as detergent-insoluble membrane (DIM) domains that resist extraction with cold non-ionic detergents such as Triton X100, Brij-96, and Lubrol WX. These microdomains are formed by lateral association of sphingolipids and cholesterol in the Golgi, and are present in the plasma membrane and other intracellular organelles such as endosomes and the TGN. As mentioned above, the BACE1 cytoplasmic tail undergoes palmitoylation, a post-translational modification that targets proteins to lipid rafts. Indeed, a significant fraction of BACE1 is localized in lipid raft microdomains in a cholesterol-dependent manner, and addition of a GPI-anchor to target BACE1 exclusively to lipid rafts increases APP processing at the β-cleavage site (Riddell et al., 2001; Cordy et al., 2003). Elegant studies by Simons and colleagues showed antibody-mediated co-patching of cell surface APP and BACE1 as well as provided evidence for amyloidogenic processing of APP in raft microdomains (Ehehalt et al., 2003). These observations are consistent with the paucity of full-length APP in raft microdomains at steady state, and the preferential accumulation of APP CTFs in adult brain and cultured cells in raft microdomains, until they can be further processed. Indeed, all four components of the γ-secretase complex (PS1 derived N- and C-terminal fragments, nicastrin, APH-1, and PEN-2) also associate with DIM fractions enriched in lipid raft markers such as caveolin, flotillin, PrP, and ganglioside GM1 (Vetrivel et al., 2004). Consistent with the typical behaviour of *bona fide* raft-resident proteins, association of γ-secretase components with DIM is also sensitive to cholesterol depletion from cellular membranes. In contrast to BACE1 and the γ-secretase complex, α-secretases have not been linked to raft microdomains based on cholesterol depletion/loading studies

(Ehehalt et al., 2003). Thus, mounting evidence suggest that lipid rafts may be the principal membrane platforms where amyloidogenic processing of APP occurs.

Detailed biochemical fractionation and magnetic immunoisolation studies indicate that active and mature components of γ-secretase complex co-reside in lipid raft microdomains with SNARES such as VAMP-4 (TGN), syntaxin 6 (TGN and vesicles) and syntaxin 13 (early endosomes) (Vetrivel et al., 2004). These findings strongly implicate lipid raft microdomains of intracellular organelles as the preferred sites of amyloidogenic processing. Interestingly, the cell-surface raft associated protein, SNAP-23 does not co-reside with mature components of γ-secretase complex (Vetrivel et al., 2004), raising the possibility that the relatively low level of active γ-secretase complex at the cell surface could be associated with non-raft membrane domains and also explain the apparent low level of activity at the cell surface. Such spatially distinct localization of the γ-secretase allows for intramembrane processing of diverse substrates. Indeed, unlike APP, CTFs derived from several other γ-secretase substrates such as Notch1, Jagged2, N-cadherin, and DCC largely remain in non-raft membranes (Vetrivel et al., 2005). Taken together, these findings are consistent with the prediction that γ-secretase cleavage of APP occurs in lipid rafts. Further investigations are needed to address how the components of the γ-secretase and APP are recruited into raft microdomains, and clarify whether genetic mutations in APP, PS1, and PS2 modulate Aβ42 production by affecting the localization and processing of APP in lipid rafts.

3. APP function

3.1 Trophic properties

While a number of physiological roles have been attributed to APP, some unique to certain isoforms, the in vivo function(s) of the molecule remain unclear. The literature covering APP function is extensive and cannot be reviewed comprehensively (Mattson, 1997). Suffice to say that a number of functional domains have since been mapped to the extra- and intracellular region of APP. These include metal (copper and zinc) binding, extracellular matrix components (heparin, collagen, and laminin), neurotrophic and adhesion domains, and protease inhibition (Kunitz protease inhibitor domain present in APP751 and APP770 isoforms). One of the earliest indication of APP function came from assessing growth pattern of fibroblasts where APP levels were decreased by expression of an antisense *APP* construct (Saitoh et al., 1989). These cells grew slowly but the growth retardation can be restored by treatment with secreted APPs. The active domain was subsequently mapped to a pentapeptide domain "RERMS" near the middle of the extracellular domain (positions 403-407)

(Ninomiya et al., 1993). The activity is not limited to fibroblasts as infusion of this pentapeptide as well as APPs into brain resulted in increased synaptic density and improved memory retention in animals (Roch et al., 1994; Meziane et al., 1998). Because, as mentioned above, APPs is constitutively released from cells following α-secretase cleavage, these findings indicated that APP has autocrine and paracrine functions in growth regulation.

In all, a trophic role for APP has been perhaps the most consistently and arguably the best established function for the molecule. APP has been shown to stimulate neurite outgrowth in a variety of experimental settings. This phenotype is compatible with the upregulation of APP expression during neuronal maturation (Hung et al., 1992). The N-terminal heparin-binding domain of APP (residues 28-123), just upstream from the "RERMS" sequence, also stimulates neurite outgrowth and promotes synaptogenesis. Interestingly, the crystal structure of this domain shows similarities to known cysteine-rich growth factors (Rossjohn et al., 1999). Conversely, injection of APP antibodies directly into the brain led to impairment in behavioral tasks in adult rat (Meziane et al., 1998). Finally, a recent report indicated the presence of binding sites for APPs in epidermal growth factor (EGF)-responsive neural stem cells in the subventricular zone in the adult rodent brain (Caille et al., 2004). In this context, APPsα acts in concert with EGF to stimulate the proliferation of these cells both in neurospheres in culture and *in vivo*. However, APPs is necessary, but not sufficient, for full activity, as it appears to act as a co-factor with EGF. If these findings are true in human brain, then the reduction in APPs levels in cerebrospinal fluid of individuals with AD may indicate the loss of additional trophic activity in AD, together with the reduction of other growth factors in brain (see Chapter 15).

3.2   Cell adhesion

An "RHDS" motif near the extralumenal portion of APP or at the C-terminus of APPs that is contained within the Aβ region appear to promote cell adhesion. It is believed that this region acts in an integrin-like manner and can, accordingly, be blocked by RGDS peptide sequence derived from the fibronectin-binding domain (Ghiso et al., 1992). Similarly, APP colocalizes with integrins on the surface of axons and at sites of adhesion (Storey et al., 1996; Yamazaki et al., 1997). Evidence of interaction with laminin and collagen provides further evidence of adhesion promoting properties. Interestingly, because the RHDS sequence is contained within the N-terminus of Aβ (residues 4-7), similar cell adhesive promoting properties have also been attributed to Aβ peptide itself. This latter property, however, is difficult to tease out in view of the cytotoxicity of Aβ peptide when tested in a variety cell systems *in vitro*. Furthermore, it is difficult to separate the cell adhesive from the neurite outgrowth promoting roles of APP. Clearly, these are probably somewhat inseparable, as neuronal migration, neurite outgrowth, and even synaptogenesis would involve substrate adhesion. The phenotype of APP and APLP-deficient animals are certainly in agreement with these proposed physiological activity of these molecules (see below).

3.3   Is APP a receptor?

Although APP was initially proposed to act as a cell surface receptor, the evidence supporting this idea has been unconvincing. Aside from interactions with extracellular matrix proteins, only recently has a candidate ligand been proposed. It was reported that F-spondin, a neuronally secreted signaling glycoprotein that may function in neuronal development and repair, binds to the extracellular domain of APP as well as APLP1 and APLP2 (Ho and Sudhof, 2004). This binding reduces β-secretase cleavage of APP and nuclear transactivation of AICD (see below), suggesting therefore that F-spondin may be a ligand that regulates APP processing.

As mentioned above, γ-secretase processing of APP also releases an intracellular domain of APP, termed AICD (Figure 1). This processing step is not unique for APP, and indeed may be a rather generalized phenomenon whereby membrane anchored proteins are cleaved to either release cytosolic fragments that participate in cell signaling, as in the Notch receptor, or for degradation. Because APP undergoes the same γ-secretase membrane proteolysis as Notch, the analogy to Notch is simply too tempting or obvious, even though the evidence that APP is itself a cofactor for transcriptional activation within the nucleus, remains to be firmly established. Using a heterologous signalling reporter system, AICD can form a transcriptionally active complex together with two other molecules, Fe65 and Tip60 (Cao and Sudhof, 2001). Although it was initially felt that AICD must enter the nucleus with Fe65, subsequent study showed that nuclear translocation of AICD is not required but may be indirect through Fe65 (Cao and Sudhof, 2004). An alternative approach to address this question is to look for AICD activated candidate genes. In this regards, two genes have been proposed to date, KAI1, a tumour suppressor gene, and neprilysin, a neutral endopeptidase with Aβ degrading activity (Baek et al., 2002; Pardossi-Piquard et al., 2005). The latter pathway is particularly interesting because it suggests that γ-secretase release of AICD can regulate the degradation of Aβ in the extracellular space. If this is true, it will be important to know the feedback pathways that modulate γ-secretase activity to regulate neprilysin expression.

3.4   APP-deficient animals

In view of the above discussion, it is perhaps a little surprising then that with so many functions attributed to APP, the initial phenotype of APP-deficient mice obtained by gene-targeting was rather unrevealing (Zheng et al., 1995). These mice were lighter in body mass and with age; there was weakness in the extremities. Examination of the brain revealed gliosis only, a rather non-specific astrocytic reaction. Postnatal growth deficit was also noted in the APLP1-deficient mice but APLP2-deficient mice demonstrated no apparent phenotype (von Koch et al., 1997; Heber et al., 2000). Interestingly, $APLP2^{-/-}/APLP1^{-/-}$ and $APP^{-/-}/APLP2^{-/-}$ double mutants, but not $APP^{-/-}/APLP1^{-/-}$ animals, showed early postnatal lethality, indicating that members of the APP gene family are essential genes that exhibit partial overlapping functions. Curiously, the histopathological phenotype of the animals that displayed early lethality was rather bland by initial descriptions. Similarly, neurons cultured from these animals were unaltered in their basal growth rates or response to excitotoxicity. However, in the peripheral nervous system, $APP^{-/-}/APLP2^{-/-}$ double knockout animals exhibited poorly formed neuromuscular junction with reduced apposition of pre- and postsynaptic elements of the junctional synapses (Wang et al., 2005). The number of synaptic vesicles at the presynaptic terminals were also reduced, a finding confirmed by defective neurotransmitter release. With knowledge of the neuromuscular junction phenotypes of $APP^{-/-}/APLP2^{-/-}$ double knockout mice in mind, examination of the parasympathetic submandibular ganglia of these animals also showed a reduction in active zone size, synaptic vesicle density, and number of docked vesicles per active zone (Yang et al., 2005). This function of APP/APLP is evolutionarily conserved, as evidenced by the decreased number of synaptic boutons in neuromuscular junction of *Drosophila* larvae lacking *APPL*, and involves interaction of APPL with the cytosolic adaptor Mint and a transmembrane cell adhesion molecule named Fasciclin II (Ashley et al., 2005).

Deficiency of all three *APP* genes led to death shortly after birth. The majority of the animals showed cortical dysplasia suggestive of migrational abnormalities of the neuroblasts and partial loss of cortical Cajal Retzius cells (Herms et al., 2004). Taken together, the recent findings presented a convincing picture that members of the *APP* gene family play essential roles in the development of the nervous system relating to synapse structure and function, as well as in neuronal migration. Whether these abnormalities underlie the early postnatal survival of the animals remain to be established. Further, whether these activities are due to mechanical properties or mediated by activating signaling pathways, or both, are interesting questions that remain to be elucidated.

4.   Phenotype of excess APP

In view of the trophic properties of APP, it would be natural to predict that overexpression of APP would demonstrate phenotypes related to the enhanced neurite outgrowth, enhanced cell growth, etc. Indeed, many studies have reported such findings (Leyssen et al., 2005). Surprisingly and more interestingly, however, convincing negative phenotypes have also been reported. Overexpression of APP in cells induced to differentiate into neurons led to cell death (Yoshikawa et al., 1992). *In vivo*, genetic engineering to overexpress APP carrying various familial AD mutations in transgenic mice resulted in the development of amyloid deposition and amyloid associated changes in brain, including loss of synaptic markers, confirming the pathogenic nature of these mutations. Careful examination also showed axonal swellings and varicosities, months before any evidence of amyloid deposition or amyloid associated pathology in brain (Stokin et al., 2005). Perhaps the best example of the consequences of APP overexpression is trisomy 21 in humans and trisomy 16 in mice, the latter containing many of the cognate human chromosome 21 orthologous genes. Individuals with Down's syndrome (DS) who live beyond the 3rd decade of life almost invariably develop histopathology indistinguishable from AD (Burger and Vogel, 1973). *APP* is present in three copies in trisomy 21 and this excess gene dosage leads to early elevation of $A\beta$ levels, even in brains of fetuses (Teller et al., 1996). Several lines of evidence support the concept that *APP* gene triplication is necessary and possibly even sufficient to cause the AD histopathology in DS individuals. First, fine mapping of genes duplicated in several individuals with partial trisomy, where some but not all chromosome 21 genes are triplicated, excluded *APP* and *SOD1* genes in generating classical features of DS (Korenberg et al., 1990). Second, a remarkable case report of a 78 year-old woman with DS features due to partial trisomy 21 who at postmortem examination did not have any of the expected AD pathological changes in brain (Prasher et al., 1998). The segment of the chromosome that was triplicated in this individual excluded the *APP* gene, thereby confirming that *APP* or possibly genes immediately adjacent to *APP* is necessary for the development of AD histopathology. Third, the segmental trisomy 16 mouse (Ts65Dn), a genetic model of DS, shows physiological and structural abnormalities that are in common with human DS. For example, vesicular enlargements in neuronal perikarya containing endosomal markers (Rab5, EEA1, etc.) present in AD and DS individuals are also seen in this mouse model. Interestingly, these changes can be reversed if the *APP* gene dosage was reduced back to the euploid state when the Ts65Dn mice were crossed to the APP-deficient animals, showing that these changes are uniquely due to *APP* gene dosage (Cataldo et al., 2003). Finally, the studies from DS cases have led to some to suggest that AD may be caused by triplication of *APP* in brain. However, generalized *APP* gene triplication appears to be excluded in AD but somatic aneuploidy remains a distinct possibility (Yang et al., 2003; Rehen et al., 2005). If true, then this intriguing idea can certainly provide a plausible mechanism for the development of sporadic AD.

5.   Summary

This review has covered some of the salient aspects of APP biology, concentrating on the recent advances in processing, trafficking, and function of APP and the related APLP1 and APLP2 members. The importance of APP in AD clearly lies in its role as precursor to the Aβ peptide that plays a central role in the amyloid hypothesis. However, APP has a number of additional biological activities, some of which impact neuronal development and function. Growing evidence suggests that perturbations of some of these activities may also contribute to AD pathogenesis and neurodegeneration. As such, it will be important to continue to investigate the normal function of APP.

References

Abad-Rodriguez J, Ledesma MD, Craessaerts K, Perga S, Medina M, Delacourte A, Dingwall C, De Strooper B, Dotti CG (2004) J Cell Biol 167:953–960.
Allinson TM, Parkin ET, Turner AJ, Hooper NM (2003) J Neurosci Res 74:342–352.
Ando K, Iijima KI, Elliott JI, Kirino Y, Suzuki T (2001) J Biol Chem 276:40353–40361.
Ashley J, Packard M, Ataman B, Budnik V (2005) J Neurosci 25:5943–5955.
Baek SH, Ohgi KA, Rose DW, Koo EH, Glass CK, Rosenfeld MG (2002) Cell 110:55–67.
Burger PC, Vogel FS (1973) Am J Pathol 73:457–476.
Buxbaum JD, Thinakaran G, Koliatsos V, O'Callahan J, Slunt HH, Price DL and Sisodia SS (1998) J Neurosci 18:9629–9637.
Caille I, Allinquant B, Dupont E, Bouillot C, Langer A, Muller U, Prochiantz A (2004) Development 131:2173–2181.
Cao X, Sudhof TC (2001) Science 293:115–120.
Cao X, Sudhof TC (2004) J Biol Chem 279:24601–24611.
Cataldo AM, Petanceska S, Peterhoff CM, Terio NB, Epstein CJ, Villar A, Carlson EJ, Staufenbiel M, Nixon RA (2003) J Neurosci 23:6788–6792.
Chyung JH, Raper DM, Selkoe DJ (2004) J Biol Chem 280:4383–4392.
Cook DG, Forman MS, Sung JC, Leight S, Kolson DL, Iwatsubo T, Lee VM, Doms RW (1997) Nature Med 3:1021–1023.
Cordy JM, Hussain I, Dingwall C, Hooper NM, Turner AJ (2003) Proc Natl Acad Sci USA 100:11735–11740.
Coulson EJ, Paliga K, Beyreuther K, Masters CL (2000) Neurochem Int 36:175–184.
Edbauer D, Winkler E, Regula JT, Pesold B, Steiner H, Haass C (2003) Nat Cell Biol 5:486–488.
Ehehalt R, Keller P, Haass C, Thiele C, Simons K (2003) J Cell Biol 160:113–123.
Ghiso J, Rostagno A, Gardella JE, Liem L, Gorevic PD, Frangione B (1992) Biochem J 288 (Pt 3):1053–1059.

Greenfield JP, Tsai J, Gouras GK, Hai B, Thinakaran G, Checler F, Sisodia SS, Greengard P, Xu H (1999) Proc Natl Acad Sci USA 96:742–747.

Haass C, Koo EH, Teplow DB, Selkoe DJ (1994) Proc Natl Acad Sci USA 91:1564–1568.

Hardy J (1997) Trends Neurosci 20:154–159.

Heber S, Herms J, Gajic V, Hainfellner J, Aguzzi A, Rulicke T, von Kretzschmar H, von Koch C, Sisodia S, Tremml P, Lipp HP, Wolfer DP, Muller U (2000) J Neurosci 20:7951–7963.

Herms J, Anliker B, Heber S, Ring S, Fuhrmann M, Kretzschmar H, Sisodia S, Muller U (2004) Embo J 23:4106–4115.

Hill K, Li Y, Bennett M, McKay M, Zhu X, Shern J, Torre E, Lah JJ, Levey AI, Kahn RA (2003) J Biol Chem 278:36032–36040.

Ho A, Sudhof TC (2004) Proc Natl Acad Sci USA 101:2548–2553.

Hung AY, Koo EH, Haass C, Selkoe DJ (1992) Proc Natl Acad Sci USA 89:9439–9443.

Iwatsubo T (2004) Curr Opin Neurobiol 14:379–383.

Kaether C, Lammich S, Edbauer D, Ertl M, Rietdorf J, Capell A, Steiner H, Haass C (2002) J Cell Biol 158:551–561.

Kaether C, Skehel P, Dotti CG (2000) Mol Biol Cell 11:1213–1224.

Kamal A, Almenar-Queralt A, LeBlanc JF, Roberts EA, Goldstein LS (2001) Nature 414:643–648.

Kamal A, Stokin GB, Yang Z, Xia CH, Goldstein LS (2000) Neuron 28:449–459.

Khvotchev M, Sudhof TC (2004) J Biol Chem 279:47101–47108.

King GD, Turner RS (2004) Exp Neurol 185:208–219.

Kojro E, Gimpl G, Lammich S, Marz W, Fahrenholz F (2001) Proc Natl Adac Sci Usa 98:5815–5820.

Koo EH, Kopan R (2004) Nature Med 10 Suppl:S26–33.

Koo EH, Sisodia SS, Archer DR, Martin LJ, Weidemann A, Beyreuther K, Fischer P, Masters CL, Price DL(1990) Proc Natl Adac Sci USA 87:1561–1565.

Koo EH, Squazzo SL (1994) J Biol Chem 269:17386–17389.

Kopan R, Ilagan MX (2004) Nat Rev Mol Cell Biol 5:499–504.

Korenberg JR, Kawashima H, Pulst SM, Allen L, Magenis E, Epstein CJ (1990) Am J Med Genet Suppl 7:91–97.

LaMonte BH, Wallace KE, Holloway BA, Shelly SS, Ascano J, Tokito M, Van Winkle T, Howland DS, Holzbaur EL (2002) Neuron 34:715–727.

Lazarov O, Lee M, Peterson DA, Sisodia SS (2002) J Neurosci 22:9785–9793.

Lazarov O, Morfini GA, Lee EB, Farah MH, Szodorai A, DeBoer SR, Koliatsos VE, Kins S, Lee VM, Wong PC, Price DL, Brady ST, Sisodia SS (2005) J Neurosci 25:2386–2395.

Lee EB, Zhang B, Liu K, Greenbaum EA, Doms RW, Trojanowski JQ, Lee VM (2005) J Cell Biol 168:291–302.

Lee MS, Kao SC, Lemere CA, Xia W, Tseng HC, Zhou Y, Neve R, Ahlijanian MK, Tsai LH (2003) J Cell Biol 163:83–95.

Levy E, Carman MD, Fernandez-Madrid IJ, Power MD, Lieberburg I, van Duinen SG, Bots GT, Luyendijk W, Frangione B (1990) Science 248:1124–1126.

Leyssen M, Ayaz D, Hebert SS, Reeve S, De Strooper B, Hassan BA (2005) Embo J 24:2944–2955.

Marquez-Sterling NR, Lo ACY, Sisodia SS, Koo EH (1997) J Neurosci 17:140–151.

Mattson MP (1997) Physiol Rev 77:1081–1132.

Meziane H, Dodart JC, Mathis C, Little S, Clemens J, Paul SM, Ungerer A (1998) ProcNatl Adac Sci Usa 95:12683–12688.

Ninomiya H, Roch JM, Sundsmo MP, Otero DA, Saitoh T (1993) J Cell Biol 121:879–886.

Pardossi-Piquard R, Petit A, Kawarai T, Sunyach C, Alves da Costa C, Vincent B, Ring S, D'Adamio L, Shen J, Muller U, St George Hyslop P, Checler F (2005) Neuron 46:541–554.

Parent AT, Barnes NY, Taniguchi Y, Thinakaran G, Sisodia SS (2005) J Neurosci 25:1540–1549.

Perez RG, Soriano S, Hayes JD, Ostaszewski B, Xia W, Selkoe DJ, Chen X, Stokin GB, Koo EH (1999) J Biol Chem 274:18851–18856.

Pietrzik CU, Yoon IS, Jaeger S, Busse T, Weggen S, Koo EH (2004) J Neurosci 24:4259–4265.

Prasher VP, Farrer MJ, Kessling AM, Fisher EM, West RJ, Barber PC, Butler AC (1998) Ann Neurol 43:380–383.

Rehen SK, Yung YC, McCreight MP, Kaushal D, Yang AH, Almeida BS, Kingsbury MA, Cabral KM, McConnell M J, Anliker B, Fontanoz M, Chun J (2005) J Neurosci 25:2176–2180.

Riddell DR, Christie G, Hussain I, Dingwall C (2001) Curr Biol 11:1288–1293.

Roch JM, Masliah E, Roch-Levecq AC, Sundsmo MP, Otero DA, Veinbergs I, Saitoh T (1994) ProcNatl Adac Sci Usa 91:7450–7454.

Rossjohn J, Cappai R, Feil SC, Henry A, McKinstry WJ, Galatis D, Hesse L, Multhaup G, Beyreuther K, Masters CL, Parker MW(1999) Nature Struct Bio 6:327–331.

Saitoh T, Sundsmo M, Roch JM, Kimura N, Cole G, Schubert D, Oltersdorf T, Schenk DB (1989) Cell 58:615–622.

Sawamura N, Ko M, Yu W, Zou K, Hanada K, Suzuki T, Gong JS, Yanagisawa K, Michikawa M (2004) J Biol Chem 279:11984–11991.

Simons K, Toomre D (2000) Nat Rev Mol Cell Biol 1:31–39.

Simons M, Keller P, De Strooper B, Beyreuther K, Dotti CG, Simons K (1998) Proc Natl Adac Sci Usa 95:6460–6464.

Sisodia SS (1992) Proc Natl Adac Sci Usa 89:6075–6079.

Sisodia SS, Kim SH, Thinakaran G (1999) Am J Hum Genet 65:7–12.

Stokin GB, Lillo C, Falzone TL, Brusch RG, Rockenstein E, Mount SL, Raman R, Davies P, Masliah E, Williams DS, Goldstein LS(2005) Science 307:1282–1288.

Storey E, Spurck T, Pickett-Heaps J, Beyreuther K, Masters CL (1996) Brain Res 735:59–66.

Takahashi RH, Milner TA, Li F, Nam EE, Edgar MA, Yamaguchi H, Beal MF, Xu H, Greengard P, Gouras GK(2002) Am J Pathol 161:1869–1879.

Teller JK, Russo C, DeBusk LM, Angelini G, Zaccheo D, Dagna-Bricarelli F, Scartezzini P, Bertolini S, Mann DM, Tabaton M, Gambetti P (1996) Nature Med 2:93–95.

Thinakaran G, Parent AT (2004) Pharmacol Res 50:411–418.

Torroja L, Chu H, Kotovsky I, White K (1999) Curr Biol 9:489–492.

Vassar R (2004) J Mol Neurosci 23:105–114.

Vetrivel, KS, Cheng H, Kim SH, Chen Y, Barnes NY, Parent AT, Sisodia SS, Thinakaran G (2005) J Biol Chem 280:25892–25900.

Vetrivel KS, Cheng H, Lin W, Sakurai T, Li T, Nukina N, Wong PC, Xu H, Thinakaran G (2004) J Biol Chem 279:44945–44954.

von Koch CS, Zheng H, Chen H, Trumbauer M, Thinakaran G, van der Ploeg LH, Price DL, Sisodia SS (1997) Neurobiol Aging 18:661–669.

Wang P, Yang G, Mosier DR, Chang P, Zaidi T, Gong YD, Zhao NM, Dominguez B, Lee KF, Gan WB, Zheng H (2005) J Neurosci 25:1219–1225.

Xu H, Sweeney D, Wang R, Thinakaran G, Lo AC, Sisodia SS, Greengard P, Gandy S (1997) Proc Natl Adac Sci Usa 94:3748–3752.

Yamazaki T, Koo EH, Selkoe DJ(1997) J Neurosci 17:1004–1010.

Yang G, Gong YD, Gong K, Jiang WL, Kwon E, Wang P, Zheng H, Zhang XF, Gan WB, Zhao NM (2005) Neurosci Lett 384:66–71.

Yang Y, Mufson EJ, Herrup K (2003) J Neurosci 23:2557–2563.

Yoshikawa K, Aizawa T, Hayashi Y (1992) Nature 359:64–67.

Zheng H, Jiang M, Trumbauer ME, Sirinathsinghji DJ, Hopkins R, Smith DW, Heavens RP, Dawson GR, Boyce S, Conner MW, Stevens KA, Slunt HH, Sisoda SS, Chen HY, Van der Ploeg LH (1995) Cell 81:525–531.

Chapter 3

The Biology of the Presenilin Complexes

Tomoko Wakabayashi[1], Takeshi Iwatsubo[2] and Bart De Strooper[1]

[1] *Laboratory for neuronal cell biology and gene transfer*
Center for Human Genetics, KU Leuven and VIB
Herestraat 49, 3000 Leuven, Belgium
Email: Bart.DeStrooper@med.kuleuven.be

[2] *Department of Neuropathology and Neuroscience*
Graduate School of Pharmaceutical Sciences
University of Tokyo
7-3-1 Hongo, Bunkyo-ku, Tokyo, Japan
Email: iwatsubo@mol.f.u-tokyo.ac.jp

1.　　Introduction

1.1　　APP proteolytic processing

Alzheimer's Disease (AD) is characterized by the deposition of two kinds of abnormal protein aggregates, senile plaques and neurofibrillary tangles, and by neuronal dysfunction and cell loss in the brain. Senile plaques are primarily composed of extracellular deposits of hydrophobic 37-43 amino acid Aβ peptides. Aβ peptides are derived by successive enzymatic cleavages of the type I membrane protein, β-amyloid precursor protein (APP) (Haass and Selkoe 1993). APP is first cleaved close to the membrane in the extracellular domain by either α- or β-secretase, resulting in a release of soluble APP ectodomains, and residual membrane-tethered C-terminal protein stubs, termed C83 or C99, respectively (The numbers indicate the length of each carboxyl-terminal fragment). C83 and C99 are substrates for γ-secretase, an activity that generates p3 and Aβ peptides, respectively. γ-Secretase processes substrates at different positions within the membrane domain and thus, both Aβ and p3 have "ragged" termini. Aβ has been best studied in this regard and species between 37 and 43 amino acid residues have been identified. γ-Secretase cleavage of APP also releases the intracellular carboxy-terminal "APP intracellular domain" or "AICD". The function of both $A\beta$ and AICD is the subject of intense investigations.

Because $A\beta42$ is the primary constituent of the amyloid fibrils deposited in the AD brains, and mutations in APP and presenilin enhance the production of this peptide, γ-secretase cleavage of APP is a pivotal step in AD pathogenesis. It is striking that this proteolytic reaction occurs within the highly hydrophobic environment of the membrane.

1.2   Identification of presenilin

Genetic studies in familial AD (FAD) cases have identified disease-linked mutations in three genes that contribute to AD. The first pathogenic mutations in early-onset FAD families were found in the *APP* gene on chromosome 21 (Chartier-Harlin et al. 1991; Goate et al. 1991; Murrell et al. 1991). However, subsequent studies indicated that mutations in *APP* account only for a small fraction of FAD cases. Several genetic studies indicated a major locus for FAD on chromosome 14 in early onset autosomal dominant AD, and in 1995, the *Presenilin1 (PS1)* gene on chromosome 14 (14q24.3) was identified by positional cloning (Sherrington et al. 1995). Shortly thereafter, it was shown that mutations in the closely related *PS2* gene on chromosome 1 (1q42.2) could cause FAD as well (Levy-Lahad et al. 1995; Rogaev et al. 1995).

Studies in transgenic mice (Borchelt et al. 1996; Duff et al. 1996) and cultured cells (Citron et al. 1997; Scheuner et al. 1996; Tomita et al. 1997) have revealed that expression of FAD-linked PS variants elevates $A\beta42/A\beta40$ ratios. Moreover, transgenic mice that co-express FAD-mutant PS1 and APP develop amyloid plaques much earlier than age-matched mutant APP mice (Borchelt et al. 1997). Therefore, *PS* mutations cause a change in the $A\beta42/40$ ratio, but whether PS is directly involved in γ-secretase processing of APP was unclear. However, in PS-deficient neurons and fibroblasts, APP processing was greatly impaired, leading to the accumulation of the C83 and C99 APP fragments, the direct substrates of γ-secretase, and inhibition of $A\beta$ (and p3) generation (De Strooper et al. 1998; Xia et al. 1998). Thus, PS are directly required for γ-secretase cleavage of APP. Overall, the findings imply that mutations in the substrate (APP) or in the proteolytic machinery (PS) result in similar changes in $A\beta42$ generation (Scheuner et al. 1996). This provides very strong support for the "amyloid cascade hypothesis".

2. Molecular and cellular biology of the presenilins

The overall gene structures of PS1 and PS2 are similar, consisting of ten translated exons that undergo tissue-specific alternative splicing (Clark et al. 1995; Levy-Lahad et al. 1996). The biological importance of differentially spliced PS is still unclear. Analysis of mRNA expression revealed that both PS1 and PS2 are present in most human and mouse tissues but expressed at different levels (Hébert et al. 2004; Lee et al. 1996). In brain, PS is expressed predominantly in neuronal cells, while lower levels of PS are also detected in glial populations (Kovacs et al. 1996; Lee et al. 1996).

Human PS1 and PS2 polypeptides display 67% identity to each other. PS homologues were found across diverse species including plants, insects and vertebrates, indicating that PS plays essential roles in all multicellular organisms. PS1 and PS2 are highly hydrophobic proteins that consist of 467 and 448 amino acid residues, respectively. The membrane topology of the PS is still a matter of ongoing debate. Theoretical predictions identify seven to ten transmembrane domains, and no signal peptide (Sherrington et al. 1995). Experimental studies all agree that the first 6 hydrophobic domains span the membrane (De Strooper et al. 1997; Doan et al. 1996; Henricson et al. 2005; Laudon et al. 2005; Li and Greenwald 1996). Based on immunocytochemistry studies combined with selective permeabilization of cells or using hybrid SEL-12 (*C.elegans* presenilin) – LacZ fusion proteins, a model with eight trans-membrane domains has been proposed (Doan et al. 1996; Li and Greenwald 1996). In this model (Fig. 1A), the amino- and carboxyl-termini are both oriented towards the cytoplasm, and a relatively large hydrophilic loop between TM (transmembrane domain) 6 and TM7 also protrude into the cytoplasm. However, alternative models have been proposed (Fig.1B). Most recently, elegant experiments were performed introducing minimal modifications (such as short glycosylation tags and biotinylation experiments) in PS and expressing those proteins in a PS-negative background. These studies have proposed a nine transmembrane topology for presenilin, which is also supported by theoretical calculations (Henricson et al. 2005; Laudon et al. 2005; Spasic et al. 2006). This model indicates that the carboxy-terminus of PS is lumenal/extracellular (Fig. 1B).

PS1 and PS2 are unglycosylated polypeptides of $\sim$43 kDa and 50 kDa, respectively (De Strooper et al. 1997; Walter et al. 1996). Shortly after synthesis, full-length PS is cleaved by an unidentified "presenilinase activity" yielding an N-terminal fragment (NTF) and a C-terminal fragment (CTF) (Thinakaran et al. 1996). Proteolytic cleavage in PS1 occurs mainly between residue 291 and 292 in the large cytoplasmic loop region between TM6 and 7 (Podlisny et al. 1997). In cultured cells and tissues, NTF and CTF are the principal PS-related polypeptides that are observed, but low levels of full length PS

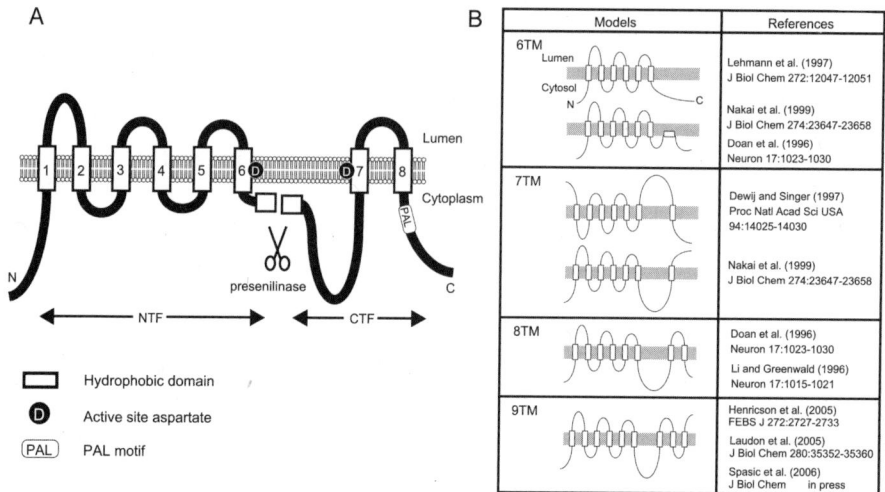

Figure 1. Transmembrane structure of presenilin. A. Eight transmembrane model of presenilin. Two putative active site aspartates in TM6 and TM7 are marked. The location of the endoproteolytic cleavage site by "presenilinase activity" within the large cytoplasmic loop is indicated. NTF: aminoterminal fragment, CTF: carboxyterminal fragment. B. Several alternative models are reported in the literature and schematically represented here.

are occasionally detected. In contrast, overexpression of PS in cultured cells results in the accumulation of full-length proteins without increasing NTF/CTF levels (De Strooper et al. 1997; Thinakaran et al. 1996). The half-life of the holoprotein species ($\sim$2 hours) is much shorter than that of the PS fragments (over 24 hours), indicating that only cleaved PS fragments become incorporated and stabilized in the γ-secretase complex. Full-length PS, on the other hand is unstable and degraded by a proteasome-mediated pathway (Kim et al. 1997; Steiner et al. 1998).

The subcellular localization of presenilins is also a matter of debate. Overexpression of presenilins leads to accumulation in ER and to a certain extent, the Golgi apparatus, and is present in "aggresomes" (De Strooper et al. 1997; Johnston et al. 1998). Overexpression data are difficult to interpret, because we now know that the localization of PS is also determined by its incorporation into the γ-secretase complex, and that this step is rate-limiting. In primary cultures of neuronal cells expressing presenilin at endogenous levels (Annaert et al. 1999), the bulk of immunoreactivity is observed in ER, and very little if any staining is observed in Golgi, or at the cell surface. However, a small pool of PS, likely incorporated in the active complex, is found at the cell surface (Kaether et al. 2002; Rechards et al. 2003) and endocytic/TGN compartments. It is believed that this latter fraction is responsible for the cleavage of APP (and probably other substrates as well). Less clear is the significance of PS in the lysosomal compartments and the mitochondria as reported in the literature (Hansson et al. 2004; Pasternak et al. 2003). γ-Secretase might also be localized in phagosomes, which could be relevant in the context of the overall function of γ-secretase in the turnover of transmembrane domains of integral membrane proteins (see below) (Jutras et al. 2005).

3.   Genetics of PS

To date, more than 140 autosomal-dominant mutations that cause early-onset AD have been found in *PS1*, while 10 have been identified in *PS2* (http://www.molgen.ua.ac.be/ADMutations/). PS1 mutations are more common than mutations in either APP or PS2. The mutations are, in general, missense mutations that result in amino acid substitutions at conserved positions in PS1 and PS2. One exception is a splice acceptor site mutation in exon 9, resulting in the in-frame deletion of exon 9 (the mutation referred to as Δ9) (Perez-Tur et al. 1995). The deletion on its own has little effect on γ-secretase, but the amino acid substitution (S290C) at the splice site is sufficient to cause an elevation in Aβ42 production (Steiner et al. 1999a).

The mutations in PS are scattered over the entire sequence, but have a slight preference for the transmembrane helices and exon 8. It is still unclear how the mutations in PS affect function: as ablating PS expression blocks Aβ generation, investigators have proposed that the clinical mutations in presenilin cause a gain of "toxic" or "abnormal" function. However, the FAD-linked PS mutants weakly rescue the PS deficient phenotypes in *C.elegans* (Baumeister et al. 1997; Levitan et al. 1996). Additional accumulating evidence in the literature indicates that, in general, FAD-linked PS variants are associated with partial loss of functions as assessed in a variety of assays (Baki et al. 2004; Leissring et al. 2000; Marambaud et al. 2003; Nishimura et al. 1999). On the other hand, it is clear that all mutations increase the Aβ42/Aβ40 ratio. However since this reflects a decrease in Aβ40 and an increase in Aβ42, these results do not necessarily imply that absolute amounts of Aβ are elevated (Walker et al. 2005). If Aβ42 is an inefficiently generated Aβ species, one could reconcile the data, at least conceptually, by proposing that PS mutations cause partial loss of function of the γ-secretase complex (hypomorphs), resulting in a protease that is biochemically inefficient. This interpretation is not necessarily in contradiction with rescue experiments in PS1 knock out mice showing that transgenic expression of FAD mutant PS1, can (partially) rescue the embryonic lethal phenotype (Qian et al. 1998; Davis et al. 1998).

Genetic studies have been also instrumental in the understanding of the biological function of PS. In *C.elegans* (Levitan and Greenwald 1995), *Drosophila melanogaster* (Struhl and Greenwald 1999; Ye et al. 1999) and *Mus musculus* (De Strooper et al. 1998; Shen et al. 1997; Wong et al. 1997), PS deficiencies cause phenotypes that are similar to those observed with Notch loss of function mutations. In fact, the *C.elegans* homologue of the human *PS* genes, *sel-12*, was identified independently from the work on the human genes, in a screen for genes capable of suppressing a *lin-12* (a member of the Notch receptor family) phenotype (Levitan and Greenwald 1995). Notch is a type I integral membrane protein that is constitutively processed by furin at site 1 (S1). The ligands of Notch receptors are members of the DSL (Delta/Serrate/Lag-1) family. While the (S1) cleavage occurs constitutively, a second (S2) and third (S3) cleavage only takes place upon binding to Notch ligands (Brou et al. 2000; Logeat et al. 1998; Mumm et al. 2000). The S3 cleavage releases the Notch intracellular domain (NICD) that is subsequently translocated to the nucleus. NICD modulates target gene expression via binding to the CSL (CBF1/Suppressor of Hairless/Lag-1) family of transcription factors (Kopan and Goate 2000).

While the genetics do not tell at what level PS influences the Notch signalling pathway, the mode of Notch processing is reminiscent of that of APP cleavage. Therefore, in studies similar to those performed with APP (De Strooper et al., 1998), Notch1 processing was investigated in neurons and other cells derived from *PS*-deficient mice (De Strooper et al. 1999). These studies revealed that S3 cleavage of Notch was blocked, indicating that PS was indeed responsible for the final cut in Notch (De Strooper et al. 1999; Struhl and Greenwald 1999). The S3 cleavage releases the Notch intracellular domain and results in signal transduction to the nucleus (Schroeter et al. 1998). These studies provided the first evidence for a direct role of PS1 in APP and Notch processing.

Knockout mice for the *PS1* homologue, *PS2*, were also generated. Unlike *PS1-/-* mice, *PS2-/-* mice were viable, fertile, and showed no abnormalities in APP metabolism. Limited pathology was observed in the lungs of aged animals (Donoviel et al. 1999; Herreman et al. 1999). However, loss of both *PS1* and *PS2* in mice led to early embryonic lethality, a severe Notch deficient phenotype, and a complete loss of Aβ production, suggesting that *PS2* and *PS1* have partially overlapping functions *in vivo* (Donoviel et al. 1999; Herreman et al. 1999; 2000; Steiner et al. 1999b).

4.   PS function in intramembrane proteolysis: Clearance of integral membrane stubs and signaling

4.1   PS is a protease

Aspartyl proteases require two catalytic aspartate residues to hydrolyze the peptide bond of a substrate. Based on this notion, Wolfe and colleagues identified two critical aspartate residues in transmembrane domain 6 and 7 (Wolfe et al. 1999a). Mutation of either or both of these aspartate residues in PS abrogated γ-secretase activity (Steiner et al. 1999b; Wolfe et al. 1999b). It is unusual to find charged amino acids within transmembrane sequences, but these two aspartates are completely conserved in PSs from all species. PS in which the aspartyl residues are mutated do not undergo endoproteolysis, cause accumulation of C83/C99 fragments, and reduce p3/Aβ production. Although the sequences around the two aspartates, YD and GLGD, do not fit with the motifs of canonical aspartic proteases (D(T/S)G(T/S)), the active site of PS is similar to the catalytic sites of the bacterial type-4 prepilin peptidase (TFPP) (Steiner et al. 2000). TFPP has an eight TM structure with active sites close to the transmembrane regions. Direct support for the hypothesis that PS were catalytical active proteases came from studies showing that several γ-secretase inhibitors bound to presenilin (Esler et al. 2000; Li et al. 2000a; Seiffert et al. 2000). In these studies, photoactivated forms of putative transition-state analogue inhibitors of γ-secretase specifically bound to NTF and CTF of presenilins, while full-length PS was not labelled, further suggesting that full length PS is an inactive precursor for the active dimeric form of PS. Thus, the catalytic centre of γ-secretase resides at the heterodimeric interface of PS fragments. Besides two aspartates within TM6 and TM7, there is also a highly conserved PAL motif in the C-terminus of all reported PSs. Mutations within this sequence abolish both the endoproteolysis and the enzymatic activity of PS (Tomita et al. 2001; Wang et al. 2004).

4.2   Presenilin is part of a larger γ-secretase complex

In detergent-solubilized membranes, γ-secretase activity resides within high molecular weight fractions (Capell et al. 1998; Yu et al. 1998). These fractions contain PS NTF/CTF, while holoprotein is detected in lower molecular weight fractions. Therefore, it seems that γ-secretase activity is mediated by a multiprotein complex. The exact size of the active complex remains an issue of debate. Early estimates based on gel filtration indicated Mr of above 2×10^6 kDa, the exclusion limit of the chromatography column used in those experiments (Li et al. 2000b). This is probably an overestimation due to incomplete solubilisation of the membranes in which the complex resides. More recent estimates are based on Blue Native gel electrophoresis and indicate

Mr ranging between 200 and 900 kDa (Edbauer et al. 2002; Evin et al. 2005; Yu et al. 1998).

Different approaches have led to the identification of three additional components of the complex. Nicastrin (Nct) was purified biochemically using antibodies against PS (Yu et al. 2000). Nct is a well conserved, type I integral membrane protein, synthesized as a ∼110 kDa immature protein and highly glycosylated to ∼130 kDa. In *C.elegans*, Nct was genetically identified as Aph-2 (anterior pharynx defective), which, like PS, is involved in the GLP-1 (*C.elegans* Notch) pathway (Goutte et al. 2000). Although the observations confirm the importance of Nct in γ-secretase activity, overexpression of both Nct and PS was not sufficient to enhance γ-secretase activity in cells.

Two additional hydrophobic membrane proteins that failed to be identified using classical mass spectophotometry methods were identified in genetic screens. Aph-1 and Pen-2 were identified as modifiers of GLP-1 signalling in *C. Elegans* (Francis et al. 2002; Goutte et al. 2002). Aph-1 is proposed to have a seven TM structure with the C-terminus located in the cytoplasm. In human, three Aph-1 isoforms (two splice variants of Aph-1a and one Aph-1b) have been identified. Pen-2 displays a hairpin-like structure with two transmembrane domains and both termini are located luminally (Crystal et al. 2003).

Antibodies against one of any four proteins (PS, Nct, Aph-1 and Pen-2) immunoisolate all four proteins, and the immunoprecipitates contain γ-secretase activity *in vivo* (Kimberly et al. 2003; Steiner et al. 2002; Takasugi et al. 2003). Furthermore, down regulation of Nct by RNA interference (RNAi) resulted in a loss of γ-secretase activity (Edbauer et al. 2002; Francis et al. 2002). Nct is also involved in the Notch signalling pathway and Aβ generation in fly and mouse (Chung and Struhl 2001; Edbauer et al. 2002; Hu et al. 2002; Li et al. 2003; Lopez-Schier and St Johnston 2002). Downregulation of Aph-1 or Pen-2 levels in cells causes a decline in γ-secretase activity, similar to that observed in cells with reduced PS or Nct levels (Francis et al. 2002; Lee et al. 2002; Takasugi et al. 2003). Thus, all four proteins are indispensable for APP and Notch processing. Indeed, expression of any combination of three proteins did not enhance γ-secretase activity, while the four proteins expressed together in mammalian or fly cells (Kimberly et al. 2003; Takasugi et al. 2003) cause increased γ-secretase cleavage of APP or Notch. In yeast cells that lack the four γ-secretase components, coexpression of all four human proteins reconstitutes γ-secretase activity (Edbauer et al. 2003). Thus, PS, Nct, Aph-1 and Pen-2 are the minimal components that constitute the core of the γ-secretase complex.

The functions of the individual components of the γ-secretase complex are not fully understood, but each polypeptide influences the maturation and stability of other molecules within the complex. For example, in the absence of PS, Nct does not leave the endoplasmic reticulum and Pen-2 protein levels decrease (Chen et al. 2003; Herreman et al. 2003; Leem et al. 2002). However,

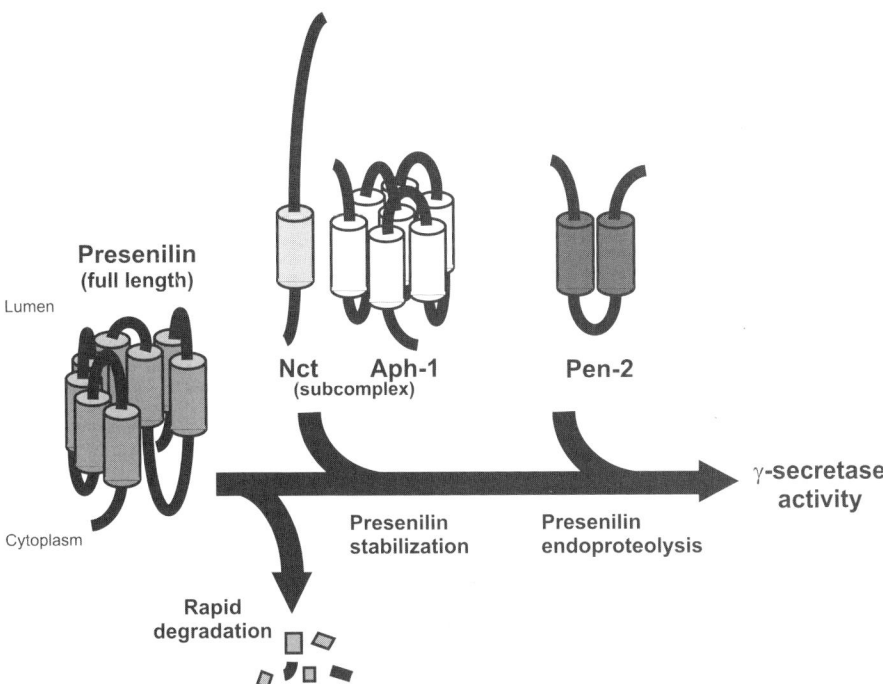

Figure 2. γ-Secretase components and the stepwise assembly and activation of the γ-secretase complex. Presenilin holoprotein (full-length) is rapidly degraded, whereas a fraction of presenilin is stabilized by binding to Nicastrin and Aph-1 (subcomplex). Subsequent binding of Pen-2 to the ternary complex induces the endoproteolysis of presenilin and confers γ-secretase activity. Cylindrical columns represent the putative transmembrane domains of each component.

Aph-1 is not markedly influenced by PS depletion (Lee et al. 2002; Gu et al. 2003). In the absence of Nct, PS and Pen-2 are destabilized (Chung and Struhl 2001; Edbauer et al. 2002; Hu et al. 2002; Steiner et al. 2002). The loss of Aph-1a results in decreased levels of PS, mature Nct and Pen-2. In addition to the decline in PS fragments and mature Nct levels, the loss of Pen-2 leads to the accumulation of full-length PS and immature Nct (Francis et al. 2002; Luo 2003; Takasugi et al. 2003). Apparently, immature Nct and Aph-1 form an intermediate complex (subcomplex) in the ER prior to the binding of PS and Pen-2 (Hu and Fortini 2003; La Voie et al. 2003). Overall, these studies suggest a hypothetical sequence for the assembly of the complex (Fig. 3).

First, immature Nct and Aph-1 form a subcomplex in the ER that binds and stabilizes full-length PS. Incorporation of Pen-2 into the PS/Nct/Aph-1 ternary complex stabilizes the whole complex and results in PS endoproteolysis, and activation of the γ-secretase complex (Takasugi et al. 2003). Nct is extensively glycosylated, but inhibitors of glycosylation do not interfere with complex assembly or γ-secretase activity (Herreman et al. 2003).

The stoichiometry of the γ-secretase components is still under discussion. The sum of the molecular weight of the individual subunits is 200–250 kDa. The higher molecular weight estimates reported in the literature suggest that either the complex forms multimers or additional proteins are in the complex. Some evidence suggests that the catalytic PS subunits can form homodimers (Hébert et al. 2003; Schroeter et al. 2003). The mode of interaction(s) of each subunit within the complex is not fully resolved, but the transmembrane domains are clearly important. A conserved GXXXG motif in Aph-1, known to facilitate transmembrane helix-helix interactions (Edbauer et al. 2004; Lee et al. 2004; Niimura et al. 2005), is important for the Aph1-Nicastrin interaction. The C-terminal residues of PS are also important for the generation of active γ-secretase (Tomita et al. 1999), probably because this region mediates interactions with the transmembrane domain of Nct (Kaether et al. 2004). Finally, deletion or replacement of the transmembrane domain of Nicastrin prevents incorporation into the complex, further corroborating the importance of hydrophobic interactions for the assembly of the complex (Capell et al. 2003; Morais et al. 2003; Shah et al. 2005).

The subcellular localization of the γ-secretase complex and its activity is also subject of intensive research. The discrepancy between the PS subcellular localization in the endoplasmic reticulum (see above) and γ-secretase activity is somewhat puzzling (Annaert et al. 1999). However, assembly and activation of the γ-secretase complex in the early secretory compartment can be demonstrated when Nct is retained in the ER with an ER retention signal (Capell et al. 2005; Kim et al. 2004). In contrast, activity-dependent inhibitors modified with fluorophores display binding at the cell surface (Tarassishin et al. 2004). Therefore, it appears likely that the four components of the γ-secretase assemble and become enzymatically active in early compartments of the biosynthetic pathway, but that the assembled complex rapidly leaves the ER, to encounter and cleave its substrates in the late secretory compartments, such as the *trans*-Golgi network, endosomes and plasma membrane. This would explain why little intact complex is present in the ER, and suggests that it is mainly unassembled PS that is observed in the ER in immunocytochemical staining experiments of untransfected cells. Somewhat opposed to this hypothesis is the observation that transition-state-analogue-inhibitors apparently only bind a small subset of the total pool of γ-secretase complex in the cell. This suggests in fact that only a fraction of γ-secretase complex is active in the cell (Beher et al. 2003; Lai et al. 2003), compatible with the idea that additional proteins or factors are involved in the activation of the protease. In conclusion, the subcellular localization of PS does therefore not necessarily correspond with that of the active γ-secretase activity and further work to precisely understand the activation of the complex is needed.

4.3   Nicastrin is a gate keeper for the γ-secretase substrates

As already stated, γ-secretase only cleaves APP and Notch after the bulk of the ectodomain has been shed. γ-secretase cleaves these remaining stubs in the transmembrane domain with an extremely relaxed specificity (a hydrophobic α-helix seems all that is required (Lichtenthaler et al. 2002)). On the other hand, the size of the ectodomain is important since bulky ectodomains of 200 to 300 amino acid residues prevent γ-secretase cleavage, whereas proteins with ectodomains smaller than about 50 amino acid residues are efficiently cleaved (Struhl and Adachi 2000). Thus, γ-secretase is able to measure the length of the ectodomain of its substrate, which is rather an unexpected property for a protease. Recent elegant work (Shah et al. 2005) provides a mechanistic insight. In these studies, it was shown that the extracellular domain of nicastrin binds specifically to the amino-terminal residue of membrane-bound protein fragments. A domain of nicastrin is reminiscent of aminopeptidase structures, and the carboxyl function of a particular glutamate residue (E333) in this region is crucial for the binding interaction. This domain also encompasses a stretch of amino acid residues (DYIGS) previously shown by Yu et al. (2000) to modulate APP processing. This substrate binding domain in nicastrin is accordingly called the DAP domain (DYIGS and peptidase homologous region). The DAP domain of nicastrin is not involved in assembly of the γ-secretase complex (see above) but, as Shah et al. confirm, the transmembrane domain of nicastrin is needed for the interaction of nicastrin with the other partners of the complex (Capell et al. 2003; Morais et al. 2003).

4.4   PS and regulated intramembrane proteolysis

As mentioned above, presenilins in the γ-secretase complex cleave substrates with an extremely relaxed specificity in the membrane, and are therefore likely very important to clear the membrane of such transmembrane stubs that remain after ectodomain shedding ("proteasome of the membrane" (Kopan and Ilagan 2004)). However presenilins are also involved in Notch signalling and thus are also molecular switches in important developmental processes. In fact, several other examples of proteases that cleave in the membrane are known to have important signalling functions as well. For instance, the active site of PS, GxGD (x = variable), represents a novel motif that is also found in other polytopic membrane aspartyl proteases like signal peptide peptidase (SPP) and SPP-like proteases (SPPLs/PSHs) (Grigorenko et al. 2002; Ponting et al. 2002; Weihofen et al. 2002). Similar to PSs, SPP and SPPLs contain two active site

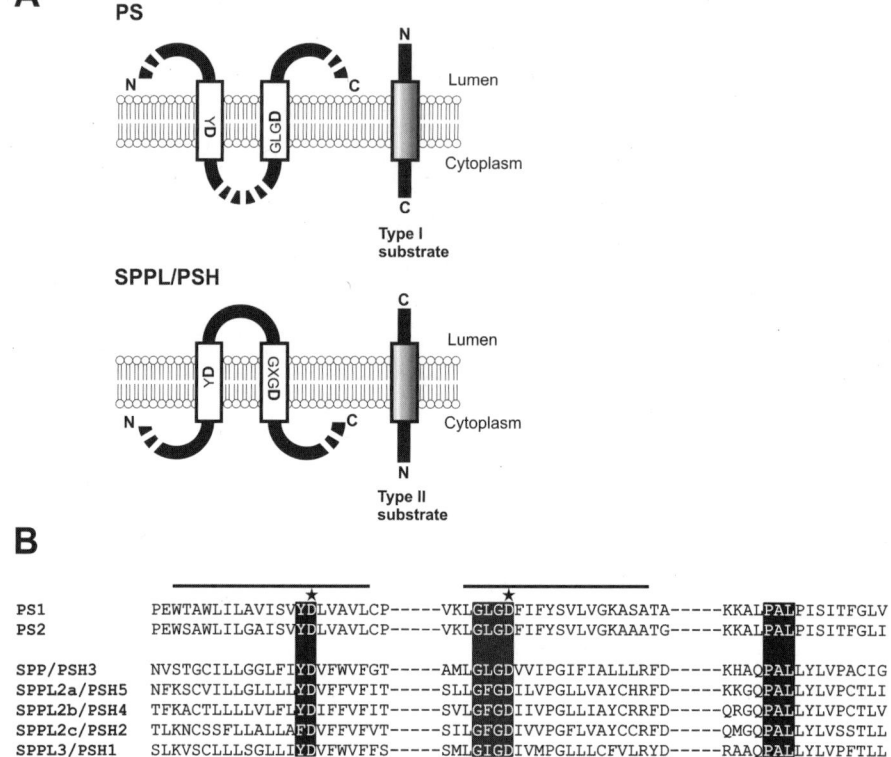

Figure 3. Conserved motifs in presenilins and SPPL/PSHs. A. Models of transmembrane domains that comprise putative active sites and its substrates. Presenilins (top) cleave the transmembrane domains of substrates with type I orientation. SPP and some related SPPL (PSH) family members (bottom) have GXGD type active sites in the opposite topology and cleave substrates with type II topology. B. Alignment of the three highly conserved domains, YD, GXGD and PAL motifs in human presenilin and SPPL/PSHs. Bars designate the transmembrane domains of human presenilins. Residues highlighted in black indicate the conserved motifs. Stars indicate the critical aspartates.

aspartates that are positioned opposite to each other in adjacent transmembrane domains (Fig. 3).

However, the transmembrane orientation of the domains containing the active site is opposite to that of PSs (Fig.3A). Accordingly, SPP cleaves peptides with type II membrane topology. Some of the SPPLs are also suggested to possess protease activity (Krawitz et al. 2005).

In addition to the polytopic membrane aspartyl proteases, rhomboid, a serine protease responsible for the processing of the ligands such as Spitz (Urban et al. 2001), and the site-2 protease (S2P), a zinc-metalloprotease responsible for the cleavage of sterol regulatory element binding proteins (SREBPs) (Rawson et al. 1997) have been identified as crucial factors in cell signalling events. These proteases are able to cleave substrates in a "regulated" manner, meaning that they only cleave after an initial cleavage in the ectodomain upon ligand recognition (presenilin/S2P), or after the specific transport of the substrate to the compartment where the protease resides (rhomboid). The released fragments can act as a transcription factor (e.g. SREBP) or as a ligand for a receptor (like Spitz) (Brown et al. 2000; Wolfe and Kopan 2004).

Given the relaxed specificity of presenilin/γ-secretase, it is not surprising that a rapidly growing list of substrates have been identified over the last several years (see Table 1).

As with the case of APP, γ-secretase cleavage of these substrates requires the prior trimming of the large ectodomain. Following γ-secretase cleavage, intracellular domains (C-terminal fragments) of several substrates are released from the membrane. In analogy with Notch, it has been suggested that many of these intracellular domains can move to the nucleus and regulate gene transcription, but further work *in vivo* is needed to make those conclusions more definitive (see Table 1). PS-dependent RIP could also mediate non-nuclear signalling, for example, by affecting the assembly of receptor complexes or adherence junctions on the plasma membrane. Several proteins such as cadherins, nectin-1α and CD44 are known to mediate cell-cell adhesion (Georgakopoulos et al. 1999; Kim et al. 2002; Marambaud et al. 2002; Nagano et al. 2004), suggesting that PSs might regulate intercellular adhesion in neurons or epithelial cells by cleaving those proteins. However, it should be noted that these proteins only become a substrate after ectodomain shedding, and thus, the regulation of these processes likely occurs at the level of the initial cleavage step rather than by the γ-secretase-mediated cleavage event.

While Nicastrin is pivotal for recognition of γ-secretase substrates, the question as to whether different γ-secretase complexes have different biochemical properties should be further considered. There are two *PS* genes and two *Aph-1* genes (that encode alternatively spliced transcripts) in human, and PS1 or PS2 and Aph-1a or Aph-1b are incorporated in a mutually exclusive way into different complexes (Hébert et al. 2004; Shirotani et al. 2004). Thus four (and maybe more) different γ-secretase complexes do exist, and likely in the same cell type (Hébert et al. 2004). Gene targeting studies of *PS1* and *PS2*, or *Aph-1a, Aph-1b* and *Aph-1c* in mice indicate that the two PS and the three (rodent) Aph-1 proteins contribute differentially to the overall γ-secretase activity in mice (Mastrangelo et al. 2005; Serneels et al. 2005) and are involved in quite different biological processes. It remains to be seen whether these differences are a reflection of specific tissue expression or variable biochemical properties of each complex.

Table 1. Reported substrates of γ-secretase.

Substrate	Predicted function	Reference
APP	Aβ generation, gene regulation with Fe65/Tip60	De Strooper et al. (1998) Nature 391:387–390
APLP-1, APLP-2	gene regulation with Fe65	Naruse et al. (1998) Neuron 21:1213–1221
		Scheinfeld et al. (2002) J Biol Chem 277:44195–44201
Notch 1, 2, 3 and 4	gene regulation with CSL family transcription factors	De Strooper et al. (1999) Nature 398:518–522
		Saxena et al. (2001) J Biol Chem 276:40268–40273
Delta	gene regulation via AP-1	Bland et al. (2003) J Biol Chem 278:13607–13610
Jagged	gene regulation via AP-1	LaVoie et al. (2003) J Biol Chem 278:34427–34437
E-cadherin	disassembly of E-cadherin/catenin complex (adherens junction)	Marambaud et al. (2003) EMBO J 21:1948–1956
N-cadherin	repression of CBP/CREB-mediated transcription	Marambaud et al. (2002) Cell 114:635–645
ErbB-4	gene regulation with YAP	Ni et al. (2001) Science 294:2179–2181
NRG-1	gene regulation	Bao et al. (2003) J Cell Biol 161:1133–1141
CD44	potentiation of CBP-mediated transactivation ?	Lammich et al. (2002) J Biol Chem 277:44754–44759
		Murakami et al. (2003) Oncogene 22:1511–1516
p75NTR	disassembly of the p75-Trk heteromeric neurotrophin receptor	Kanning et al. (2003) J Neurosci 23:5425–5436
	Rho activation (inhibition of axonal growth)	Domeniconi et al. (2005) Neuron 46:849–855
DCC	gene regulation ?	Taniguchi et al. (2003) J Biol Chem 278:30425–30428
Nectin-1	regulation of the stability of adherens junction ?	Kim et al. (2002) J Biol Chem 277:49976–49981
syndecan-3	CASK signaling	Schulz et al. (2003) J Biol Chem 278:48651–48657
LRP	gene regulation with Tip60	May et al. (2002) J Biol Chem 277:18736–18743
CSF-1R	gene regulation ?	Wilhelmsen et al. (2004) Mol Cell Biol 24:454–464
CD43	gene regulation ?	Andersson et al. (2005) Biochem J 387:377–384
Alcadein	supression of Fe65-dependent gene transactivity of AICD	Araki et al. (2004) J Biol Chem 279:24343–24354
GluR subunit 3	regulation of receptor function and mobility within the cell ?	Meyer et al. (2003) J Biol Chem 278:23786–23796
voltage gated sodium channel b subunits	?	Wong et al. (2005) J Biol Chem 280:23009–23017; Kim et al., J. Biol. Chem. 280, 23251–23261
Megalin	?	Zou et al. (2004) J Biol Chem 279:34302–34310
apoER2	?	May et al. (2003) J Biol Chem 278:37386–37392

4.5 γ-Secretase inhibitors and AD treatment

Besides general aspartyl protease inhibitors such as pepstatin, several small compounds inhibit γ-secretase activity both *in vitro* and *in vivo*. A series of transition state analogue inhibitors, such as the hydroxyethylene dipeptide isostere, L-685,458, and the peptidomimetics designed on the basis of the γ-secretase cleavage site in APP and modified with difluoro ketone/alcohol groups, specifically bind to the γ-secretase compex and inhibit activity (Esler et al. 2002; Li et al. 2000b; Wolfe et al. 1999b). Modified versions of these inhibitors have identified PS as the binding site, suggesting that PS is responsible for γ-secretase activity. The dipeptide type γ-secretase inhibitor, DAPT (N-[N-(3,5-difluorophenacetyl)-L-alanyl]-S-phenylglycine t-butyl ester), was identified from a large scale compound screening (Dovey et al. 2001). In this report, dose-dependent decreases in cortical Aβ levels were observed after acute administration of DAPT in mutant APP transgenic mice. Benzo-diazepine analogue compounds have been shown to be potent inhibitors with low IC$_{50}$ (half-maximal inhibitory concentration). The most potent compound reported, compound E, inhibits Aβ production in cell culture with an IC$_{50}$ of 0.3 nM (Seiffert et al. 2000). A series of fenylamine sulfonamides (Rishton et al. 2000), benzocaprolactams, certain low molecular weight hydroxyethylenes and sulfoneamides (Anderson et al. 2005) that inhibit γ-secretase activity have also been disclosed by several companies. Short helical peptides with α-aminoisobutyric acid based on the APP transmembrane domain have been demonstrated to inhibit γ-secretase activity (Das et al. 2003) as well.

Quite unexpectedly, cell-based studies have shown that compounds of the class of the non-steroidal anti-inflammatory drugs (NSAIDs) can modulate γ-secretase (Weggen et al. 2001). Some commonly used NSAIDs selectively inhibit Aβ42 generation, and in turn increase Aβ38 production, a less fibrillogenic species. Importantly, these compounds do not affect Notch signalling. The effect is independent of the inhibitory effect on cyclooxygenases (COX) 1 and 2. Based on epidemiological evidence that patients with arthritis taking NSAIDs show reduced incidence and slower progression of AD, NSAIDs are potential anti-AD drugs. However, this area needs further exploration in prospective studies and clinical trials. It has been suggested that Aβ42-lowering NSAIDs bind to PS and change its conformation, resulting in small changes in the way APP (and possibly other substrates) are presented to the catalytic site (Lleó et al. 2004).

An important down side of these studies is that many drugs now used in the clinic actually significantly raise the Aβ42/Aβ40 ratio (Kukar et al. 2005). It remains to be seen whether this is a risk factor for the development of AD.

While γ-secretase inhibitors are being developed with the aim to cure AD, they have also contributed to our understanding of the γ-secretase complex.

For instance, immobilized transition state analogue inhibitors were used to purify the active γ-secretase complex. As a result, the substrates of γ-secretase were copurified with the protease subunits, suggesting that the initial substrate binding site is different from the active site (Beher et al. 2003; Esler et al. 2002), in accordance with the role of Nct in substrate binding (see above).

5.   Other functions of PS

The role of PS/γ-secretase in clearance of membrane stubs of type I integral membrane proteins and in cellular signalling via the RIP mechanism is well supported by the *in vivo* evidence in different species. However, PSs have been implicated in several other functions as well. Two hybrid screenings have yielded a vast amount of candidate proteins potentially interacting with PS. One particular well studied example is the interaction of presenilin with β-catenin and other members of the armadillo-repeat protein family, such as neural plakophilin related armadillo protein (NPRAP or β-catenin) and p0071 (Levesque et al. 1999; Stahl et al. 1999; Zhou et al. 1997). The cell surface pool of β-catenin is bound to the cadherin-cell adhesion complex and provides a link to the actin cytoskeleton. The cytoplasmic pool is in complex with Axin, APC and GSK-3β and can translocate to the nucleus, regulating gene expression. This pool is activated by Wnt signaling. PS1 appears to contribute to the sequential β-catenin phosphorylation, independent of the Wnt-regulated Axin complex. In the absence of PS1, β-catenin accumulates and causes malignant transformation (Kang et al. 2002). Mice deficient in PS1 expression in the skin develop skin tumors, correlating with β-catenin stimulated LEF/TCF transcription (Xia et al. 2001). However, since deficient Notch signalling in the skin causes a similar phenotype, deficient γ-secretase could contribute to the phenotype as well (Nicolas et al. 2003). Since PS1 dependent β-catenin regulation is rescued by presenilin in which the catalytic aspartate residues are mutated, the role of PS in β-catenin signaling is apparently γ-secretase activity independent (Meredith et al. 2002). Apparently, the presence of Cadherins (see above) is essential to mediate the molecular and functional interactions of PS and β-catenin (Serban et al. 2005).

Strong evidence also points to a role of PS in Ca^{2+} homeostasis (LaFerla 2002). Both γ-secretase dependent (possibly via AICD (Leissring et al. 2002)) and independent mechanisms have been invoked. Inositol 1,4,5-triphosphate (IP$_3$) mediated intracellular Ca^{2+} release in cells expressing FAD-linked PS mutants is potentiated (Leissring et al. 1999a, b). Finally, it has been shown that PSs facilitate capacitative calcium entry (CCE) (Leissring et al. 2000; Yoo et al. 2000). CCE is a process in which influx of Ca^{2+} through store-operated Ca^{2+} channels on the plasma membranes are activated when intracellular Ca^{2+}

stores are depleted. PS1 deficiency enhances CCE while FAD-linked mutations attenuate CCE. For a further discussion of the relevance for Alzheimer's Disease, please see the review by LaFerla (LaFerla 2002).

Independent of these studies, the C-terminal portion of PS2 had been identified in a screen for genes involved in T cell apoptosis (Vito et al. 1996). While the PS2 CTF is antiapoptotic, overexpression of full-length PS2 promotes apoptosis after trophic factor withdrawal or induced by β-amyloid peptide (Wolozin et al. 1996). Some of PS binding proteins, such as Bcl-X_L, PSAP and calsenilin, relate to apoptotic pathways, which indirectly support the possible role for PSs in apoptosis. On the other hand, mice deficient in PS do not display significant signs of enhanced or decreased apoptosis.

More recent in depth studies on the physiological functions of presenilin have been performed in different genetic *PS1* and *PS2* gene knockout animals. These were recently reviewed and further references are available in Marjaux et al. (2004). PS null mice that lack PSs specifically in the postnatal forebrain exhibit age-related impairments in LTP and in hippocampal memory, followed by synaptic and neuronal degeneration. Synaptic levels of the NMDA receptor and CaMKII are selectively decreased, and decreased levels of CBP cause reduced expression of CREB/CBP target genes, such as *c-fos* and *BDNF* (Saura et al. 2004). This severe phenotype is only observed when PS is completely inactivated (Feng et al. 2001; Yu et al. 2001). The authors suggested that part of the phenotype could be explained by PS function in molecular trafficking of, for instance, the NMDA receptor, in line with previous *in vitro* experiments suggesting a role for PS in protein trafficking (Naruse et al. 1998).

Finally, recent publications indicate that PS are involved in autophagic degradative pathways, but whether these functions are related to trafficking or turnover of transmembrane proteins remains to be clarified (Esselens et al. 2004; Wilson et al. 2004).

6. Challenges in presenilin research

Enormous progress has been made since the presenilins were identified in 1995 (Levy-Lahad et al. 1995; Rogaev et al. 1995). We now know that they are catalytic subunits of a larger γ-secretase complex, the components of which are well defined. We have reasonable notions regarding the role of γ-secretase in the turn-over of membrane anchored protein stubs, and in regulated intramembrane proteolysis signalling, particularly in the Notch pathway. A major challenge is to provide more *in vivo* support for the many other functions in which presenilin/γ-secretase has been implied. Does presenilin also have functions independent of γ-secretase, and does it form other complexes that are responsible for these alternative functions?

The principal goal in Alzheimer's Disease research is to develop a cure. γ-Secretase remains a very interesting target in that regard, since it executes the final step in the production of the Aβ peptide. Worries that blocking γ-secretase could result in major side effects (mainly via disturbing Notch signalling) have seriously delayed clinical trials with the inhibitors that have been developed over the last years. However, it becomes more and more clear that we have still to learn a lot about this complex. For examples: how is the substrate, after docking to the Nicastrin binding site, transferred to the putative hydrophilic pore of the presenilin catalytic subunit?; are there additional binding sites that contribute to the recognition of the substrates?; why do at least four different γ-secretases exist?; what is the exact structure of the complex; and finally, can we decipher the mechanism by which this protease recognizes and cleaves its substrates? It will take a long time before all these questions are answered, but it is also likely that by answering them we will find new ways to subtly influence its activity, so that we decrease the amyloid burden in patients without significantly impacting upon the important biology of this fascinating complex.

References

Anderson JJ, Holtz G, Baskin PP, Turner M, Rowe B, Wang B, Kounnas MZ, Lamb BT, Barten D, Felsenstein K, McDonald I, Srinivasan K, Munoz B, Wagner SL (2005) Biochem Pharmacol 69:689–698.

Annaert WG, Levesque L, Craessaerts K, Dierinck I, Snellings G, Westaway D, George-Hyslop PS, Cordell B, Fraser P, De Strooper B (1999) J Cell Biol 147:277–294.

Baki L, Shioi J, Wen P, Shao Z, Schwarzman A, Gama-Sosa M, Neve R, Robakis NK (2004) EMBO J 23:2586–2596.

Baumeister R, Leimer U, Zweckbronner I, Jakubek C, Grunberg J, Haass C (1997) Genes Funct 1:149–159.

Beher D, Fricker M, Nadin A, Clarke EE, Wrigley JD, Li YM, Culvenor JG, Masters CL, Harrison T, Shearman MS (2003) Biochemistry 42:8133–8142.

Borchelt DR, Thinakaran G, Eckman CB, Lee MK, Davenport F, Ratovitsky T, Prada CM, Kim G, Seekins S, Yager D, Slunt HH, Wang R, Seeger M, Levey AI, Gandy SE, Copeland NG, Jenkins NA, Price DL, Younkin SG, Sisodia SS (1996) Neuron 17:1005–1013.

Borchelt DR, Ratovitski T, van Lare J, Lee MK, Gonzales V, Jenkins NA, Copeland NG, Price DL, Sisodia SS (1997) Neuron 19:939–945.

Brown MS, Ye J, Rawson RB, Goldstein JL (2000) Cell 100:391–398.

Brou C, Logeat F, Gupta N, Bessia C, LeBail O, Doedens JR, Cumano A, Roux P, Black RA, Israel A (2000) Mol Cell 5:207–216.

Capell A, Grunberg J, Pesold B, Diehlmann A, Citron M, Nixon R, Beyreuther K, Selkoe DJ, Haass C (1998) J Biol Chem 273:3205–3211.

Capell A, Kaether C, Edbauer D, Shirotani K, Merkl S, Steiner H, Haass C (2003) J Biol Chem 278:52519–52523.

Capell A, Beher D, Prokop S, Steiner H, Kaether C, Shearman MS, Haass C (2005) J Biol Chem 280:6471–6478.

Chartier-Harlin MC, Crawford F, Houlden H, Warren A, Hughes D, Fidani L, Goate A, Rossor M, Roques P, Hardy J, Mullan M (1991) Nature 353:844–846.

Chen F, Tandon A, Sanjo N, Gu YJ, Hasegawa H, Arawaka S, Lee FJ, Ruan X, Mastrangelo P, Erdebil S, Wang L, Westaway D, Mount HT, Yankner B, Fraser PE, St George-Hyslop P (2003) J Biol Chem 278:19974–19979.

Chung HM, Struhl G (2001) Nat Cell Biol 3:1129–1132.

Citron M, Westaway D, Xia W, Carlson G, Diehl T, Levesque G, Johnson-Wood K, Lee M, Seubert P, Davis A, Kholodenko D, Motter R, Sherrington R, Perry B, Yao H, Strome R, Lieberburg I, Rommens J, Kim S, Schenk D, Fraser P, St George Hyslop P, Selkoe DJ (1997) Nat Med 3:67–72.

Clark RF, Hutton M, and Alzheimer's Disease Collaborative Group (1995) Nat Genet 11:219–222.

Crystal AS, Morais VA, Pierson TC, Pijak DS, Carlin D, Lee VM, Doms RW (2003) J Biol Chem 278:20117–20123.

Das C, Berezovska O, Diehl TS, Genet C, Buldyrev I, Tsai JY, Hyman BT, Wolfe MS (2003) J Am Chem Soc 125:11794–11795.

Davis JA, Naruse S, Chen H, Eckman C, Younkin S, Price DL, Borchelt DR, Sisodia SS, Wong PC (1998) Neuron 20, 603–609.

De Strooper B, Beullens M, Contreras B, Levesque L, Craessaerts K, Cordell B, Moechars D, Bollen M, Fraser P, George-Hyslop PS, Van Leuven F (1997) J Biol Chem 272:3590–3598.

De Strooper B, Saftig P, Craessaerts K, Vanderstichele H, Guhde G, Annaert W, von Figura K, Van Leuven F (1998) Nature 391:387–390.

De Strooper B, Annaert W, Cupers P, Saftig P, Craessaerts K, Mumm JS, Schroeter EH, Schrijvers V, Wolfe MS, Ray WJ, Goate A, Kopan R (1999) Nature 398:518–522.

Doan A, Thinakaran G, Borchelt DR, Slunt HH, Ratovitsky T, Podlisny M, Selkoe DJ, Seeger M, Gandy SE, Price DL, Sisodia SS (1996) Neuron 17:1023–1030.

Donoviel DB, Hadjantonakis AK, Ikeda M, Zheng H, Hyslop PS, Bernstein A (1999) Genes Dev 13:2801–2810.

Dovey HF, John V, Anderson JP, Chen LZ, de Saint Andrieu P, Fang LY, Freedman SB, Folmer B, Goldbach E, Holsztynska EJ, Hu KL, Johnson-Wood KL, Kennedy SL, Kholodenko D, Knops JE, Latimer LH, Lee M, Liao Z, Lieberburg IM, Motter RN, Mutter LC, Nietz J, Quinn KP, Sacchi KL, Seubert PA, Shopp GM, Thorsett ED, Tung JS, Wu J, Yang S, Yin CT, Schenk DB, May PC, Altstiel LD, Bender MH, Boggs LN, Britton TC, Clemens JC, Czilli DL, Dieckman-McGinty DK, Droste JJ, Fuson KS, Gitter BD, Hyslop PA, Johnstone EM, Li WY, Little SP, Mabry TE, Miller FD, Audia JE (2001) J Neurochem 76:173–181.

Duff K, Eckman C, Zehr C, Yu X, Prada CM, Perez-Tur J, Hutton M, Buee L, Harigaya Y, Yager D, Morgan D, Gordon MN, Holcomb L, Refolo L, Zenk B, Hardy J, Younkin S (1996) Nature 383:710–713.

Edbauer D, Winkler E, Haass C, Steiner H (2002) Proc Natl Acad Sci USA 99:8666–8671.

Edbauer D, Winkler E, Regula JT, Pesold B, Steiner H, Haass C (2003) Nat Cell Biol 5:486–488.

Edbauer D, Kaether C, Steiner H, Haass C (2004) J Biol Chem 279:37311–37315.

Esler WP, Kimberly WT, Ostaszewski BL, Diehl TS, Moore CL, Tsai JY, Rahmati T, Xia W, Selkoe DJ, Wolfe MS (2000) Nat Cell Biol 2:428–434.

Esler WP, Kimberly WT, Ostaszewski BL, Ye W, Diehl TS, Selkoe DJ, Wolfe MS (2002) Proc Natl Acad Sci USA 99:2720–2725.

Esselens C, Oorschot V, Baert V, Raemaekers T, Spittaels K, Serneels L, Zheng H, Saftig P, De Strooper B, Klumperman J, Annaert W (2004) J Cell Biol 166:1041–1054.

Evin G, Canterford LD, Hoke DE, Sharples RA, Culvenor JG, Masters CL (2005) Biochemistry 44:4332–4341.

Feng R, Rampon C, Tang YP, Shrom D, Jin J, Kyin M, Sopher B, Miller MW, Ware CB, Martin GM, Kim SH, Langdon RB, Sisodia SS, Tsien JZ (2001) Neuron 32:911–926.

Francis R, McGrath G, Zhang J, Ruddy DA, Sym M, Apfeld J, Nicoll M, Maxwell M, Hai B, Ellis MC, Parks AL, Xu W, Li J, Gurney M, Myers RL, Himes CS, Hiebsch R, Ruble C, Nye JS, Curtis D (2002) Dev Cell 3:85–97.

Georgakopoulos A, Marambaud P, Efthimiopoulos S, Shioi J, Cu W, Li HC, Schutte M, Gordon R, Holstein GR, Martinelli G, Mehta P, Friedrich VL Jr, Robakis NK (1999) Mol Cell 4:893–902.

Goate AM, Chartier-Harlin M-C, Mullan MJ, Brown J, Crawford F, Fidani L, Giuffra L, Haynes A, Irving N, James L, Mant R, Newton P, Rooke K, Roques P, Talbot C, Pericak-Vance M, Roses A, Williamson R, Rossor M, Owen M, Hardy J (1991) Nature 349:704–706.

Goutte C, Hepler W, Mickey KM, Priess JR (2000) Development 127:2481–2492.

Goutte C, Tsunozaki M, Hale VA, Priess JR (2002) Proc Natl Acad Sci USA 99:775–779.

Grigorenko AP, Moliaka YK, Korovaitseva GI, Rogaev EI (2002) Biochemistry (Mosc) 67:826–835.

Gu Y, Chen F, Sanjo N, Kawarai T, Hasegawa H, Duthie M, Li W, Ruan X, Luthra A, Mount HT, Tandon A, Fraser PE, St George-Hyslop P (2003) J Biol Chem 278:7374–7380.

Haass C, Selkoe DJ (1993) Cell 75:1039–1042.

Hansson CA, Frykman S, Farmery MR, Tjernberg LO, Nilsberth C, Pursglove SE, Ito A, Winblad B, Cowburn RF, Thyberg J, Ankarcrona M (2004) J Biol Chem 279:51654–51660.

Hébert SS, Godin C, Tomiyama T, Mori H, Levesque G (2003) Biochem Biophys Res Commun 301:119–126.

Hébert SS, Serneels L, Dejaegere T, Horre K, Dabrowski M, Baert V, Annaert W, Hartmann D, De Strooper B (2004) Neurobiol Dis 17:260–272.

Henricson A, Kall L, Sonnhammer EL (2005) FEBS J 272:2727–2733.

Herreman A, Hartmann D, Annaert W, Saftig P, Craessaerts K, Serneels L, Umans L, Schrijvers V, Checler F, Vanderstichele H, Baekelandt V, Dressel R, Cupers P, Huylebroeck D, Zwijsen A, Van Leuven F, De Strooper B (1999) Proc Natl Acad Sci USA 96:11872–11877.

Herreman A, Serneels L, Annaert W, Collen D, Schoonjans L, De Strooper B (2000) Nat Cell Biol 2:461–462.

Herreman A, Van Gassen G, Bentahir M, Nyabi O, Craessaerts K, Mueller U, Annaert W, De Strooper B (2003) J Cell Sci 116:1127–1136.

Hu Y, Ye Y, Fortini ME (2002) Dev Cell 2:69–78.

Hu Y, Fortini ME (2003) J Cell Biol 161:685–690.

Johnston JA, Ward CL, Kopito RR (1998) J Cell Biol 143:1883–1898.

Jutras I, Laplante A, Boulais J, Brunet S, Thinakaran G, Desjardins M (2005) J Biol Chem 280:36310–36317.

Kaether C, Lammich S, Edbauer D, Ertl M, Rietdorf J, Capell A, Steiner H, Haass C (2002) J Cell Biol 158:551–561.

Kaether C, Capell A, Edbauer D, Winkler E, Novak B, Steiner H, Haass C (2004) EMBO J 23:4738–4748.

Kang DE, Soriano S, Xia X, Eberhart CG, De Strooper B, Zheng H, Koo EH (2002) Cell 110:751–762.

Kim DY, Ingano LA, Kovacs DM (2002) J Biol Chem 277:49976–49981.

Kim SH, Yin YI, Li YM, Sisodia SS (2004) J Biol Chem 279:48615–48619.

Kim TW, Pettingell WH, Hallmark OG, Moir RD, Wasco W, Tanzi RE (1997) J Biol Chem 272:11006–11010.

Kimberly WT, LaVoie MJ, Ostaszewski BL, Ye W, Wolfe MS, Selkoe DJ (2003) Proc Natl Acad Sci USA 100:6382–6387.

Kopan R, Goate A (2000) Genes Dev 14:2799–2806.

Kopan R, Ilagan MX (2004) Nat Rev Mol Cell Biol 5:499–504.

Kovacs DM, Fausett HJ, Page KJ, Kim TW, Moir RD, Merriam DE, Hollister RD, Hallmark OG, Mancini R, Felsenstein KM, Hyman BT, Tanzi RE, Wasco W (1996) Nat Med 2:224–229.

Krawitz P, Haffner C, Fluhrer R, Steiner H, Schmid B, Haass C (2005) J Biol Chem 280:39515–39523.

Kukar T, Murphy MP, Eriksen JL, Sagi SA, Weggen S, Smith TE, Ladd T, Khan MA, Kache R, Beard J, Dodson M, Merit S, Ozols VV, Anastasiadis PZ, Das P, Fauq A, Koo EH, Golde TE (2005) Nat Med 11, 545–550.

LaFerla FM (2002) Nat Rev Neurosci 3:862–872.

Lai MT, Chen E, Crouthamel MC, DiMuzio-Mower J, Xu M, Huang Q, Price E, Register RB, Shi XP, Donoviel DB, Bernstein A, Hazuda D, Gardell SJ, Li YM (2003) J Biol Chem 278:22475–22481.

Laudon H, Hansson EM, Melen K, Bergman A, Farmery MR, Winblad B, Lendahl U, von Heijne G, Naslund J (2005) J Biol Chem 280:35352–35360.

LaVoie MJ, Fraering PC, Ostaszewski BL, Ye W, Kimberly WT, Wolfe MS, Selkoe DJ (2003) J Biol Chem 278:37213–37222.

Lee MK, Slunt HH, Martin LJ, Thinakaran G, Kim G, Gandy SE, Seeger M, Koo E, Price DL, Sisodia SS (1996) J Neurosci 16:7513–7525.

Lee SF, Shah S, Li H, Yu C, Han W, Yu G (2002) J Biol Chem 277:45013–45019.

Lee SF, Shah S, Yu C, Wigley WC, Li H, Lim M, Pedersen K, Han W, Thomas P, Lundkvist J, Hao YH, Yu G (2004) J Biol Chem 279:4144–4152.

Leem JY, Vijayan S, Han P, Cai D, Machura M, Lopes KO, Veselits ML, Xu H, Thinakaran G (2002) J Biol Chem 277:19236–19240.

Leissring MA, Paul BA, Parker I, Cotman CW, LaFerla FM (1999a) J Neurochem 72:1061–1068.

Leissring MA, Parker I, LaFerla FM (1999b) J Biol Chem 274:32535–32538.

Leissring MA, Akbari Y, Fanger CM, Cahalan MD, Mattson MP, LaFerla FM (2000) J Cell Biol 149:793–798.

Leissring MA, Murphy MP, Mead TR, Akbari Y, Sugarman MC, Jannatipour M, Anliker B, Muller U, Saftig P, De Strooper B, Wolfe MS, Golde TE, LaFerla FM (2002) Proc Natl Acad Sci USA 99:4697–4702.

Levesque G, Yu G, Nishimura M, Zhang DM, Levesque L, Yu H, Xu D, Liang Y, Rogaeva E, Ikeda M, Duthie M, Murgolo N, Wang L, VanderVere P, Bayne ML, Strader CD, Rommens JM, Fraser PE, St George-Hyslop P (1999) J Neurochem 72:999–1008.

Levitan D, Greenwald I (1995) Nature 377:351–354.

Levitan D, Doyle TG, Brousseau D, Lee MK, Thinakaran G, Slunt HH, Sisodia SS, Greenwald I (1996) Proc Natl Acad Sci USA 93:14940–14944.

Levy-Lahad E, Wasco W, Poorkaj P, Romano DM, Oshima J, Pettingell WH, Yu CE, Jondro PD, Schmidt SD, Wang K (1995) Science 269:973–977.

Levy-Lahad E, Poorkaj P, Wang K, Fu YH, Oshima J, Mulligan J, Schellenberg GD (1996) Genomics 34:198–204.

Li T, Ma G, Cai H, Price DL, Wong PC (2003) J Neurosci 23:3272–3277.

Li X, Greenwald I (1996) Neuron 17:1015–1021.

Li YM, Xu M, Lai MT, Huang Q, Castro JL, DiMuzio-Mower J, Harrison T, Lellis C, Nadin A, Neduvelil JG, Register RB, Sardana MK, Shearman MS, Smith AL, Shi XP, Yin KC, Shafer JA, Gardell SJ (2000a) Nature 405:689–694.

Li YM, Lai MT, Xu M, Huang Q, DiMuzio-Mower J, Sardana MK, Shi XP, Yin KC, Shafer JA, Gardell SJ (2000b) Proc Natl Acad Sci USA 97:6138–6143.

Lichtenthaler SF, Beher D, Grimm HS, Wang R, Shearman MS, Masters CL, Beyreuther K (2002) Proc Natl Acad Sci USA 99:1365–1370.

Lleó A, Berezovska O, Herl L, Raju S, Deng A, Bacskai BJ, Frosch MP, Irizarry M, Hyman BT (2004) Nat Med 10:1065–1066.

Logeat F, Bessia C, Brou C, LeBail O, Jarriault S, Seidah NG, Israel A (1998) Proc Natl Acad Sci USA 95:8108–8112.

Lopez-Schier H, St Johnston D (2002) Dev Cell 2:79–89.

Luo WJ, Wang H, Li H, Kim BS, Shah S, Lee HJ, Thinakaran G, Kim TW, Yu G, Xu H (2003) J Biol Chem 278:7850–7854.

Marambaud P, Shioi J, Serban G, Georgakopoulos A, Sarner S, Nagy V, Baki L, Wen P, Efthimiopoulos S, Shao Z, Wisniewski T, Robakis NK (2002) EMBO J 21:1948–1956.

Marambaud P, Wen PH, Dutt A, Shioi J, Takashima A, Siman R, Robakis NK (2003) Cell 114:635–645.

Marjaux E, Hartmann D, De Strooper B (2004) Neuron 42:189–192.

Mastrangelo P, Mathews PM, Chishti MA, Schmidt SD, Gu Y, Yang J, Mazzella MJ, Coomaraswamy J, Horne P, Strome B, Pelly H, Levesque G, Ebeling C, Jiang Y, Nixon RA, Rozmahel R, Fraser PE, St George-Hyslop P, Carlson GA, Westaway D (2005) Proc Natl Acad Sci USA 102:8972–8977.

Meredith JE Jr, Wang Q, Mitchell TJ, Olson RE, Zaczek R, Stern AM, Seiffert D (2002) Biochem Biophys Res Commun 299:744–750.

Morais VA, Crystal AS, Pijak DS, Carlin D, Costa J, Lee VM, Doms RW (2003) J Biol Chem 278:43284–43291.

Mumm JS, Schroeter EH, Saxena MT, Griesemer A, Tian X, Pan DJ, Ray WJ, Kopan R (2000) Mol Cell 5:197–206.

Murrell J, Farlow M, Ghetti B, Benson MD (1991) Science 254:97–99.

Nagano O, Murakami D, Hartmann D, De Strooper B, Saftig P, Iwatsubo T, Nakajima M, Shinohara M, Saya H (2004) J Cell Biol 165:893–902.

Naruse S, Thinakaran G, Luo JJ, Kusiak JW, Tomita T, Iwatsubo T, Qian X, Ginty DD, Price DL, Borchelt DR, Wong PC, Sisodia SS(1998) Neuron 21:1213–1221.

Nicolas M, Wolfer A, Raj K, Kummer JA, Mill P, van Noort M, Hui CC, Clevers H, Dotto GP, Radtke F (2003) Nat Genet 33:416–421.

Niimura M, Isoo N, Takasugi N, Tsuruoka M, Ui-Tei K, Saigo K, Morohashi Y, Tomita T, Iwatsubo T (2005) J Biol Chem 280:12967–12975.

Nishimura M, Yu G, Levesque G, Zhang DM, Ruel L, Chen F, Milman P, Holmes E, Liang Y, Kawarai T, Jo E, Supala A, Rogaeva E, Xu DM, Janus C, Levesque L, Bi Q, Duthie M, Rozmahel R, Mattila K, Lannfelt L, Westaway D, Mount HT, Woodgett J, Fraser PE, St George-Hyslop P (1999) Nat Med 5:164–169.

Pasternak SH, Bagshaw RD, Guiral M, Zhang S, Ackerley CA, Pak BJ, Callahan JW, Mahuran DJ (2003) J Biol Chem 278:26687–26694.

Perez-Tur J, Froelich S, Prihar G, Crook R, Baker M, Duff K, Wragg M, Busfield F, Lendon C, Clark RF, Roques P, Fuldner RA, Johnston JA, Cowburn R, Forsell C, Axelman K, Lilius L, Houlden H, Karran E, Roberts GW, Rossor M, Adams MD, Hardy J, Goate A, Lannfelt L, Hutton M (1995) Neuro Report 7:297–301.

Podlisny MB, Citron M, Amarante P, Sherrington R, Xia W, Zhang J, Diehl T, Levesque G, Fraser P, Haass C, Koo EH, Seubert P, St George-Hyslop P, Teplow DB, Selkoe DJ (1997) Neurobiol Dis 3:325–337.

Ponting CP, Hutton M, Nyborg A, Baker M, Jansen K, Golde TE (2002) Hum Mol Genet 11:1037–1044.

Qian S, Jiang P, Guan XM, Singh G, Trumbauer ME, Yu H, Chen HY, Van de Ploeg LH, Zheng H (1998) Neuron 20:611–617.

Rawson RB, Zelenski NG, Nijhawan D, Ye J, Sakai J, Hasan MT, Chang TY, Brown MS, Goldstein JL (1997) Mol Cell 1:47–57.

Rechards M, Xia W, Oorschot VM, Selkoe DJ, Klumperman J (2003) Traffic 4:553–565.

Rishton GM, Retz DM, Tempest PA, Novotny J, Kahn S, Treanor JJ, Lile JD, Citron M (2000) J Med Chem 43:2297–2299.

Rogaev EI, Sherrington R, Rogaeva EA, Levesque G, Ikeda M, Liang Y, Chi H, Lin C, Holman K, Tsuda T (1995) Nature 376:775–778.

Saura CA, Choi SY, Beglopoulos V, Malkani S, Zhang D, Shankaranarayana Rao BS, Chattarji S, Kelleher RJ 3rd, Kandel ER, Duff K, Kirkwood A, Shen J (2004) Neuron 42:23–36.

Scheuner D, Eckman C, Jensen M, Song X, Citron M, Suzuki N, Bird TD, Hardy J, Hutton M, Kukull W, Larson E, Levy-Lahad E, Viitanen M, Peskind E, Poorkaj P, Schellenberg G, Tanzi R, Wasco W, Lannfelt L, Selkoe D, Younkin S (1996) Nat Med 2:864–870.

Schroeter EH, Kisslinger JA, Kopan R (1998) Nature 393:382–386.

Schroeter EH, Ilagan MX, Brunkan AL, Hecimovic S, Li YM, Xu M, Lewis HD, Saxena MT, De Strooper B, Coonrod A, Tomita T, Iwatsubo T, Moore CL, Goate A, Wolfe MS, Shearman M, Kopan R (2003) Proc Natl Acad Sci USA 100:13075–13080.

Seiffert D, Bradley JD, Rominger CM, Rominger DH, Yang F, Meredith JE Jr, Wang Q, Roach AH, Thompson LA, Spitz SM, Higaki JN, Prakash SR, Combs AP, Copeland RA, Arneric SP, Hartig PR, Robertson DW, Cordell B, Stern AM, Olson RE, Zaczek R (2000) J Biol Chem 275:34086–34091.

Serban G, Kouchi Z, Baki L, Georgakopoulos A, Litterst CM, Shioi J, Robakis NK (2005) J Biol Chem 280:36007–36012.

Serneels L, Dejaegere T, Craessaerts K, Horre K, Jorissen E, Tousseyn T, Hebert S, Coolen M, Martens G, Zwijsen A, Annaert W, Hartmann D, De Strooper B (2005) Proc Natl Acad Sci USA 102:1719–1724.

Shah S, Lee SF, Tabuchi K, Hao YH, Yu C, Laplant Q, Ball H, Dann CE 3rd, Sudhof T, Yu G (2005) Cell 122:435–447.

Shen J, Bronson RT, Chen DF, Xia W, Selkoe DJ, Tonegawa S (1997) Cell 89:629–639.

Sherrington R, Rogaev EI, Liang Y, Rogaeva EA, Levesque G, Ikeda M, Chi H, Lin C, Li G, Holman K, Tsuda T, Mar L, Foncin JF, Bruni AC, Montesi MP, Sorbi S, Rainero I, Pinessi L, Nee L, Chumakov I, Pollen D, Brookes A, Sanseau P, Polinsky RJ, Wasco W, Da Silva HAR, Haines JL, Pericak-Vance MA, Tanzi RE, Roses AD, Fraser PE, Rommens JM, St George-Hyslop PH (1995) Nature 375:754–760.

Shirotani K, Edbauer D, Prokop S, Haass C, Steiner H (2004) J Biol Chem 279:41340–41345.

Spasic D, Tolia A, Dillen K, Baert V, De Strooper B, Vrijens S, Annaert W (2006) J Biol Chem *in press.*

Stahl B, Diehlmann A, Sudhof TC (1999) J Biol Chem 274:9141–9148.

Steiner H, Capell A, Pesold B, Citron M, Kloetzel PM, Selkoe DJ, Romig H, Mendla K, Haass C (1998) J Biol Chem 273:32322–32331.

Steiner H, Romig H, Grim MG, Philipp U, Pesold B, Citron M, Baumeister R, Haass C (1999a) J Biol Chem 274:7615–7618.

Steiner H, Duff K, Capell A, Romig H, Grim MG, Lincoln S, Hardy J, Yu X, Picciano M, Fechteler K, Citron M, Kopan R, Pesold B, Keck S, Baader M, Tomita T, Iwatsubo T, Baumeister R, Haass C (1999b) J Biol Chem 274:28669–28673.

Steiner H, Kostka M, Romig H, Basset G, Pesold B, Hardy J, Capell A, Meyn L, Grim ML, Baumeister R, Fechteler K, Haass C (2000) Nat Cell Biol 2:848–851.

Steiner H, Winkler E, Edbauer D, Prokop S, Basset G, Yamasaki A, Kostka M, Haass C (2002) J Biol Chem 277:39062–39065.

Struhl G, Greenwald I (1999) Nature 398:522–525.

Struhl G, Adachi A (2000) Mol Cell 6:625–636.

Takasugi N, Tomita T, Hayashi I, Tsuruoka M, Niimura M, Takahashi Y, Thinakaran G, Iwatsubo T (2003) Nature 422:438–441.

Tarassishin L, Yin YI, Bassit B, Li YM (2004) Proc Natl Acad Sci USA 101:17050–17055.

Thinakaran G, Borchelt DR, Lee MK, Slunt HH, Spitzer L, Kim G, Ratovitsky T, Davenport F, Nordstedt C, Seeger M, Hardy J, Levey AI, Gandy SE, Jenkins NA, Copeland NG, Price DL, Sisodia SS (1996) Neuron 17:181–190.

Tomita T, Maruyama K, Saido TC, Kume H, Shinozaki K, Tokuhiro S, Capell A, Walter J, Grunberg J, Haass C, Iwatsubo T, Obata K (1997) Proc Natl Acad Sci USA 94:2025–2030.

Tomita T, Takikawa R, Koyama A, Morohashi Y, Takasugi N, Saido TC, Maruyama K, Iwatsubo T (1999) J Neurosci 19:10627–10634.

Tomita T, Watabiki T, Takikawa R, Morohashi Y, Takasugi N, Kopan R, De Strooper B, Iwatsubo T (2001) J Biol Chem 276:33273–33281.

Urban S, Lee JR, Freeman M (2001) Cell 107:173–182.

Vito P, Lacana E, D'Adamio L (1996) Science 271:521–525.

Walker ES, Martinez M, Brunkan AL, Goate A (2005) J Neurochem 92:294–301.

Walter J, Capell A, Grunberg J, Pesold B, Schindzielorz A, Prior R, Podlisny MB, Fraser P, Hyslop PS, Selkoe DJ, Haass C (1996) Mol Med 2:673–691.

Wang J, Brunkan AL, Hecimovic S, Walker E, Goate A (2004) Neurobiol Dis 15:654–666.

Weggen S, Eriksen JL, Das P, Sagi SA, Wang R, Pietrzik CU, Findlay KA, Smith TE, Murphy MP, Bulter T, Kang DE, Marquez-Sterling N, Golde TE, Koo EH (2001) Nature 414:212–216.

Weihofen A, Binns K, Lemberg MK, Ashman K, Martoglio B (2002) Science 296:2215–2218.

Wilson CA, Murphy DD, Giasson BI, Zhang B, Trojanowski JQ, Lee VM (2004) J Cell Biol 165:335–346.

Wolfe MS, Xia W, Ostaszewski BL, Diehl TS, Kimberly WT, Selkoe DJ (1999a) Nature 398:513–517.

Wolfe MS, Xia W, Moore CL, Leatherwood DD, Ostaszewski B, Rahmati T, Donkor IO, Selkoe DJ (1999b) Biochemistry 38:4720–4727.

Wolfe MS, Kopan R (2004) Science 305:1119–1123.

Wolozin B, Iwasaki K, Vito P, Ganjei JK, Lacana E, Sunderland T, Zhao B, Kusiak JW, Wasco W, D'Adamio L (1996) Science 274:1710–1713.

Wong PC, Zheng H, Chen H, Becher MW, Sirinathsinghji DJ, Trumbauer ME, Chen HY, Price DL, Van der Ploeg LH, Sisodia SS (1997) Nature 387:288–292.

Xia W, Zhang J, Ostaszewski BL, Kimberly WT, Seubert P, Koo EH, Shen J, Selkoe DJ (1998) Biochemistry 37:16465–16471.

Xia X, Qian S, Soriano S, Wu Y, Fletcher AM, Wang XJ, Koo EH, Wu X, Zheng H (2001) Proc Natl Acad Sci USA 98:10863–10868.

Ye Y, Lukinova N, Fortini ME (1999) Nature 398:525–529.

Yoo AS, Cheng I, Chung S, Grenfell TZ, Lee H, Pack–Chung E, Handler M, Shen J, Xia W, Tesco G, Saunders AJ, Ding K, Frosch MP, Tanzi RE, Kim TW (2000) Neuron 27:561–572.

Yu G, Chen F, Levesque G, Nishimura M, Zhang DM, Levesque L, Rogaeva E, Xu D, Liang Y, Duthie M, St George-Hyslop PH, Fraser PE (1998) J Biol Chem 273:16470–16475.

Yu G, Nishimura M, Arawaka S, Levitan D, Zhang L, Tandon A, Song YQ, Rogaeva E, Chen F, Kawarai T, Supala A, Levesque L, Yu H, Yang DS, Holmes E, Milman P, Liang Y, Zhang DM, Xu DH, Sato C, Rogaev E, Smith M, Janus C, Zhang Y, Aebersold R, Farrer LS, Sorbi S, Bruni A, Fraser P, St George-Hyslop P (2000) Nature 407:48–54.

Yu H, Saura CA, Choi SY, Sun LD, Yang X, Handler M, Kawarabayashi T, Younkin L, Fedeles B, Wilson MA, Younkin S, Kandel ER, Kirkwood A, Shen J (2001) Neuron 31:713–726.

Zhou J, Liyanage U, Medina M, Ho C, Simmons AD, Lovett M, Kosik KS (1997) Neuroreport 8:1489–1494.

Chapter 4

β-secretase: Physiological Role and Target Validation[*]

Fiona M. Laird[1,5], Mohamed H. Farah[1,5], Hey-Kyoung Lee[2],
Alena V. Savonenko[1,3,5], Donald L. Price[1,3-5] and
Philip C. Wong[1,4,5]

Departments of [1]Pathology, [3]Neurology and [4]Neuroscience and
[5]Division of Neuropathology
The Johns Hopkins University School of Medicine
Baltimore, Maryland 21205

[2] Department of Biology
University of Maryland
College Park, Maryland 20742

1.   Introduction

Alzheimer's Disease (AD), the most common cause of dementia in the elderly (Morris and Price, 2001; Mayeux, 2003; Petersen, 2003), selectively damages the brain but does not involve other tissues/organs. Thus, AD is a classical example of the mystery of "selective vulnerability", a term that refers to the predilection of certain neurological and psychiatric diseases that involve the nervous system and certain regions/circuits and cells in the brain and spinal cord. Therefore, the clinical signs of AD are the consequence of the degeneration of selected populations of neurons in brain regions/neural circuits critical for memory, cognitive performance, and personality (Price and Sisodia, 1998; Mesulam, 1999). In AD, amyloid-β (Aβ) neuritic plaques and neurofibrillary pathology are preferentially localized to the central nervous system (CNS), particularly the cortex, hippocampus, and amygdala. Although the biological basis for the selective vulnerability of the brain to Aβ amyloidosis is not well

---
[*] **Correspondence to:** Philip C. Wong, Department of Pathology, The Johns Hopkins University School of Medicine, 558 Ross Research Building, 720 Rutland Avenue, Baltimore, MD 21205-2196, Email: wong@jhmi.edu

understood, it has been postulated that levels of APP in concert with levels and activities of the pro-amyloidogenic enzyme β-secretase β-site APP cleavage enzyme1 (BACE1) play critical roles in the susceptibility of the CNS and in the pathogenesis of this common brain disease of the elderly (Wong et al., 2002).

In the CNS, Aβ peptides are generated by sequential endoproteolytic cleavages of neuronal APP by two membrane-bound enzyme activities; BACE1 cleaves APP to generate APP-β carboxyl terminal fragments (APP-βCTFs) (Hussain et al., 1999; Sinha et al., 1999; Vassar et al., 1999; Yan et al., 1999; Lin et al., 2000) and cleavage of APP-βCTFs by the γ-secretase complex (Francis et al., 2002; De Strooper, 2003; Selkoe and Kopan, 2003; Takasugi et al., 2003) leads to secretion of Aβ peptides. In other cells, APP can also be cleaved endoproteolytically within the Aβ sequence through an alternative, non-amyloidogenic pathway involving α-secretase thought to be TNF-alpha Converting Enzyme (TACE) (Allinson et al., 2003), or BACE2 (Farzan et al., 2000; Yan et al., 2001); subsequently, γ-secretase cleavage of these APP-αCTFs generates p3 fragments. These α-secretase and BACE2 cleavages, which occur in non-neural tissues, preclude the formation of Aβ peptides and thus are thought to protect these organs from Aβ amyloidosis. The physiological roles of APP and its derivatives (APPs, APP-CTFs, Aβ, and AICD) are not well understood (Zheng et al., 1995; Dawson et al., 1999). APP is a member of a gene family including the APLPs (APLP1 and 2) (von Koch et al., 1997; Heber et al., 2000) and functional redundancy has made interpretations difficult. However, emerging evidence indicates that the APP-βCTF or αCTF that are cleaved by γ-secretase lead to the release of a cytosolic fragment termed the APP intracellular domain (AICD), which forms a multimeric complex with the nuclear adaptor protein, Fe65, and the histone acetyltransferase, Tip60, to regulate gene transcription (Cao and Sudhof, 2001; Baek et al., 2002). The complex formed between APP and Fe65 has also been implicated in neurite growth and synapse modification (Sabo et al., 2003).

A variety of studies demonstrate that mutations in *APP* and *presenilins* (*PS1* and *PS2*) cause a subset of familial AD (FAD) (Sisodia and St George-Hyslop, 2002) and do so by increasing the production, length and/or aggregation propensities/toxicity of Aβ peptides. Multiple lines of evidence support the view that excessive accumulation of Aβ peptides, particularly Aβ42, and oligomeric Aβ species may be particularly significant in the pathogenesis of AD (Hardy and Selkoe, 2002). Recent studies indicate that APP is transported rapidly anterograde in axons in both the PNS and CNS (Kamal et al., 2001; Lazarov et al., 2002; Sheng et al., 2002, 2003). In the CNS, the actions of β- and γ-secretases, lead to the normal release of Aβ peptides at axon terminals where they may modulate synaptic activity, perhaps by interacting with NMDA receptors at glutaminergic synapses (Kamenetz et al., 2003). The observation that Aβ might be involved in NMDA receptor trafficking supports the view that abnormal accumulation of Aβ can cause synaptic dysfunction (Snyder et al., 2005). Thus, elevated concentrations of Aβ, particularly oligomeric species, at these sites appear to interfere with synaptic communication and in circuits of brain responsible for memory, leading to impairments in performance on memory tasks (Price and Sisodia, 1998; Chapman et al., 1999; Hsia et al., 1999; Lansbury, 1999; Serpell et al., 2000; Selkoe, 2002; Walsh et al., 2002). Because BACE1 cleavage of APP is critical in Aβ amyloidosis it has been suggested that pharmacological inhibition of BACE1 will be an attractive strategy to ameliorate Aβ deposition in AD (Citron, 2002). In this chapter, we will review recent advances towards our understanding of the physiological role of BACE1 as well as validating BACE1 as a therapeutic target for anti-amyloid therapy. Specifically, we will summarize evidence to support: 1) BACE1 is a major determinant of Aβ amyloidosis in the brain; 2) BACE1 and APP processing pathways play critical roles in cognitive, synaptic and emotional functions; 3) deletion of BACE1 prevents age-associated cognitive deficits occurring in a mouse model of Aβ amyloidosis; and 4) Aβ deposition is sensitive to *BACE1* dosage and is efficiently cleared from brains of aged mouse models of Aβ amyloidosis.

2. BACE1 is a major determinant of Aβ amyloidosis in the CNS

In AD, why should the brain and not other organs, such as the pancreas (with a high level of BACE1 mRNA), be particularly prone to develop Aβ deposits? The relative differences between BACE1 and BACE2 mRNA expression in various tissues (Vassar et al., 1999; Bennett et al., 2000) suggested that BACE1 may be a major factor in selective predisposition of Aβ accumulation in the CNS (Wong et al., 2001). Thus, the distributions and levels of BACE1 and BACE2/TACE, along with APP, are proposed to be key determinants of susceptibility of the CNS to Aβ amyloidosis (Wong et al., 2001). Summarized below is direct evidence that supports this hypothesis. First, BACE1 protein is much more abundant in the brain as compared to other organs. With the exception of the relatively higher levels of BACE1 in the olfactory bulb, the levels of BACE1 in most regions of the brain appear to be uniform (Laird et al., 2005). Second, although BACE1 protein is present at comparable levels in most brain regions, neurons are enriched in BACE1 (and APP), both of these proteins are conspicuous in presynaptic terminals in the hippocampus (Laird et al., 2005), a region that is critical for learning and memory and one that is vulnerable in AD. Notably, strong BACE1 immunoreactivity was observed in the hilus of dentate gyrus and stratum lucidum of CA3 region (terminal field of mossy fiber pathway) (Laird et al., 2005). Moreover, BACE1 immunoreactivities were localized to the giant boutons of the mossy fibers that form synapses with proximal dendrites of CA3 pyramidal cells; these giant boutons were readily labeled by anti-synaptophysin antibody (Laird et al., 2005). Taken together, these results are consistent with the view that BACE1 is enriched in presynaptic terminals. Because hippocampal granule cells are continuously undergoing turnover throughout the life of the animal, the enrichment of BACE1 in these highly plastic cells suggests that BACE1 may play a role in either synaptic development or plasticity. While these findings are consistent with reports documenting that APP is anterogradely transported in the CNS (Buxbaum et al., 1998; Lazarov et al., 2002; Sheng et al., 2002) initial reports indicated that BACE1 is anterogradely transported in CNS to a greater extent than in the PNS (Kamal et al., 2001; Sheng et al., 2003). However, recent studies failed to detect the transport of BACE1 in the PNS (Lazarov et al., 2005). Importantly, BACE1 and APP-βCTF are present in terminal fields of some populations of neurons and lesions of inputs to these regions reduce levels of APP, β-CTFs and Aβ (Lazarov et al., 2002; Sheng et al., 2002). Third, physiologically high levels of BACE1 in neurons are coupled with low levels of α-secretase and BACE2 activities. In contrast, non-neuronal cells exhibit low levels of active BACE1 and high levels of BACE2 and/or α-secretase activities (Laird et al., 2005). Although a high level of BACE1 mRNA is observed in the pancreas (Vassar et al., 1999), it appears that some of the pancreatic mRNAs are alternatively spliced to generate a BACE1 isoform that is incapable of cleaving APP (Bodendorf et al., 2001) and full length BACE1 cannot be detected in mouse pancreas (Laird et al., 2005). Taken together, these results support the hypothesis that a high ratio of BACE1 to BACE2/α-secretase activities is a biological explanation as to why neurons form amyloid in the CNS while other cells do not. This interpretation raises the interesting possibility that there exist polymorphisms that might predispose individuals to elevated levels of BACE1. The recent reports that show elevated BACE1 activities in cohorts of individuals with AD are consistent with this view (Fukumoto et al., 2002; Li and Sudhof, 2004). Genetic studies suggest that an association of C to G polymorphism in exon 5 of *BACE1* might serve as a risk factor for AD and that this effect is most pronounced in carriers of the ApoE4 allele (Kirschling et al., 2003). However, it remains to be determined whether polymorphisms in regulatory regions of *BACE1* are associated with AD.

3. Deletion of BACE1 prevents neuropathological abnormalities and age-associated cognitive deficits occurring in a mouse model of Aβ amyloidosis

Transgenic mice expressing mutant *APP* or mutant *APP* in conjunction with mutant *PS1* exhibit many, but not all of the features of AD (Borchelt et al., 1997; Holcomb et al., 1998; Ashe, 2001; Chapman et al., 2001; Wong et al., 2002). These animals have been employed to test potential therapeutic strategies by a variety of approaches (Dewachter and Van Leuven, 2002; Kotilinek et al., 2002; DeMattos et al., 2004). Since previous studies demonstrated that BACE1 is the neuronal β-secretase required for the secretion of Aβ peptides in the CNS (Cai et al., 2001; Luo et al., 2001; Roberds et al., 2001), it was anticipated that the ablation of BACE1 would impact upon the deposition of Aβ in the brains in mouse models of amyloidosis. Indeed, this prediction is confirmed in studies documenting that the deletion of *BACE1* completely prevented Aβ deposits in brains of younger (12 months of age) *APPswe* mice lacking *BACE1* (Luo et al., 2003) as well as in aged (18 months of age) *APPswe;PS1ΔE9* mice (Laird et al., 2005). In addition, deletion of *BACE1* prevents neuritic alterations, astrocytosis, and microgliosis in *APPswe;PS1ΔE9* mice, observations that are consistent with the idea that accumulation of Aβ induces microgliosis and astrocytosis in the brain and deletion of *BACE1* prevents both of these glial inflammatory responses (Laird et al., 2005). Given that the levels of p3 fragments observed *in vitro* in *BACE1* null cells (Cai et al., 2001) does not seem to contribute to development of Aβ deposition *in vivo*, these results are consistent with the view that no compensatory mechanisms exist for BACE1-driven Aβ production and indicate that BACE1 might be an excellent therapeutic target for amelioration of Aβ amyloidosis in AD.

Although deletion of BACE1 prevents both Aβ deposition and neuropathological abnormalities occurring in *APPswe;PS1ΔE9* mice, an important question is whether these mutant mice develop age-associated deficits in the acquisition and retention of spatial memories in the absence of BACE1 (Jankowsky et al., 2002; Savonenko et al., 2003). Initial studies showing that deficits in a Y-maze alternation task observed in young *APPswe* mice (at a time before Aβ deposits were observed) can be prevented in the absence of BACE1 (Ohno et al., 2004) are consistent with the hypothesis that oligomeric Aβ (McLean et al., 1999; Walsh et al., 2002), or other Aβ-derived diffusible ligands (ADDL) (Gong et al., 2003), might be important for cognitive deficits in these mouse models. To address the question as to whether in aged animals the absence of BACE1 prevents Aβ amyloidosis and age-dependent cognitive deficits and whether partial reduction of BACE 1 ameliorates cognitive impairments in aged mutant *APP;PS1* mice, the Morris water maze task which is sensitive to age-associate cognitive deficits in different mutant APP transgenic models

(Ashe, 2001) including *APPswe;PS1ΔE9* mice (Savonenko et al., 2005), is employed. Analysis of cognitive effects of *BACE1* ablation in 16–18 month-old *APPswe;PS1ΔE9* mice with 0, 50, or 100 % of BACE1 activity showed as expected: *APPswe;PS1ΔE9* mice swam a significantly longer distance to find the platform and spent less time in the vicinity of the platform than control non-transgenic mice (Laird et al., 2005). However, *APPswe;PS1ΔE9; BACE1$^{-/-}$* mice performed as well as non-transgenic mice in both platform and probe trials, indicating that deletion of *BACE1* prevented age-dependent cognitive deficits observed in *APPswe/PS1ΔE9* mice (Laird et al., 2005). However, partial decrease of BACE1 to 50% of normal level was not sufficient to ameliorate cognitive deficits since *APPswe;PS1ΔE9;BACE1$^{+/-}$* mice were significantly impaired as compared to non-transgenic controls and indistinguishable from *APPswe;PS1ΔE9* mice (Laird et al., 2005). These results demonstrate that while complete deletion of BACE1 prevents age-related cognitive deficits occurring in a mouse model of Aβ amyloidosis, a 50% decrease of BACE1 is not sufficient to significantly ameliorate cognitive deficits. Thus, these studies demonstrating that the ablation of BACE1 prevents the neuropathological abnormalities and age-associated cognitive deficits occurring in a mouse model of amyloidosis favor the view that BACE1 inhibitors may prove beneficial in efforts to attenuate Aβ amyloidosis in AD. However, caution is necessary as complete inhibition of BACE1 could lead to memory deficits as occurs in *BACE1$^{-/-}$* mice (see below). Analysis of conditional *BACE1* knockout mouse models will be of value in clarifying this issue.

Because these behavioral analyses were performed with aged mutant *APP;PS1* mice lacking *BACE1* in a hybrid strain background (C57/Bl6J and 129Sv), it was conceivable that the genetic background was a confounding factor in the Morris water maze study. To address this issue, it was important to verify that baseline performance was intact or to exclude the possibility that strain background impacts significantly on the effect of *APP;PS1* transgenes. Comparative Morris water maze analysis of mutant *APP;PS1* mice on congenic C57Bl6/J background showed a similar effect as compared to mutant *APP;PS1* mice on an F2 C57Bl6/J;129Sv background (Laird et al., 2005). Moreover, one of the well-known behavioral characteristics of the 129Sv strain that may affect the performance in the Morris Water maze is a relatively high floating time (Wolfer et al., 1997). However, a comparison of the time which cohorts of F2 C57B6/129sv mice spent floating, showed no significant differences between the various genotypes (Laird et al., 2005). Thus, these observations are consistent with the view that the genetic background used in these studies did not affect the sensitivity of the Morris Water maze task.

4. BACE1 and APP processing pathway play critical roles in cognition and emotion

Although the deletion of BACE1 did not lead to overt developmental abnormalities, an observation that BACE1 null mice exhibit mild cognitive deficits in the Y-maze task (Ohno et al., 2004) raised the question as to whether BACE1 plays a critical role in learning and memory. To assess hippocampus-dependent cognitive function in BACE1 null mice, three different behavioral tasks were used. First, using the Morris water maze task to assess spatial reference memory, it was observed that 3 month-old $BACE1^{-/-}$ mice were able to learn and remember the hidden platform location as efficiently as $BACE1^{+/-}$ or non-transgenic littermates. However, 16 month-old $BACE1^{-/-}$ mice exhibited a significant impairment in spatial reference memory during probe trial measures as compared to littermate controls (Laird et al., 2005). During platform trials, the performances of aged $BACE1^{-/-}$ mice were not significantly different from non-transgenic mice indicating that they were able to adopt non-spatial strategies in finding the hidden platform (Laird et al., 2005). These results indicated that $BACE1^{-/-}$ mice exhibit an age-dependent deficit in spatial reference memory. Secondly, the radial water maze tasks was employed to analyze the impact of BACE1 on spatial working memory (Arendash et al., 2001; Dudchenko, 2004). $BACE1^{-/-}$ mice were significantly impaired in the radial water maze; $BACE1^{-/-}$ mice made significantly more errors before finding the platform as compared to $BACE1^{+/-}$ mice or $BACE1^{+/+}$ mice (Laird et al., 2005). This deficit cannot be attributed to differences in visual acuity since 3 month-old $BACE1^{-/-}$ mice were not impaired on the classical Morris water maze where the visual demands are equivalent. One critical aspect of the radial water maze task is a variable start position randomly assigned to a different arm for each trial. This variability in start position ensures that the task is solved using allocentric, hippocampus-dependent, rather than egocentric, hippocampus-independent, strategies (Eichenbaum et al., 1990; King et al., 2002). Interestingly, when a start position was made invariant across the trials, performances of the $BACE1^{-/-}$ mice in the radial water maze were normal (Laird et al., 2005). Therefore, by reducing the hippocampal demands of the task, the $BACE1^{-/-}$ mice were able to perform as well as littermate controls, possibly by utilizing hippocampus-independent strategies to complete the task. Thirdly, the Y-maze task was employed as an independent method to assess spatial working memory in $BACE1^{-/-}$ mice. Similar to the radial water maze, a deficit in cognitive function was seen in $BACE1^{-/-}$ mice as judged by the Y-maze task. Although $BACE1^{-/-}$ mice visited a similar number of arms as the $BACE1^{+/-}$ or $BACE1^{+/+}$ mice, $BACE1^{-/-}$ mice showed significant deficits in arm alternation, indicating that the ablation of *BACE1* resulted in the early cognitive deficits in spatial working memory (Laird et al., 2005). Taken together, these findings support the view that BACE1 plays a critical role in both spatial reference memory and working memory.

Interestingly, $BACE1^{-/-}$ mice exhibited a reduced speed of swimming compared to littermate controls (Laird et al., 2005) in the Morris water maze task suggesting that anxiety levels in these mice might be reduced since swim speed in rodents may reflect a stress/anxiety reaction to the placement in cold water (Winocur and Hasher, 2004). To test this hypothesis, an open field task, a popular model of anxiety-like behaviors (Crawley, 1999; Prut and Belzung, 2003), was used to confront animals with an unknown environment. In such a situation, mice spontaneously prefer the periphery of the apparatus (thigmotaxis) due to an anxiety-induced inhibition of exploration in the aversive central parts of the open field. An increase in time spent, distance traveled and number of entries to the central part, as well as a decreased latency to enter the central part would indicate anxiolysis. In the open field task, $BACE1^{-/-}$ mice traveled a significantly longer distance, and showed a significantly higher proportion of activity and number of visits in the central parts of the open field as compared to $BACE1^{+/-}$ or $BACE1^{+/+}$ mice (Laird et al., 2005). This "low anxiety" phenotype was also confirmed in the plus maze where 3 month-old $BACE1^{-/-}$ mice visited open arms of the maze more often as compared to $BACE1^{+/-}$ mice or littermate controls. Together, these findings are consistent with the view that $BACE1$ knockout mice exhibit a lower level of anxiety as compared to control littermates, implicating an important role for BACE1 in emotion (Laird et al., 2005).

The findings that $BACE1$ deficient mice exhibit significant impairments in a number of hippocampus-dependent cognitive tasks provides direct evidence that BACE1 plays a critical role in learning and memory. Because a variety of possible substrates of BACE1 have been identified (Kitazume et al., 2001; Gruninger-Leitch et al., 2002; Lichtenthaler et al., 2003; Li and Sudhof, 2004), it is difficult to ascertain at this juncture which pathways may account for the cognitive deficits observed in $BACE1$ null mice. However, the observation that transgenic expression of $APPswe$ and $PS1\Delta E9$ is sufficient to restore the memory deficits, but not emotional alterations occurring in $BACE1^{-/-}$ mice suggests the APP processing pathway is a candidate for critical signaling mechanisms influencing cognitive functions (Laird et al., 2005). Consistent with this view is the finding that APP-deficient mice appear to exhibit age-associated deficits in tasks that measure spatial reference memory (Dawson et al., 1999).

Although it is not known precisely how APP signaling impacts on learning and memory, we postulate that a deficiency in neuronal AICD might contribute to the cognitive deficits observed in $BACE1$ null animals. The demonstration that a multimeric complex comprised of AICD, Fe65 and Tip60 is capable of stimulating heterologous reporter constructs (Cao and Sudhof, 2001) coupled with the identification of a subset of NF-kB genes regulated by the AICD-mediated transcription complex (Baek et al., 2002) are consistent with the view that AICD plays a signaling role analogous to that of NICD signaling. Recent studies indicate that AICD may play a role in neuronal differentiation through the c-Jun N-terminal kinase pathway (Kimberly et al., 2005). In this model, we suggest that the majority of AICD produced in neurons is BACE1-dependent due to the relative abundance of BACE1 enzymatic activities found in these cells, we hypothesize that only a minor proportion is derived from non-amyloidogenic pathways, i.e. via BACE2 or α-secretase activities. Thus, we envision that $BACE1^{-/-}$ mice have low levels of AICD in neurons, leading to misregulation of AICD/Fe65/Tip60-dependent transcription which results by as yet to be characterized processes leading to memory deficits observed in $BACE1^{-/-}$ mice. Consistent with this concept are observations that cognitive impairments are prevented in $APPswe;PS1\Delta E9;BACE1^{-/-}$ mice because the levels of AICD can be compensated by increases in levels of APP, by greater efficiency of ε-cleavage mediated by PS1ΔE9-dependent γ-secretase activity, or by both mechanisms. While future investigations will be required to distinguish these possibilities, studies showed that increased expression of APPswe is capable of preventing mild memory deficits in young $BACE1^{-/-}$ mice (Ohno et al., 2004). Thus increased levels of APP appear to be sufficient to compensate for the postulated impaired signaling underlying cognitive deficits. Intriguingly, recent studies demonstrated that isoform-specific deletion of *Fe65* is associated with memory deficits (Wang et al., 2004). One interpretation of this finding is that interactions of AICD with Fe65-dependent pathways are important for normal cognitive function.

While it is clear that the cognitive deficits observed in *BACE1* null mice can be prevented in $APPswe;PS1\Delta E9;BACE1^{-/-}$ mice, the emotional deficits occurring in *BACE1* null mice persisted in $APPswe;PS1\Delta E9;BACE1^{-/-}$ mice (Laird et al., 2005). Consistent with this view are demonstrations that increases in motor activity found in *BACE1* null mice as assessed by a Y-maze paradigm also occurred in $APPswe;BACE1^{-/-}$ mice (Ohno et al., 2004). Moreover, isoform-specific Fe65 knockout mice showed cognitive but not emotional deficits (Wang et al., 2004). Taken together, these findings strongly support the idea that other potential substrates of BACE1 (other than APP family members) may play critical roles in neuronal circuits that impact on aspects of emotion in $BACE1^{-/-}$ mice. These discoveries raise issues regarding mechanism-based alterations in physiology and behavior that may be associated with use of BACE1 inhibitors.

5.   BACE1 modulates synaptic function in hippocampus

To examine the role of BACE1 in synaptic transmission and plasticity, a series of electrophysiological parameters on area CA1, one of the main integral outputs from the hippocampus, were assessed in $BACE1^{-/-}$ mice. While basal synaptic transmission (which predominantly measure the AMPA receptor mediated component) and pharmacologically isolated NMDA receptor-mediated responses were similar between wild-type and $BACE1^{-/-}$ mice there was a significant increase in paired-pulse facilitation (PPF) ratio in $BACE1^{-/-}$ mice as compared to littermate controls at 50 msec interstimulus interval (Laird et al., 2005). Since changes in PPF ratio have been attributed to alterations in presynaptic release probability (Manabe et al., 1993; Saura et al., 2004), the increased PPF ratio observed in $BACE1^{-/-}$ mice may indicate a deficit in presynaptic release. However, recent data indicate that PPF can also arise from postsynaptic modifications, namely subunit composition of AMPA receptors (Rozov et al., 1998; Shin et al., 2004). Thus, a postsynaptic role for BACE1 cannot be excluded.

To determine whether synaptic plasticity is altered in the $BACE1^{-/-}$ mice, long-term potentiation (LTP) was examined using four trains of theta-burst stimulation (TBS). While no differences were observed in the magnitude of LTP up to 2 hours following the TBS protocol or in long-term depression (LTD) induced with one train of paired-pulse 1 Hz protocol, there was a significantly larger LTD reversal (de-depression) in $BACE1^{-/-}$ mice (when 4 trains of TBS were delivered after LTD saturation) as compared to control mice (Laird et al., 2005). Since both LTP and de-depression are induced by the same TBS protocol, these results indicate that $BACE1^{-/-}$ mice show a selective increase in de-depression. To determine if this is due to differential summation of responses during TBS, the area under the field potentials during the TBS was compared and it was observed that a significant increase in responses during TBS in the $BACE1^{-/-}$ only occurred during the de-depression (Laird et al., 2005). Thus, these outcomes indicate that $BACE1^{-/-}$ mice display specific deficits in paired-pulse facilitation and de-depression implicating significant alterations in mechanisms of pre-synaptic release and synaptic plasticity. Regardless of the exact role BACE1 plays in synaptic functions, the increase in PPF ratios indicates that a lack of BACE1 can increase synaptic transmission at higher frequencies. This observation may explain the larger de-depression in $BACE1^{-/-}$ mice as compared to control mice. However, contrary to expectations, no increase in LTP in $BACE1^{-/-}$ mice, which is induced by the same TBS protocol was observed. One possibility is that the expression of LTP and de-depression utilize different molecular mechanisms (Abeliovich et al., 1993; Lee et al., 2000). In this case, BACE1 dependent pathways may act as a negative regulator of de-depression mechanisms, while leaving LTP mechanisms intact. That an enhanced summation of responses was observed in $BACE1^{-/-}$ mice only during LTD reversal suggests that BACE1 acts selectively on the induction mechanism of de-depression. Regardless of the mechanism, the enhanced transmission of information at high frequencies (due to increase in ratio of PPF) suggests that altered BACE1 functions lead to interference with information transmission across synapses. Additionally, the enhanced de-depression (LTD reversal) observed in $BACE1^{-/-}$ mice may prevent stable expression of LTD, and hence would interfere with information storage/memory. It is plausible that the enhanced PPF and/or de-depression may be responsible for the behavioral deficits seen in $BACE1$ null mice by either preventing effective transmission and/or storage of relevant information. In future studies, it will be interesting to clarify whether abnormalities in synaptic plasticity reported here in $BACE1^{-/-}$ mice are dependent on APP processing. Crosses of APP transgenic mice to $BACE1^{-/-}$ mice should be of value in clarifying this issue. Finally, as BACE1 also accumulates in CA3 in the hippocampus, it will be important to define whether deficits in synaptic transmission and plasticity also occur in this area in $BACE1^{-/-}$ mice.

6.   Aβ deposition is sensitive to $BACE1$ dosage and efficiently cleared from brains of aged mouse models of Aβ amyloidosis

Because of the potential value of inhibiting BACE1 in efforts to ameliorate Aβ deposition in AD, it would be important to address whether Aβ burden is sensitive to BACE1 dosage and whether Aβ deposits can be efficiently cleared from the CNS by silencing BACE1. To determine whether Aβ burden is sensitive to dosage of $BACE1$ in the CNS, Aβ production in young $APPswe$; $PS1\Delta E9$; $BACE1^{+/+}$ and $APPswe$; $PS1\Delta E9$; $BACE1^{+/-}$ mice before the onset of Aβ deposition was assessed. In 3-month-old $APPswe$; $PS1\Delta E9$; $BACE1^{+/-}$ mice as compared to age-matched $APPswe$; $PS1\Delta E9$ mice, both APP β-CTF, and Aβ1-40/1-42 were significantly reduced (Laird et al., 2005). To quantify the reduction in Aβ burden after the onset of Aβ deposition, filter trap and unbiased stereological approaches were employed to measure the amount of aggregated Aβ and percentage of the brain occupied by Aβ deposits, respectively. Importantly, no Aβ aggregation was detected from the filter trap assay in 12- or 20-month old $APPswe;PS1\Delta E9$ transgenic mice in the $BACE1$ null background (Laird et al., 2005). The filter trap assay reveals a 27% reduction of Aβ aggregates in brains of 12 month-old $APPswe$; $PS1\Delta E9$ transgenic mice as compared to those in the $BACE1^{+/-}$ background (Laird et al., 2005). However, no significant differences were observed in 20-month-old animals. Unbiased stereological analysis of brain sections of the same sets of mice revealed a 37% decrease in percentages of brain volume occupied by Aβ plaques in 12-month-old *APPswe; PS1ΔE9; BACE1$^{+/-}$* as compared to *APPswe; PS1ΔE9; BACE1$^{+/+}$* mice (Laird et al., 2005). However, no significant reduction of Aβ deposition was found in brains of 20 month-old *APPswe; PS1ΔE9; BACE1$^{+/-}$* mice paralleling the result from filter trap assays (Laird et al., 2005). Taken together, these results demonstrate that Aβ burden is sensitive to BACE1 dosage in young but not aged mice suggesting that BACE1 is no longer a limiting factor in this aged mouse model, possibly due to compromised Aβ clearance mechanisms including those mediated by ApoE, ApoJ (clustrin), insulin degrading enzyme (IDE), and neprilysin (Iwata et al., 2001; Leissring et al., 2003; Marr et al., 2003; DeMattos, 2004; Iwata et al., 2004). Future studies will be necessary to clarify the ways by which Aβ clearance mechanisms may be compromised with age.

To determine whether Aβ deposits can be efficiently cleared from the brain by reducing the level of BACE1, RNA interference (RNAi) strategies to silence BACE1 in the hippocampus of aged mouse models of amyloidosis were employed (Laird et al., 2005; Singer et al., 2005). First, lentivirus expressing shRNA to silence *BACE1* was demonstrated to be effective in reducing levels of BACE1 mRNA and BACE1 protein, as well as secretion of Aβ_{1-40} (Laird et al., 2005; Singer et al., 2005). Subsequently, stereotaxically injected lentivirus expressing shRNA to silence *BACE1* selectively in the hippocampus of mutant *APP* mice (Singer et al., 2005) or *APPswe;PS1ΔE9* mice (Laird et al., 2005) showed a significant reduction in Aβ burden (as assessed by unbiased stereological methods) in injected hippocampus as compared to the hippocampus of transgenic mice injected with an irrelevant lentivirus. However, no differences on Aβ burden were seen in the cortex. Furthermore, silencing BACE1 through RNAi in the hippocampus also attenuated neuropathology and cognitive deficits occurring in mutant *APP* mice (Singer et al., 2005). Taken together, these findings demonstrate that Aβ deposits can be efficiently cleared from the brain by silencing *BACE1* and further validate BACE1 as a target for anti-amyloid therapy for AD.

7.   Summary

Over the past several years, significant advances have been made towards our understanding of the physiological role of BACE1 and APP signaling pathway. Moreover, target validation studies indicate BACE1 to be a high priority anti-amyloid therapeutic target for the treatment of AD. However, inhibition of BACE1 activity may not be completely free of mechanism-based consequences related to possible roles of BACE1-dependent APP/AICD signaling in cognitive functions. It is anticipated that novel mechanism-based treatments such as BACE1 inhibitors will become available in the future. Therefore, it will be critical to understand some of the BACE1/APP mechanisms that influence cognition and emotional circuits in the CNS. BACE1 conditional knockout mouse models will be invaluable for clarifying whether cognitive deficits seen in *BACE1*$^{-/-}$ mice are related to development or to aging, and whether BACE1 is involved directly in memory formation, as well as for evaluating to what extent Aβ peptide associated synaptic abnormalities are reversible following reductions of BACE1 activity. Finally, studies summarized in this review emphasize the pivotal roles that BACE1 plays in both health and disease, findings that are pertinent to the development of effective and safe anti-amyloid therapies for the treatment of this devastating disease of the elderly.

Acknowledgments

The authors wish to thank the many colleagues who have worked at JHMI as well as those at other institutions for their contributions to some of the original work cited in this review and for their helpful discussions. Aspects of this work were supported by grants from the U.S. Public Health Service (AG05146, NS41438, NS45150 and NS047308) as well as the Metropolitan Life Foundation, Adler Foundation, CART Foundation, and Bristol-Myers Squibb Foundation. MHF is supported by an NRSA (AG02556) and the John Douglas French Foundation.

References

Abeliovich A, Chen C, Goda Y, Silva AJ, Stevens CF, Tonegawa S (1993) Modified hippocampal long-term potentiation in PKC gamma-mutant mice. Cell 75:1253–1262.

Allinson TM, Parkin ET, Turner AJ, Hooper NM (2003) ADAMs family members as amyloid precursor protein alpha-secretases. J Neurosci Res 74:342–352.

Arendash GW, King DL, Gordon MN, Morgan D, Hatcher JM, Hope CE, Diamond DM (2001) Progressive, age-related behavioral impairments in transgenic mice carrying both mutant amyloid precursor protein and presenilin-1 transgenes. Brain Res 891:42–53.

Ashe KH (2001) Learning and memory in transgenic mice modeling Alzheimer's disease. Learn Mem 8:301–308.

Baek SH, Ohgi KA, Rose DW, Koo EH, Glass CK, Rosenfeld MG (2002) Exchange of N-CoR corepressor and Tip60 coactivator complexes links gene expression by NF-kappaB and beta-amyloid precursor protein. Cell 110:55–67.

Bennett BD, Babu-Khan S, Loeloff R, Louis JC, Curran E, Citron M, Vassar R (2000) Expression analysis of BACE2 in brain and peripheral tissues. J Biol Chem 275:20647–20651.

Bodendorf U, Fischer F, Bodian D, Multhaup G, Paganetti P (2001) A splice variant of beta-secretase deficient in the amyloidogenic processing of the amyloid precursor protein. J Biol Chem 276:12019–12023.

Borchelt DR, Ratovitski T, van Lare J, Lee MK, Gonzales V, Jenkins NA, Copeland NG, Price DL, Sisodia SS (1997) Accelerated amyloid deposition in the brains of transgenic mice coexpressing mutant presenilin 1 and amyloid precursor proteins. Neuron 19:939–945.

Buxbaum JD, Thinakaran G, Koliatsos V, O'Callahan J, Slunt HH, Price DL, Sisodia SS (1998) Alzheimer amyloid protein precursor in the rat hippocampus: transport and processing through the perforant path. J Neurosci 18:9629–9637.

Cai H, Wang Y, McCarthy D, Wen H, Borchelt DR, Price DL, Wong PC (2001) BACE1 is the major beta-secretase for generation of Abeta peptides by neurons. Nat Neurosci 4:233–234.

Cao X, Sudhof TC (2001) A transcriptionally [correction of transcriptively] active complex of APP with Fe65 and histone acetyltransferase Tip60. Science 293:115–120.

Chapman PF, Falinska AM, Knevett SG, Ramsay MF (2001) Genes, models and Alzheimer's disease. Trends Genet 17:254–261.

Chapman PF, White GL, Jones MW, Cooper-Blacketer D, Marshall VJ, Irizarry M, Younkin L, Good MA, Bliss TV, Hyman BT, Younkin SG, Hsiao KK (1999) Impaired synaptic plasticity and learning in aged amyloid precursor protein transgenic mice. Nat Neurosci 2:271–276.

Citron M (2002) Alzheimer's disease: treatments in discovery and development. Nat Neurosci 5 Suppl:1055–1057.

Crawley JN (1999) Behavioral phenotyping of transgenic and knockout mice: experimental design and evaluation of general health, sensory functions, motor abilities, and specific behavioral tests. Brain Res 835:18–26.

Dawson GR, Seabrook GR, Zheng H, Smith DW, Graham S, O'Dowd G, Bowery BJ, Boyce S, Trumbauer ME, Chen HY, Van der Ploeg LH, Sirinathsinghji DJ (1999) Age-related cognitive deficits, impaired long-term potentiation and reduction in synaptic marker density in mice lacking the beta-amyloid precursor protein. Neuroscience 90:1–13.

De Strooper B (2003) Aph-1, Pen-2, and Nicastrin with Presenilin generate an active gamma-Secretase complex. Neuron 38:9–12.

DeMattos RB (2004) Apolipoprotein E dose-dependent modulation of beta-amyloid deposition in a transgenic mouse model of Alzheimer's disease. J Mol Neurosci 23:255–262.

DeMattos RB, Cirrito JR, Parsadanian M, May PC, O'Dell MA, Taylor JW, Harmony JA, Aronow BJ, Bales KR, Paul SM, Holtzman DM (2004) ApoE and clusterin cooperatively suppress Abeta levels and deposition: evidence that ApoE regulates extracellular Abeta metabolism in vivo. Neuron 41:193–202.

Dewachter I, Van Leuven F (2002) Secretases as targets for the treatment of Alzheimer's disease: the prospects. Lancet Neurol 1:409–416.

Dudchenko PA (2004) An overview of the tasks used to test working memory in rodents. Neurosci Biobehav Rev 28:699–709.

Eichenbaum H, Stewart C, Morris RG (1990) Hippocampal representation in place learning. J Neurosci 10:3531–3542.

Farzan M, Schnitzler CE, Vasilieva N, Leung D, Choe H (2000) BACE2, a beta-secretase homolog, cleaves at the beta site and within the amyloid-beta region of the amyloid-beta precursor protein. Proc Natl Acad Sci USA 97:9712–9717.

Francis R, McGrath G, Zhang J, Ruddy DA, Sym M, Apfeld J, Nicoll M, Maxwell M, Hai B, Ellis MC, Parks AL, Xu W, Li J, Gurney M, Myers RL, Himes CS, Hiebsch R, Ruble C, Nye JS, Curtis D (2002) aph-1 and pen-2 are required for Notch pathway signaling, gamma-secretase cleavage of betaAPP, and presenilin protein accumulation. Dev Cell 3:85–97.

Fukumoto H, Cheung BS, Hyman BT, Irizarry MC (2002) Beta-secretase protein and activity are increased in the neocortex in Alzheimer disease. Arch Neurol 59:1381–1389.

Gong Y, Chang L, Viola KL, Lacor PN, Lambert MP, Finch CE, Krafft GA, Klein WL (2003) Alzheimer's disease-affected brain: presence of oligomeric A beta ligands (ADDLs) suggests a molecular basis for reversible memory loss. Proc Natl Acad Sci USA 100:10417–10422.

Gruninger-Leitch F, Schlatter D, Kung E, Nelbock P, Dobeli H (2002) Substrate and inhibitor profile of BACE (beta-secretase) and comparison with other mammalian aspartic proteases. J Biol Chem 277:4687–4693.

Hardy J, Selkoe DJ (2002) The amyloid hypothesis of Alzheimer's disease: progress and problems on the road to therapeutics. Science 297:353–356.

Heber S, Herms J, Gajic V, Hainfellner J, Aguzzi A, Rulicke T, von Kretzschmar H, von Koch C, Sisodia S, Tremml P, Lipp HP, Wolfer DP, Muller U (2000) Mice with combined gene knock-outs reveal essential and partially redundant functions of amyloid precursor protein family members. J Neurosci 20:7951–7963.

Holcomb L, Gordon MN, McGowan E, Yu X, Benkovic S, Jantzen P, Wright K, Saad I, Mueller R, Morgan D, Sanders S, Zehr C, O'Campo K, Hardy J, Prada CM, Eckman C, Younkin S, Hsiao K, Duff K (1998) Accelerated Alzheimer-type phenotype in transgenic mice carrying both mutant amyloid precursor protein and presenilin 1 transgenes. Nat Med 4:97–100.

Hsia AY, Masliah E, McConlogue L, Yu GQ, Tatsuno G, Hu K, Kholodenko D, Malenka RC, Nicoll RA, Mucke L (1999) Plaque-independent disruption of neural circuits in Alzheimer's disease mouse models. Proc Natl Acad Sci USA 96:3228–3233.

Hussain I, Powell D, Howlett DR, Tew DG, Meek TD, Chapman C, Gloger IS, Murphy KE, Southan CD, Ryan DM, Smith TS, Simmons DL, Walsh FS, Dingwall C, Christie G (1999) Identification of a novel aspartic protease (Asp 2) as beta-secretase. Mol Cell Neurosci 14:419–427.

Iwata N, Tsubuki S, Takaki Y, Shirotani K, Lu B, Gerard NP, Gerard C, Hama E, Lee HJ, Saido TC (2001) Metabolic regulation of brain Abeta by neprilysin. Science 292:1550–1552.

Iwata N, Mizukami H, Shirotani K, Takaki Y, Muramatsu S, Lu B, Gerard NP, Gerard C, Ozawa K, Saido TC (2004) Presynaptic localization of neprilysin contributes to efficient clearance of amyloid-beta peptide in mouse brain. J Neurosci 24:991–998.

Jankowsky JL, Savonenko A, Schilling G, Wang J, Xu G, Borchelt DR (2002) Transgenic mouse models of neurodegenerative disease: opportunities for therapeutic development. Curr Neurol Neurosci Rep 2:457–464.

Kamal A, Almenar-Queralt A, LeBlanc JF, Roberts EA, Goldstein LS (2001) Kinesin-mediated axonal transport of a membrane compartment containing beta-secretase and presenilin-1 requires APP. Nature 414:643–648.

Kamenetz F, Tomita T, Hsieh H, Seabrook G, Borchelt D, Iwatsubo T, Sisodia S, Malinow R (2003) APP processing and synaptic function. Neuron 37:925–937.

Kimberly WT, Zheng JB, Town T, Flavell RA, Selkoe DJ (2005) Physiological regulation of the beta-amyloid precursor protein signaling domain by c-Jun N-terminal kinase JNK3 during neuronal differentiation. J Neurosci 25:5533–5543.

King JA, Burgess N, Hartley T, Vargha-Khadem F, O'Keefe J (2002) Human hippocampus and viewpoint dependence in spatial memory. Hippocampus 12:811–820.

Kirschling CM, Kolsch H, Frahnert C, Rao ML, Maier W, Heun R (2003) Polymorphism in the BACE gene influences the risk for Alzheimer's disease. Neuroreport 14:1243–1246.

Kitazume S, Tachida Y, Oka R, Shirotani K, Saido TC, Hashimoto Y (2001) Alzheimer's beta-secretase, beta-site amyloid precursor protein-cleaving enzyme, is responsible for cleavage secretion of a Golgi-resident sialyltransferase. Proc Natl Acad Sci USA 98:13554–13559.

Kotilinek LA, Bacskai B, Westerman M, Kawarabayashi T, Younkin L, Hyman BT, Younkin S, Ashe KH (2002) Reversible memory loss in a mouse transgenic model of Alzheimer's disease. J Neurosci 22:6331–6335.

Laird FM, Cai H, Savonenko AV, Farah MH, He K, Melnikova T, Wen H, Chiang H, Xu G, Koliatsos V, Borchelt DR, Price DL, Lee H, Wong PC (2005) BACE1, a Major Determinant of Selective Vulnerability of the Brain to Ab Amyloidogenesis, is Essential for Cognitive, Emotional and Synaptic Functions. J Neurosci 25:11693–11709.

Lansbury PT, Jr. (1999) Evolution of amyloid: what normal protein folding may tell us about fibrillogenesis and disease. Proc Natl Acad Sci USA 96:3342–3344.

Lazarov O, Lee M, Peterson DA, Sisodia SS (2002) Evidence that synaptically released beta-amyloid accumulates as extracellular deposits in the hippocampus of transgenic mice. J Neurosci 22:9785–9793.

Lazarov O, Morfini GA, Lee EB, Farah MH, Szodorai A, DeBoer SR, Koliatsos VE, Kins S, Lee VM, Wong PC, Price DL, Brady ST, Sisodia SS (2005) Axonal transport, amyloid precursor protein, kinesin-1, and the processing apparatus: revisited. J Neurosci 25:2386–2395.

Lee HK, Barbarosie M, Kameyama K, Bear MF, Huganir RL (2000) Regulation of distinct AMPA receptor phosphorylation sites during bidirectional synaptic plasticity. Nature 405:955–959.

Leissring MA, Farris W, Chang AY, Walsh DM, Wu X, Sun X, Frosch MP, Selkoe DJ (2003) Enhanced proteolysis of beta-amyloid in APP transgenic mice prevents plaque formation, secondary pathology, and premature death. Neuron 40:1087–1093.

Li Q, Sudhof TC (2004) Cleavage of amyloid-beta precursor protein and amyloid-beta precursor-like protein by BACE 1. J Biol Chem 279:10542–10550.

Lichtenthaler SF, Dominguez DI, Westmeyer GG, Reiss K, Haass C, Saftig P, De Strooper B, Seed B (2003) The cell adhesion protein P-selectin glycoprotein ligand-1 is a substrate for the aspartyl protease BACE1. J Biol Chem 278:48713–48719.

Lin X, Koelsch G, Wu S, Downs D, Dashti A, Tang J (2000) Human aspartic protease memapsin 2 cleaves the beta-secretase site of beta-amyloid precursor protein. Proc Natl Acad Sci USA 97:1456–1460.

Luo Y, Bolon B, Damore MA, Fitzpatrick D, Liu H, Zhang J, Yan Q, Vassar R, Citron M (2003) BACE1 (beta-secretase) knockout mice do not acquire compensatory gene expression changes or develop neural lesions over time. Neurobiol Dis 14:81–88.

Luo Y, Bolon B, Kahn S, Bennett BD, Babu-Khan S, Denis P, Fan W, Kha H, Zhang J, Gong Y, Martin L, Louis JC, Yan Q, Richards WG, Citron M, Vassar R (2001) Mice deficient in BACE1, the Alzheimer's beta-secretase, have normal phenotype and abolished beta-amyloid generation. Nat Neurosci 4:231–232.

Manabe T, Wyllie DJ, Perkel DJ, Nicoll RA (1993) Modulation of synaptic transmission and long-term potentiation: effects on paired pulse facilitation and EPSC variance in the CA1 region of the hippocampus. J Neurophysiol 70:1451–1459.

Marr RA, Rockenstein E, Mukherjee A, Kindy MS, Hersh LB, Gage FH, Verma IM, Masliah E (2003) Neprilysin gene transfer reduces human amyloid pathology in transgenic mice. J Neurosci 23:1992–1996.

Mayeux R (2003) Apolipoprotein E, Alzheimer disease, and African Americans. Arch Neurol 60:161–163.

McLean CA, Cherny RA, Fraser FW, Fuller SJ, Smith MJ, Beyreuther K, Bush AI, Masters CL (1999) Soluble pool of Abeta amyloid as a determinant of severity of neurodegeneration in Alzheimer's disease. Ann Neurol 46:860–866.

Mesulam MM (1999) Neuroplasticity failure in Alzheimer's disease: bridging the gap between plaques and tangles. Neuron 24:521–529.

Morris JC, Price AL (2001) Pathologic correlates of nondemented aging, mild cognitive impairment, and early–stage Alzheimer's disease. J Mol Neurosci 17:101–118.

Ohno M, Sametsky EA, Younkin LH, Oakley H, Younkin SG, Citron M, Vassar R, Disterhoft JF (2004) BACE1 Deficiency Rescues Memory Deficits and Cholinergic Dysfunction in a Mouse Model of Alzheimer's Disease. Neuron 41:27–33.

Petersen RC (2003) Mild cognitive impairment clinical trials. Nat Rev Drug Discov 2:646–653.

Price DL, Sisodia SS (1998) Mutant genes in familial Alzheimer's disease and transgenic models. Annu Rev Neurosci 21:479–505.

Prut L, Belzung C (2003) The open field as a paradigm to measure the effects of drugs on anxiety-like behaviors: a review. Eur J Pharmacol 463:3–33.

Roberds SL, Anderson J, Basi G, Bienkowski MJ, Branstetter DG, Chen KS, Freedman SB, Frigon NL, Games D, Hu K, Johnson-Wood K, Kappenman KE, Kawabe TT, Kola I, Kuehn R, Lee M, Liu W, Motter R, Nichols NF, Power M, Robertson DW, Schenk D, Schoor M, Shopp GM, Shuck ME, Sinha S, Svensson KA, Tatsuno G, Tintrup H, Wijsman J, Wright S, McConlogue L (2001) BACE knockout mice are healthy despite lacking the primary beta-secretase activity in brain: implications for Alzheimer's disease therapeutics. Hum Mol Genet 10:1317–1324.

Rozov A, Zilberter Y, Wollmuth LP, Burnashev N (1998) Facilitation of currents through rat Ca2+-permeable AMPA receptor channels by activity-dependent relief from polyamine block. J Physiol 511 (Pt 2):361–377.

Sabo SL, Ikin AF, Buxbaum JD, Greengard P (2003) The amyloid precursor protein and its regulatory protein, FE65, in growth cones and synapses in vitro and in vivo. J Neurosci 23:5407–5415.

Saura CA, Choi SY, Beglopoulos V, Malkani S, Zhang D, Shankaranarayana Rao BS, Chattarji S, Kelleher RJ, 3rd, Kandel ER, Duff K, Kirkwood A, Shen J (2004) Loss of presenilin function causes impairments of memory and synaptic plasticity followed by age-dependent neurodegeneration. Neuron 42:23–36.

Savonenko A, Xu GM, Melnikova T, Morton JL, Gonzales V, Wong MP, Price DL, Tang F, Markowska AL, Borchelt DR (2005) Episodic-like memory deficits in the APPswe/PS1dE9 mouse model of Alzheimer's disease: Relationships to beta-amyloid deposition and neurotransmitter abnormalities. Neurobiol Dis 18:602–617.

Savonenko AV, Xu GM, Price DL, Borchelt DR, Markowska AL (2003) Normal cognitive behavior in two distinct congenic lines of transgenic mice hyperexpressing mutant APP SWE. Neurobiol Dis 12:194–211.

Selkoe D, Kopan R (2003) Notch and Presenilin: regulated intramembrane proteolysis links development and degeneration. Annu Rev Neurosci 26:565–597.

Selkoe DJ (2002) Alzheimer's disease is a synaptic failure. Science 298:789–791.

Serpell LC, Blake CC, Fraser PE (2000) Molecular structure of a fibrillar Alzheimer's A beta fragment. Biochemistry 39:13269–13275.

Sheng JG, Price DL, Koliatsos VE (2002) Disruption of corticocortical connections ameliorates amyloid burden in terminal fields in a transgenic model of Abeta amyloidosis. J Neurosci 22:9794–9799.

Sheng JG, Price DL, Koliatsos VE (2003) The beta-amyloid-related proteins presenilin 1 and BACE1 are axonally transported to nerve terminals in the brain. Exp Neurol 184:1053–1057.

Shin J, Shen F, Huguenard JR (2004) Polyamines modulate AMPA receptor dependent synaptic responses in immature layer V pyramidal neurons. J Neurophysiol.

Singer O, Marr RA, Rockenstein E, Crews L, Coufal NG, Gage FH, Verma IM, Masliah E (2005) Targeting BACE1 with siRNAs ameliorates Alzheimer disease neuropathology in a transgenic model. Nat Neurosci 8:1343–1349.

Sinha S, Anderson JP, Barbour R, Basi GS, Caccavello R, Davis D, Doan M, Dovey HF, Frigon N, Hong J, Jacobson-Croak K, Jewett N, Keim P, Knops J, Lieberburg I, Power M, Tan H, Tatsuno G, Tung J, Schenk D, Seubert P, Suomensaari SM, Wang S, Walker D, Zhao J, McConlogue L, John V (1999) Purification and cloning of amyloid precursor protein beta-secretase from human brain. Nature 402:537–540.

Sisodia SS, St George-Hyslop PH (2002) gamma-Secretase, Notch, Abeta and Alzheimer's disease: where do the presenilins fit in? Nat Rev Neurosci 3:281–290.

Snyder EM, Nong Y, Almeida CG, Paul S, Moran T, Choi EY, Nairn AC, Salter MW, Lombroso PJ, Gouras GK, Greengard P (2005) Regulation of NMDA receptor trafficking by amyloid-beta. Nat Neurosci 8:1051–1058.

Takasugi N, Tomita T, Hayashi I, Tsuruoka M, Niimura M, Takahashi Y, Thinakaran G, Iwatsubo T (2003) The role of presenilin cofactors in the gamma-secretase complex. Nature 422:438–441.

Vassar R, Bennett BD, Babu-Khan S, Kahn S, Mendiaz EA, Denis P, Teplow DB, Ross S, Amarante P, Loeloff R, Luo Y, Fisher S, Fuller J, Edenson S, Lile J, Jarosinski MA, Biere AL, Curran E, Burgess T, Louis JC, Collins F, Treanor J, Rogers G, Citron M (1999) Beta-secretase cleavage of Alzheimer's amyloid precursor protein by the transmembrane aspartic protease BACE. Science 286:735–741.

von Koch CS, Zheng H, Chen H, Trumbauer M, Thinakaran G, van der Ploeg LH, Price DL, Sisodia SS (1997) Generation of APLP2 KO mice and early postnatal lethality in APLP2/APP double KO mice. Neurobiol Aging 18:661–669.

Walsh DM, Klyubin I, Fadeeva JV, Cullen WK, Anwyl R, Wolfe MS, Rowan MJ, Selkoe DJ (2002) Naturally secreted oligomers of amyloid beta protein potently inhibit hippocampal long-term potentiation in vivo. Nature 416:535–539.

Wang B, Hu Q, Hearn MG, Shimizu K, Ware CB, Liggitt DH, Jin LW, Cool BH, Storm DR, Martin GM (2004) Isoform-specific knockout of FE65 leads to impaired learning and memory. J Neurosci Res 75:12–24.

Winocur G, Hasher L (2004) Age and time-of-day effects on learning and memory in a non-matching-to-sample test. Neurobiol Aging 25:1107–1115.

Wolfer DP, Muller U, Stagliar M, Lipp HP (1997) Assessing the effects of the 129/Sv genetic background on swimming navigation learning in transgenic mutants: a study using mice with a modified beta-amyloid precursor protein gene. Brain Res 771:1–13.

Wong PC, Price DL, Cai H (2001) The brain's susceptibility to amyloid plaques. Science 293:1434.

Wong PC, Cai H, Borchelt DR, Price DL (2002) Genetically engineered mouse models of neurodegenerative diseases. Nat Neurosci 5:633–639.

Yan R, Munzner JB, Shuck ME, Bienkowski MJ (2001) BACE2 functions as an alternative alpha-secretase in cells. J Biol Chem 276:34019–34027.

Yan R, Bienkowski MJ, Shuck ME, Miao H, Tory MC, Pauley AM, Brashier JR, Stratman NC, Mathews WR, Buhl AE, Carter DB, Tomasselli AG, Parodi LA, Heinrikson RL, Gurney ME (1999) Membrane-anchored aspartyl protease with Alzheimer's disease beta-secretase activity. Nature 402:533–537.

Zheng H, Jiang M, Trumbauer ME, Sirinathsinghji DJ, Hopkins R, Smith DW, Heavens RP, Dawson GR, Boyce S, Conner MW, et al. (1995) beta-Amyloid precursor protein-deficient mice show reactive gliosis and decreased locomotor activity. Cell 81:525–531.

Chapter 5

Cognitive Impairment in Transgenic Aβ and Tau Models of Alzheimer's Disease

Karen H. Ashe

Departments of Neurology and Neuroscience
University of Minnesota Medical School
420 Delaware Street SE, MMC# 295, Minneapolis, MN 55455
Geriatric Research, Clinical Education Center
Minneapolis Veterans Administration Medical Center
Minneapolis, MN
Email: hsiao005@umn.edu

1.   Introduction

Alzheimer's disease (AD) is the most common cause of dementia in the United States. It currently affects over four million patients and this prevalence is predicted to quadruple in the next five decades, unless effective treatments are developed. The major clinical and neuropathological features of AD have been known for nearly a century. Alois Alzheimer, a German psychiatrist, described the progressive dementia associated with amyloid plaques, neurofibrillary tangles and neuronal loss, in 1907 (Alzheimer, 1907). Biochemical studies performed in the 1980's led to identification of the insoluble aggregates of two peptides known as amyloid-β (Aβ) and tau, which compose the amyloid plaques and neurofibrillary tangles, respectively (Glenner and Wong, 1984; Grundke-Iqbal et al., 1986; Kosik et al., 1986; Wong et al., 1985). In the 1990's molecular genetic investigations linked the amyloid precursor protein (APP), which is the polypeptide precursor to Aβ, and the tau protein to a set of autosomal dominant human neurodegenerative diseases, including familial AD, hereditary Congophilic angiopathy, and fronto-temporal dementia with Parkinsonism (Goate et al., 1991; Hutton et al., 1998; Levy et al., 1990). Mutations in some APP cleavage enzymes, which increase the production of a more amyloidogenic form of Aβ, were found to be linked to AD (Levy-Lehad et al., 1995; Sherrington et al., 1995), and provided the firmest evidence yet for the primacy of Aβ in the pathogenesis of AD. Mutations in tau are linked to familial tauopathies, but not to AD. The discovery of these mutations established tau as an important pathogenic molecule, rather than an inconsequential product of upstream pathological cascades. Yet, despite impressive advances in our understanding of APP trafficking and tau protein chemistry, APP and tau molecular genetics, Aβ and tau production and clearance, and the molecular pathology of the structural lesions created by the Aβ and tau peptides, the mechanisms by which Aβ and tau proteins disrupt brain function are poorly understood.

Understanding the molecular basis of memory loss, cognitive dysfunction and neurodegeneration in AD is crucial for developing effective AD therapies. Studies addressing these problems necessarily involve the creation of transgenic models of AD. We have focused Aβ and tau, the peptides that accumulate and neuropathologically define AD. *The work has shown that the aggregation of Aβ and tau into plaques and neurofibrillary tangles does not cause cognitive deficits in mice.* This astonishing result has redirected AD research toward the *functional* properties of Aβ and tau peptides that precede the insoluble aggregates.

This chapter is written in two parts. The first section of this chapter will focus on data examining the role of Aβ in causing cognitive deficits. The second section will discuss the role of tau in disrupting memory function.

2.   The role of Aβ in impairing cognition

2.1   APP transgenic mice

Animal models are essential for elucidating the molecular mechanisms leading to dementia in AD. Three scientific breakthroughs made the creation of the first transgenic mouse models of AD possible. First was the isolation and sequencing of the Aβ peptide in 1984 (Glenner and Wong, 1984). Second was the cloning of APP and the elucidation of its role in generating the Aβ peptide (Goldgaber et al., 1987; Kang et al., 1987; Robakis et al., 1987; Tanzi et al., 1987). Third was the discovery of the first mutation in autosomal dominant AD in APP (Goate et al., 1991) and the subsequent realization that all autosomal dominant mutations in AD appear to enhance the aggregation potential of Aβ, which may occur by one of three mechanisms. The Swedish mutation facilitating APP cleavage near the β-secretase site (Mullan et al., 1992) increases the overall production of all forms of Aβ. A mutation within Aβ, called the Arctic mutation, enhances protofibril formation (Nilsberth et al., 2001). Several mutations near the γ-secretase site increase the relative production of the more amyloidogenic Aβ42 (Chartier-Harlin et al., 1991; Goate et al., 1991; Murrell et al., 1991; Murrell et al., 2000).

This information enabled investigators to create the first APP transgenic mice modeling Alzheimer's disease. Altogether, more than 20 such mice models have been published; many models show age-related plaque deposition and memory loss. For example, the Tg2576 mouse model, which harbors human APP with the Swedish mutation, recapitulates many of the neuropathological features of AD, including amyloid plaques (Hsiao et al., 1996), oxidative stress (Smith et al., 1998) (Pappolla et al., 1998), astrogliosis (Irizarry et al., 1997), microgliosis (Benzing et al., 1999; Frautschy et al., 1998), cytokine production (Benzing et al., 1999), and dystrophic neurites (Irizarry et al., 1997). Tg2576 also show progressive deterioration in spatial reference memory (Hsiao et al., 1996; Westerman et al., 2002).

However, within a year or so of the creation of the first APP transgenic mouse models, it became apparent that many important features of AD were conspicuously absent, including neurofibrillary tangles and gross atrophy, and that there was variable neuronal and synaptic loss in the different models. Tg2576 mice were virtually devoid of such neurodegenerative changes (Irizarry et al., 1997), while synaptic loss was present in J20 and PDAPP mice (Games et al., 1995; Hsia et al., 1999), and there was some neuronal loss in APP23 mice (Calhoun et al., 1998). We do not understand the factors that account for the variations in synapses and neurons between mice, but it is clear that the differences are not entirely attributable to variations in the expression levels of Aβ. Recently, using highly sophisticated statistical analyses, it was shown that neurons are killed within the compact plaques of Tg2576 mice (Urbanc et al., 2002), but their numbers are too small to be detected using unbiased stereological estimates of overall neuron counts (Irizarry et al., 1997). Thus, some of the differences may be due to the amount of compact amyloid in the different models.

2.2   Dissociation of memory impairment from amyloid plaques

A large effort in developing AD modifying therapies has been directed at reducing the amyloid deposits in the brain (Cherny et al., 2001; Citron, 2004; Hock et al., 2003; Lim et al., 2001; Lim et al., 2000; Nicoll et al., 2003; Schenk et al., 1999). Imaging brain amyloid in living patients has been proposed as a method for following the therapeutic efficacy of amyloid modifying treatments (Klunk et al., 2004). However, amyloid deposits correlate weakly with cognitive impairment in humans (Arriagada et al., 1992; Katzman et al., 1988; Terry et al., 1991). Therefore, amyloid plaques may not be the most appropriate target against which to direct potential interventions. Identifying and isolating the forms of Aβ that specifically induce memory loss and cognitive dysfunction may greatly facilitate the development of effective AD treatments.

Multiple studies in APP transgenic mice have supported the idea that CNS dysfunction and Aβ are related (Chen et al., 2000; Hsiao et al., 1996; Hsiao et al., 1995; Janus et al., 2000; Morgan et al., 2000; Westerman et al., 2002). As early as 1995, Hsiao and colleagues showed that the expression of APP in the brains of mice in the FVB/N background strain produced an age-related CNS disorder that developed in the absence of amyloid plaques (Hsiao et al., 1995). Impaired animals showed decreased glucose utilization preferentially in the parietal and temporal lobes, similar to the pattern observed in AD patients. Interestingly, a fraction of aged non-transgenic FVB/N mice developed an identical CNS disorder, suggesting that APP expression accelerated a naturally occurring, age-related CNS abnormality. These observations provided a link between age-related cognitive decline and APP, but did not delineate the particular forms of APP or APP cleavage products that were responsible for the functional brain abnormalities. Importantly, this study was the first to demonstrate that brain dysfunction related to APP overexpression could be dissociated from amyloid formation in these mice.

Over the past decade, descriptive and experimental studies have examined the relationship between amyloid plaques and cognitive impairment in APP transgenic mice. A major emphasis has been placed on correlating amyloid plaque load with memory loss in various lines of APP transgenic mice. These studies were done to provide support for the amyloid cascade hypothesis, which stipulates that amyloid plaques cause neurodegeneration and thereby impair brain function. Multiple studies have been published on the subject (Chen et al., 2000; Chishti et al., 2001; Gordon et al., 2001; Koistinaho et al., 2001; Van Dam et al., 2003; Westerman et al., 2002). A rigorous examination of the descriptive studies along with the experimental results from Aβ immunization experiments leads to the resounding conclusion that amyloid plaques are not a major cause of cognitive impairment in plaque-forming mice.

The results of Aβ immunization studies reported by two independent groups in two different APP transgenic models completely severed the connection between amyloid plaques, insoluble Aβ and memory impairment (Dodart et al., 2002; Kotilinek et al., 2002). Memory function in an object recognition task was fully restored in aged PDAPP mice following six weekly intraperitoneal injections of m266 monoclonal antibodies (raised against Aβ(13-28)) (Dodart et al., 2002). Importantly, there was no accompanying change in amyloid plaque load (Dodart et al., 2002). Even a single injection of m266 antibodies improved performance in a hole-board exploration task, within the short interval of a few days (Dodart et al., 2002). In Tg2576 mice, spatial reference memory deficits in the Morris water maze that were present prior to plaque deposition were completely reversed following intraperitoneal injections of two full doses and one half dose of BAM-10 monoclonal antibodies (raised against Aβ(1-10)) administered within 11 days of testing (Kotilinek et al.,

2002). The reversal of memory deficits in Tg257 mice was not accompanied by changes in either SDS-soluble or SDS-insoluble, formic-acid extractable Aβ, measured by standard enzyme-linked immunosorbent assays (ELISAs) (Kotilinek et al., 2002). These two studies clearly dissociate memory deficits from amyloid plaques and gross Aβ measurements in APP transgenic mice, but failed to identify the form of Aβ that is responsible for memory deficits.

One Aβ immunization study showed no significant improvement in cognitive function in Tg2576 mice until the amyloid load was significantly reduced (Wilcock et al., 2004). Weekly intraperitoneal administration of 2286 monoclonal antibodies (raised against Aβ(28-40)) in 19- to 22-month Tg2576 mice failed to improve spontaneous alternation significantly in the Y-maze significantly after one or two months, but significantly improved performance after three months of treatment. In this study, plaque loads were unaffected after one month of treatment, but were significantly lower after two and three months. Thus, the data show an imperfect correspondence between improved behavioral performance and lower amyloid burden. Yet, if the improvement in behavioral performance was due to the elimination of plaques, then Tg2576 mice treated for two months, which showed significant reductions in amyloid burden, should have shown significantly improved spontaneous alternation, but they did not. The study is limited by small sample sizes (N = 3 in control group, and N = 6, 9 and 4 in the 1-, 2- and 3-month treatment groups, respectively).

It is also instructive to compare this study with the previous studies examining effects of passive immunization studies on cognitive function. The 2286 antibodies were directed against the carboxyl-terminus of Aβ, in contrast to the studies described above, which used m266 (Dodart et al., 2002) and BAM-10 (Kotilinek et al., 2002) directed against the amino-terminus and the mid-region of Aβ, respectively. While 2286 antibodies failed to reverse cognitive dysfunction after two months, m266 and BAM-10 antibodies reversed memory deficits within two weeks. Thus, carboxyl-terminus Aβ antibodies appear incapable of rapidly reversing cognitive dysfunction. In contrast, antibodies that led to rapid recovery of memory function have epitope specificities for the amino-terminus and mid-region of Aβ, an observation which may shed light on the potential structure of the form of Aβ that impairs cognitive function independently of amyloidosis.

The striking improvement in memory following passive immunization with m266 and BAM-10 antibodies, taken together with the lack of change in plaque load or the levels of SDS-soluble or SDS-insoluble Aβ levels, implies the existence of a form of Aβ in the brain that impairs memory independently of plaques and is not readily detectable by ELISA assays of gross brain homogenates. This form of Aβ is referred to here as Aβ star (Aβ*). However, the identity of Aβ* has been elusive.

Identifying Aβ* has been hampered by the complexities surrounding the interpretation of cognitive studies in APP transgenic mice. It has been important in the study of behavioral deficits in transgenic mice overexpressing APP to distinguish between age-independent and age-dependent abnormalities (Ashe, 2001; Chen et al., 2000; Westerman et al., 2002). The former appear to result from the overexpression of APP during brain development, while the latter are related to biochemical or structural changes that occur as the mice age. The most rigorous method for distinguishing between age-independent and age-dependent behavioral deficits is to directly compare transgenic mice overexpressing equivalent levels of wild type and mutant APP, which has been done in only one study (Westerman et al., 2002). This study showed that age-dependent memory deficits begin in Tg2576 mice at 6 months, at least four months prior to the appearance of amyloid plaques. There is no correlation between memory function and insoluble Aβ, the main component of amyloid plaque fibrils, across the lifetime of Tg2576 mice. A critical observation that emerged from this study was that memory deficits beginning at 6 months of age remain unchanged for the next 7 to 8 months, an interval during which there is an increase of between 10- and 1000-fold in the levels of water-soluble, detergent-soluble, detergent-insoluble, intracellular and lipid-raft associated Aβ and Aβ-derived diffusible ligands (Chang et al., 2003; Kawarabayashi et al., 2004; Kawarabayashi et al., 2001; Takahashi et al., 2004; Takahashi et al., 2002). The prominent dissociation between memory function and all known forms of Aβ provided additional support for the existence of Aβ*.

The descriptive studies of spatial reference memory in Tg2576 mice enabled the delineation of two criteria that Aβ* must satisfy. One, Aβ* must appear at 6 months of age, and not before. Two, the levels of Aβ* must remain stable from 6 to 13 months of age. These criteria, along with newly developed techniques for fractionating Aβ from well-defined, sub-cellular compartments of Tg2576 brain and measuring Aβ oligomers using highly sensitive immunoblotting methods, have laid the foundation for identifying and isolating Aβ* (Lesne et al., 2005). A specific 56 kilodalton Aβ assembly called Aβ*56 impairs memory in Tg2576 mice (Lesne et al., 2006). Aβ*56 correlates with memory loss in Tg2576 mice. Purified from brains of impaired Tg2576 mice, Aβ*56 impairs memory when administered to young, healthy rats (Lesne et al., 2006). Thus, Aβ*56 is a particular form of Aβ*. Whether there are other endogenous forms of Aβ* is unknown.

Recent data support the hypothesis that naturally secreted soluble oligomers of Aβ are both necessary and sufficient for Aβ to disrupt cognitive function (Cleary et al., 2005). Soluble Aβ oligomers secreted by Chinese-hamster ovary cells in culture were isolated by size-exclusion chromatography and injected into the lateral ventricles of young rats that had been trained in an operant task, the ALCR (alternating lever cyclic ratio), which measures executive function and reference memory (MacNabb et al., 1999; MacNabb et al., 2000). The ALCR task was used because it is approximately 10 times more sensitive to subtle cognitive effects of very small doses of psychoactive drugs than other more commonly used behavioral tests, such as the conventional Morris water maze (Weldon et al., 1996). In the ALCR task, substances that disrupt cognitive function produce increases in switching and perseveration errors. The detrimental effects of soluble Aβ oligomers on cognitive function occurred independently of effects on motivation or activity, because the increases in error rates occurred in the absence of changes in either absolute response rates (total responses per session length) or running response rates (responses per second during active response periods) (Cleary et al., 2005). Fractions containing femtomole amounts of oligomers, but not monomers, of Aβ increased the error rates of the rats performing the ALCR test, within 2 hours of injection. Performance returned to normal within one day of injection, and repeated weekly injections did not disrupt the ability of the rats to continue learning to improve in the ALCR task, indicating that the rats suffered no permanent neurological damage resulting from periodic exposure to the soluble Aβ oligomers. Thus, naturally secreted Aβ oligomers disrupt cognitive function independently of neurodegeneration when exogenously introduced into healthy, young rats. In summary, naturally produced soluble Aβ oligomers disrupt cognitive function in a manner that is rapid, potent and transient, and impair cognitive function without producing permanent neurological deficits.

3. The role of tau in disrupting memory function

3.1 Tau transgenic mice

Neurofibrillary tangles have long been associated with neurodegeneration in hereditary and sporadic tauopathies, AD being the most common of these disorders. However, whether and how neurofibrillary tangles disrupt cognitive function and the extent to which the structural and functional abnormalities caused by neurofibrillary pathology are capable of being modulated were unknown until quite recently. These questions are addressed more easily in transgenic mouse models than in other experimental systems.

Four landmark findings in tau biology made the creation of transgenic mice producing neurofibrillary pathology possible. First was the isolation and characterization of tau (Weingarten et al., 1975), a protein involved in promoting the aggregation and polymerization of tubulin to form microtubules. Second was the cloning of tau (Neve et al., 1986). Third was the recognition that tau was the principal protein forming the core of the paired helical filaments of neurofibrillary tangles (Grundke-Iqbal et al., 1986; Kosik et al., 1986). Fourth was the discovery of mutations in tau linked to familial tauopathies (Hutton et al., 1998).

This knowledge led to the generation of transgenic mice with neurofibrillary pathology, which were developed using both mutant and wild-type tau genes (Ishihara et al., 1999; Lewis et al., 2000). The 3xTgAD mouse carrying mutant tau, mutant APP and mutant presenilin-1 develops age-related amyloid deposits, neurofibrillary tangles and memory decline (Billings et al., 2005; Oddo et al., 2003). The htau mouse expressing all human wild-type tau isoforms but no mouse tau is the only one that produces true paired helical filaments identical to those in AD (Andorfer et al., 2003). The R406W mouse develops neurofibrillary pathology and memory deficits when mice are nearly two years old (Tatebayashi et al., 2002).

3.2   Neurofibrillary tangles do not impair memory function in mice

Cognitive function often declines with age and is believed to deteriorate initially because of changes in synaptic function rather than loss of neurons (Craik, 1977; Gallagher and Rapp, 1997; Morrison and Hof, 1997). Some individuals progress to develop AD with neurodegeneration. There is a prodrome often referred to as Mild Cognitive Impairment (MCI) or Clinical Dementia Rating (CDR) 0.5 in which individuals have subjective complaints, mild clinical abnormalities, some plaques and tangles, and neuronal loss mostly restricted to the entorhinal cortex and the hippocampus (Morris et al., 2001; Petersen et al., 1999). However, there is a large degree of overlap between the healthy non-demented cases (CDR 0) and the CDR 0.5 cases. Some individuals with CDR 0.5 have no overt loss of neurons, indicating that neuronal loss does not occur invariably prior to cognitive dysfunction (Price et al., 2001). A similar argument applies to tau pathology and neuritic degeneration, where there is extensive overlap between CDR 0 and CDR 0.5 cases, arguing also against these features causing early cognitive symptoms (Price and Morris, 1999). *The discrepancies between neuropathology and cognitive function in individual cases cannot be explained entirely on the basis of specific structural lesions in the brain.*

Recently, we created rTg4510 mice to determine the role of tau in memory impairment (Ramsden et al., 2005a). The tau transgene driving tau$_{P301L}$ expression in rTg4510 mice is regulatable (hence the "r"), and can be suppressed with the antibiotic doxycycline. rTg4510 mice develop progressive age-dependent neurofibrillary tangles, neuronal loss and cognitive deficits (Ramsden et al., 2005a). Spatial reference memory measured in the Morris water maze, a test that is sensitive to hippocampal lesions (Morris et al., 1982), becomes impaired prior to the appearance of mature, argyrophilic neurofibrillary tangles in the hippocampus (Ramsden et al., 2005a; SantaCruz et al., 2005). Following the suppression of transgenic tau, memory function recovers and neuron numbers stabilize but surprisingly, neurofibrillary tangles continue to accumulate

(SantaCruz et al., 2005). The data show that neurofibrillary tangles are neither necessary nor sufficient to cause cognitive decline or neuronal death. Moreover, these findings imply that memory function in rTg4510 mice is impaired by, as yet unidentified, tau star (tau*) molecules, which are located among the soluble tau species. Identifying and isolating tau* is a focus of current work (Ramsden et al., 2005b).

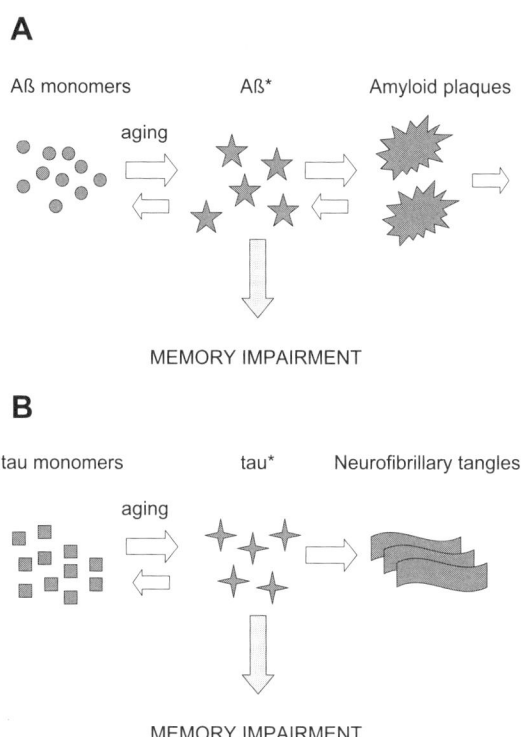

Figure 1. The Aβ* and tau* hypotheses of cognitive impairment in Alzheimer's disease. Models summarizing the relationship between memory impairment and Aβ in plaque-forming APP transgenic mice and tangle-forming tau transgenic mice. (A) The hypothetical amyloid cascade involves the conversion of monomeric Aβ (circles) to Aβ star (Aβ*), which are as yet unidentified soluble Aβ assemblies in the brain that disrupt cognitive function independently of amyloidosis or neurodegeneration. There may be a dynamic equilibrium between Aβ*, Aβ monomers and amyloid plaques. Amyloid plaques have not been shown to induce cognitive dysfunction directly. Amyloid plaques may be cleared by cells and molecules involved in the inflammatory system. (B) The hypothetical neurofibrillary tangle cascade entails the conversion of monomeric tau (squares) to tau star (tau*), which are as yet unidentified tau species in the brain that disrupt cognitive function independently of tangle formation or neurodegeneration. Neurofibrillary tangles contain a form of tau that appears to be a stable protein fate, as it accumulates even when tau protein production is greatly reduced. Neurofibrillary tangles do not appear to impair memory function.

4.   Summary

Our research addressing the molecular basis of dementia in AD has involved the creation of Aβ and tau transgenic mice. Our work and that of other investigators has shown that plaques and tangles do not cause cognitive deficits in mice. These exciting findings have redirected research away from the structural consequences of the Aβ and tau peptides and turned attention toward studies of their *functional* consequences. We refer to the specific species of Aβ and tau that underlie memory deficits as Aβ star (Aβ*) and tau star (tau*) (see Figure 1).

Aβ* and tau* are potential early causative agents in AD. Investigations of Aβ* and tau* have been made possible by the creation and study of memory and cognitive function in transgenic mice and in non-transgenic rat models of Alzheimer's disease. Identifying Aβ* and tau* is the most logical first step to developing therapies to modulate Aβ- and tau-related brain dysfunction. Successful approaches to the prevention and treatment of AD will depend upon isolating Aβ* and tau* and understanding the cellular mechanisms by which they disrupt brain function.

References

Alzheimer A (1907) A characteristic disease of the cerebral cortex. Allgemeine Zeitschrift für Psychiatrie und Psychisch-Gerichtliche Medizin 64:146–148.

Andorfer C, Kress Y, Espinoza M, de Silva R, Tucker KL, Barde YA, Duff K, Davies P (2003) Hyperphosphorylation and aggregation of tau in mice expressing normal human tau isoforms. J Neurochem 86:582–590.

Arriagada PV, Growdon JH, Hedley-Whyte ET, Hyman BT (1992) Neurofibrillary tangles but not senile plaques parallel duration and severity of Alzheimer's disease. Neurology 42:631–639.

Ashe KH (2001) Learning and memory in transgenic mice modeling Alzheimer's disease. Learn Mem 8:301–308.

Benzing WC, Wujek JR, Ward EK, Shaffer D, Ashe KH, Younkin SG, Brunden KR (1999) Evidence for glial-mediated inflammation in aged APP(SW) transgenic mice. Neurobiol Aging 20:581–589.

Billings LM, Oddo S, Green KN, McGaugh JL, Laferla FM (2005) Intraneuronal Abeta causes the onset of early Alzheimer's disease-related cognitive deficits in transgenic mice. Neuron 45:675–688.

Calhoun ME, Wiederhold KH, Abramowski D, Phinney AL, Probst A, Sturchler-Pierrat C, Staufenbiel M, Sommer B, Jucker M (1998) Neuron loss in APP transgenic mice. Nature 395:755–756.

Chang L, Bakhos L, Wang Z, Venton DL, Klein WL (2003) Femtomole immunodetection of synthetic and endogenous amyloid-beta oligomers and its application to Alzheimer's disease drug candidate screening. J Mol Neurosci 20:305–313.

Chartier-Harlin M-C, Crawford F, Houlden H, Warren A, Hughes D, Fidani L, Goate A, Rossor M, Roques P, Hardy J, Mullan M (1991) Early-onset Alzheimer's disease caused by mutation at codon 717 of the b-amyloid precursor protein gene. Nature 353:844–846.

Chen G, Chen KS, Knox J, Inglis J, Bernard A, Martin SJ, Justice A, McConlogue L, Games D, Freedman SB, Morris RG (2000) A learning deficit related to age and beta-amyloid plaques in a mouse model of Alzheimer's disease. Nature 408:975–979.

Cherny RA, Atwood CS, Xilinas ME, Gray DN, Jones WD, McLean CA, Barnham KJ, Volitakis I, Fraser FW, Kim Y, et al. (2001) Treatment with a copper-zinc chelator markedly and rapidly inhibits beta-amyloid accumulation in Alzheimer's disease transgenic mice. Neuron 30:665–676.

Chishti MA, Yang DS, Janus C, Phinney AL, Horne P, Pearson J, Strome R, Zucker N, Loukides J, French J, et al. (2001) Early-onset amyloid deposition and cognitive deficits in transgenic mice expressing a double mutant form of APP695. J Biol Chem 276:21562–21570.

Citron M (2004) Strategies for disease modification in Alzheimer's disease. Nature Reviews Neuroscience 5:677–685.

Cleary JP, Walsh DM, Hofmeister JJ, Shankar GM, Kuskowski MA, Selkoe DJ, Ashe KH (2005) Natural oligomers of the amyloid-beta protein specifically disrupt cognitive function. Nat Neurosci 8:79–84.

Craik FI (1977) Age differences in human memory. In Handbook of the Psychology of Aging, Birren JE, Schall K, eds. (New York, Van Nostrand-Reinhold), pp. 384–420.

Dodart JC, Bales KR, Gannon KS, Greene SJ, DeMattos RB, Mathis C, DeLong CA, Wu S, Wu X, Holtzman DM, Paul SM (2002) Immunization reverses memory deficits without reducing brain Abeta burden in Alzheimer's disease model. Nat Neurosci 5:452–457.

Frautschy SA, Yang F, Irizarry M, Hyman B, Saido TC, Hsiao K, Cole GM (1998) Microglial response to amyloid plaques in APPsw transgenic mice. Am J Pathol 152:307–317.

Gallagher M, Rapp PR (1997) The use of animal models to study the effects of aging on cognition. Annu Rev Psychol 48:339–370.

Games D, Adams D, Alessandrini R, Barbour R, Berthelette P, Blackwell C, Carr T, Clemens J, Donaldson T, Gillespie F, et al. (1995) Alzheimer-type neuropathology in transgenic mice overexpressing V717F beta-amyloid precursor protein. Nature 373:523–527.

Glenner GG, Wong CW (1984) Alzheimer's disease and Down's syndrome: sharing of a unique cerebrovascular amyloid fibril protein. Biochem Biophys Res Commun 122:1131–1135.

Goate AM, Chartier-Harlin CM, Mullan M, Brown J, Crawford F, Fidani L, Giuffra L, Haynes A, Irving N, James L, et al. (1991) Segregation of a missense mutation in the amyloid precursor protein gene with familial Alzheimer's disease. Nature 349:704–706.

Goldgaber D, Lerman MI, McBride OW, Saffiotti U, Gajdusek DC (1987) Characterization and chromosomal localization of a cDNA encoding brain amyloid of Alzheimer's disease. Science 235:877–880.

Gordon MN, King DL, Diamond DM, Jantzen PT, Boyett KV, Hope CE, Hatcher JM, DiCarlo G, Gottschall WP, Morgan D, Arendash GW (2001) Correlation between cognitive deficits and Abeta deposits in transgenic APP+PS1 mice. Neurobiol Aging 22:377–385.

Grundke-Iqbal I, Iqbal K, Quinlan M, Tung YC, Zaidi MS, Wisniewski HM (1986) Microtubule-associated protein tau. A component of Alzheimer paired helical filaments. J Biol Chem 261:6084–6089.

Hock C, Konietzko U, Streffer JR, Tracy J, Signorell A, Muller-Tillmanns B, Lemke U, Henke K, Moritz E, Garcia E, et al. (2003) Antibodies against beta-amyloid slow cognitive decline in Alzheimer's disease. Neuron 38:547–554.

Hsia AY, Masliah E, McConlogue L, Yu GQ, Tatsuno G, Hu K, Kholodenko D, Malenka RC, Nicoll RA, Mucke L (1999) Plaque-independent disruption of neural circuits in Alzheimer's disease mouse models. Proc Natl Acad Sci USA 96:3228–3233.

Hsiao K, Chapman P, Nilsen S, Eckman C, Harigaya Y, Younkin S, Yang F, Cole G (1996) Correlative memory deficits, Aβ elevation, and amyloid plaques in transgenic mice. Science 274:99–102.

Hsiao KK, Borchelt DR, Olson K, Johannsdottir R, Kitt C, Yunis W, Xu S, Eckman C, Younkin S, Price D, et al. (1995) Age-related CNS disorder and early death in transgenic FVB/N mice overexpressing Alzheimer amyloid precursor proteins. Neuron 15:1203–1218.

Hutton M, Lendon CL, Rizzu P, Baker M, Froelich S, Houlden H, Pickering-Brown S, Chakraverty S, Isaacs A, Grover A, et al. (1998) Association of missense and 5'-splice-site mutations in tau with the inherited dementia FTDP-17. Nature 393:702–705.

Irizarry MC, McNamara M, Fedorchak K, Hsiao K, Hyman BT (1997) APPSw transgenic mice develop age-related A beta deposits and neuropil abnormalities, but no neuronal loss in CA1. J Neuropathol Exp Neurol 56:965–973.

Ishihara T, Hong M, Zhang B, Nakagawa Y, Lee MK, Trojanowski JQ, Lee VM (1999) Age-dependent emergence and progression of a tauopathy in transgenic mice overexpressing the shortest human tau isoform. Neuron 24:751–762.

Janus C, Pearson J, McLaurin J, Mathews PM, Jiang Y, Schmidt SD, Chishti MA, Horne P, Heslin D, French J, et al. (2000) A beta peptide immunization reduces behavioural impairment and plaques in a model of Alzheimer's disease. Nature 408:979–982.

Kang J, Lemaire H-G, Unterbeck A, Salbaum JM, Masters CL, Grzeschik K-H, Multhaup G, Beyreuther K (1987) The precursor of Alzheimer's disease amyloid A4 protein resembles a cell-surface receptor. Nature 325:733–736.

Katzman R, Terry R, DeTeresa R, Brown T, Davies P, Fuld P, Renbing S, Peck A (1988) Clinical, pathological, and neurochemical changes in dementia: a subgroup with preserved mental status and numerous neocortical plaques. Ann Neurol 23:138–144.

Kawarabayashi T, Shoji M, Younkin LH, Wen-Lang L, Dickson DW, Murakami T, Matsubara E, Abe K, Ashe KH, Younkin SG (2004) Dimeric amyloid beta protein rapidly accumulates in lipid rafts followed by apolipoprotein E and phosphorylated tau accumulation in the Tg2576 mouse model of Alzheimer's disease. J Neurosci 24:3801–3809.

Kawarabayashi T, Younkin LH, Saido TC, Shoji M, Ashe KH, Younkin SG (2001) Age-dependent changes in brain, CSF, and plasma amyloid β protein in the Tg2576 transgenic mouse model of Alzheimer's disease. J Neurosci 21:372–381.

Klunk W, Engler H, Nordberg A, Wang Y, Blomqvist G, Holt DP, Bergstrom M, Savitcheva I, Huang GF, Estrada S, et al. (2004) Imaging brain amyloid in Alzheimer's disease with Pittsburgh Compound-B. Ann Neurol 55:306–319.

Koistinaho M, Ort M, Cimadevilla JM, Vondrous R, Cordell B, Koistinaho J, Bures J, Higgins LS (2001) Specific spatial learning deficits become severe with age in beta -amyloid precursor protein transgenic mice that harbor diffuse beta -amyloid deposits but do not form plaques. Proc Natl Acad Sci USA 98:14675–14680.

Kosik KS, Joachim CL, Selkoe DJ (1986) Microtubule-associated protein tau (tau) is a major antigenic component of paired helical filaments in Alzheimer disease. Proc Natl Acad Sci USA 83:4044–4048.

Kotilinek LA, Bacskai B, Westerman M, Kawarabayashi T, Younkin L, Hyman BT, Younkin S, Ashe KH (2002) Reversible memory loss in a mouse transgenic model of Alzheimer's disease. J Neurosci 22:6331–6335.

Lesne S, Kotilinek L, Kayed R, Glabe CC, Ashe KH (2005) Specific amyloid-β assemblies disrupt memory without neurodegeneration in a mouse model of Alzheimer's disease. Poster presented at: Molecular Mechanisms of Neurodegeneration: a Joint Biochemical Society/Neuroscience Ireland Focused Meeting (Dublin, Biochemical Society) http://www.biochemistry.org/meetings/abstracts/SA039/SA039P025.pdf

Lesne S, Koh MT, Kotilinek L, Kayed R, Glabe CG, Yang A, Gallagher M, Ashe KH (2006) A specific amyloid-beta protein assembly in the brain impairs memory. Nature 440:352–357.

Levy E, Carman MD, Fernandez-Madrid IJ, Power MD, Lieberburg I, van Duinen SG, Bots GT, Luyendijk W, Frangione B (1990) Mutation of the Alzheimer's disease amyloid gene in hereditary cerebral hemorrhage, Dutch type. Science 248:1124–1126.

Levy-Lehad E, Wasco W, Poorkaj P, Romano DM, Oshima J, Pettingell WH, Yu C, Jondro PD, Schmidt SD, Wang K, et al. (1995) Candidate gene for the chromosome 1 familial Alzheimer's disease locus. Science 269:973–977.

Lewis J, McGowan E, Rockwood J, Melrose H, Nacharaju P, Van Slegtenhorst M, Gwinn-Hardy K, Paul Murphy M, Baker M, Yu X, et al. (2000) Neurofibrillary tangles, amyotrophy and progressive motor disturbance in mice expressing mutant (P301L) tau protein. Nat Genet 25:402–405.

Lim GP, Chu T, Yang F, Beech W, Frautschy SA, Cole GM (2001) The curry spice curcumin reduces oxidative damage and amyloid pathology in an Alzheimer transgenic mouse. J Neurosci 21:8370–8377.

Lim GP, Yang F, Chu T, Chen P, Beech W, Teter B, Tran T, Ubeda O, Ashe KH, Frautschy SA, Cole GM (2000) Ibuprofen suppresses plaque pathology and inflammation in a mouse model for Alzheimer's disease. J Neurosci 20:5709–5714.

MacNabb C, O'Hare E, Cleary J, Georgopoulos AP (1999) Congenital hypothyroidism impairs response alternation discrimination behavior. Brain Res 847:231–239.

MacNabb C, O'Hare E, Cleary J, Georgopoulos AP (2000) Varied duration of congenital hypothyroidism potentiates perseveration in a response alternation discrimination task. Neurosci Res 36:121–127.

Morgan D, Diamond DM, Gottschall PE, Ugen KE, Dickey C, Hardy J, Duff K, Jantzen P, DiCarlo G, Wilcock D, et al. (2000) A beta peptide vaccination prevents memory loss in an animal model of Alzheimer's disease. Nature 408:982–985.

Morris JC, Storandt M, Miller JP, McKeel DW, Price JL, Rubin EH, Berg L (2001) Mild cognitive impairment represents early-stage Alzheimer disease. Arch Neurol 58:397–405.

Morris RGM, Garrud P, Rawlins JNP, O'Keefe J (1982) Place navigation impaired in rats with hippocampal lesions. Nature 297:681–683.

Morrison JH, Hof PR (1997) Life and death of neurons in the aging brain. Science 278:412–419.

Mullan M, Crawford F, Axelman K, Houlden H, Lilius L, Winblad B, Lannfelt L (1992) A pathogenic mutation for probable Alzheimer's disease in the APP gene at the N-terminus of beta-amyloid. Nat Genet 1:345–347.

Murrell J, Farlow M, Ghetti B, Benson MD (1991) A mutation in the amyloid precursor protein associated with hereditary Alzheimer's disease. Science 254:97–99.

Murrell JR, Hake AM, Quaid KA, Farlow MR, Ghetti B (2000) Early-onset Alzheimer disease caused by a new mutation (V717L) in the amyloid precursor protein gene. Arch Neurol 57:885–887.

Neve RL, Harris P, Kosik KS, Kurnit DM, Donlon TA (1986) Identification of cDNA clones for the human microtubule-associated protein tau and chromosomal localization of the genes for tau and microtubule-associated protein 2. Brain Res 387:271–280.

Nicoll JA, Wilkinson D, Holmes C, Steart P, Markham H, Weller RO (2003) Neuropathology of human Alzheimer disease after immunization with amyloid-beta peptide: a case report. Nat Med 9:448–452.

Nilsberth C, Westlind-Danielsson A, Eckman CB, Condron MM, Axelman K, Forsell C, Stenh C, Luthman J, Teplow DB, Younkin SG, et al. (2001) The 'Arctic' APP mutation (E693G) causes Alzheimer's disease by enhanced Abeta protofibril formation. Nat Neurosci 4:887–893.

Oddo S, Caccamo A, Shepherd JD, Murphy MP, Golde TE, Kayed R, Metherate R, Mattson MP, Akbari Y, LaFerla FM (2003) Triple-transgenic model of Alzheimer's disease with plaques and tangles: intracellular Abeta and synaptic dysfunction. Neuron 39:409–421.

Pappolla MA, Chyan YJ, Omar RA, Hsiao K, Perry G, Smith MA, Bozner P (1998) Evidence of oxidative stress and in vivo neurotoxicity of beta-amyloid in a transgenic mouse model of Alzheimer's disease: a chronic oxidative paradigm for testing antioxidant therapies in vivo. Am J Pathol 152:871–877.

Petersen RC, Smith GE, Waring SC, Ivnik RJ, Tangalos EG, Kokmen E (1999) Mild cognitive impairment: clinical characterization and outcome. Arch Neurol 56:303–308.

Price JL, Ko AI, Wade MJ, Tsou SK, McKeel DW, Morris JC (2001) Neuron number in the entorhinal cortex and CA1 in preclinical Alzheimer disease. Arch Neurol 58:1395–1402.

Price JL, Morris JC (1999) Tangles and plaques in nondemented aging and "preclinical" Alzheimer's disease. Ann Neurol 45:358–368.

Ramsden M, Kotilinek L, Forster C, Paulson J, McGowan E, SantaCruz K, Guimaraes A, Yue M, Lewis J, Carlson G, et al. (2005a) Age-dependent neurofibrillary tangle formation, neuron loss and memory impairment in a mouse model of human tauopathy (P301L). Journal of Neuroscience in press.

Ramsden M, Kotilinek L, Paulson J, Guimaraes A, Forster C, SantaCruz K, Ashe KH (2005b) Memory impairment in a novel mouse model of human tauopathy. Poster presented at: Molecular Mechanisms of Neurodegeneration: a Joint Biochemical Society/Neuroscience Ireland Focused Meeting (Dublin, Biochemical Society) http://www.biochemistry.org/meetings/abstracts/SA039/SA039P024.pdf

Robakis NK, Ramakrishna N, Wolfe G, Wisniewski HM (1987) Molecular cloning and characterization of a cDNA encoding the cerebrovascular and the neuritic plaque amyloid peptides. Proc Natl Acad Sci 84:4190–4194.

SantaCruz K, Lewis J, Spires T, Paulson J, Kotilinek L, Ingelsson M, Guimaraes A, DeTure M, Ramsden M, McGowan E, et al. (2005) Tau suppression in a neurodegenerative mouse model improves memory function. Science 309:476–481.

Schenk D, Barbour R, Dunn W, Gordon G, Grajeda H, Guido T, Hu K, Huang J, Johnson-Wood K, Khan K, et al. (1999) Immunization with amyloid-beta attenuates Alzheimer-disease-like pathology in the PDAPP mouse. Nature 400:173–177.

Sherrington R, Rogaev EI, Liang Y, Rogaeva EA, Levesque G, Ikeda M, Chi H, Lin C, Li G, Holman K, et al. (1995) Cloning of a gene bearing missense mutations in early-onset familial Alzheimer's disease. Nature 375:754–760.

Smith MA, Hirai K, Hsiao K, Pappolla MA, Harris PL, Siedlak SL, Tabaton M, Perry G (1998) Amyloid-beta deposition in Alzheimer transgenic mice is associated with oxidative stress. J Neurochem 70:2212–2215.

Takahashi RH, Almeida CG, Kearney PF, Yu F, Lin MT, Milner TA, Gouras GK (2004) Oligomerization of Alzheimer's beta-amyloid within processes and synapses of cultured neurons and brain. J Neurosci 24:3592–3599.

Takahashi RH, Milner TA, Li F, Nam EE, Edgar MA, Yamaguchi H, Beal MF, Xu H, Greengard P, Gouras GK (2002) Intraneuronal Alzheimer abeta42 accumulates in multivesicular bodies and is associated with synaptic pathology. Am J Pathol 161:1869–1879.

Tanzi RE, Gusella JF, Watkins PC, Bruns GAP, George-Hyslop PS, Keuren MLV, Patterson D, Pagan S, Kurnit DM, Neve RL (1987) Amyloid β protein gene: cDNA, mRNA distribution, and genetic linkage near the Alzheimer locus. Science 235:880–884.

Tatebayashi Y, Miyasaka T, Chui DH, Akagi T, Mishima K, Iwasaki K, Fujiwara M, Tanemura K, Murayama M, Ishiguro K, et al. (2002) Tau filament formation and associative memory deficit in aged mice expressing mutant (R406W) human tau. Proc Natl Acad Sci USA 99:13896–13901.

Terry RD, Masliah E, Salmon DP, Butters N, DeTeresa R, Hill R, Hansen LA, Katzman R (1991) Physical basis of cognitive alterations in Alzheimer's disease: synapse loss is the major correlate of cognitive impairment. Ann Neurol 30:572–580.

Urbanc B, Cruz L, Le R, Sanders J, Ashe KH, Duff K, Stanley HE, Irizarry MC, Hyman BT (2002) Neurotoxic effects of thioflavin S-positive amyloid deposits in transgenic mice and Alzheimer's disease. Proc Natl Acad Sci USA 99:13990–13995.

Van Dam D, D'Hooge R, Staufenbiel M, Van Ginneken C, Van Meir F, De Deyn PP (2003) Age-dependent cognitive decline in APP23 model precedes amyloid deposition. Eur J Neurosci 17:388–396.

Weingarten MD, Lockwood AH, Hwo SY, Kirschner MW (1975) A protein factor essential for microtubule assembly. Proc Natl Acad Sci USA 72:1858–1862.

Weldon DT, O'Hare E, Kuskowski MA, Cleary J, Mach JR, Jr (1996) Alternating lever cyclic-ratio schedule analysis of the effects of atropine sulfate. Pharmacol Biochem Behav 54:753–757.

Westerman MA, Cooper-Blacketer D, Mariash A, Kotilinek L, Kawarabayashi T, Younkin LH, Carlson GA, Younkin SG, Ashe KH (2002) The relationship between A beta and memory in the Tg2576 mouse model of Alzheimer's disease. J Neurosci 22:1858–1867.

Wilcock D, Rojiani A, Rosenthal A, Levkowitz G, Subbarao S, Alamed J, Wilson D, Wilson N, Freeman M, Gordon M, Morgan D (2004) Passive amyloid immunotherapy clears amyloid and transiently activates microglia in a transgenic mouse model of amyloid deposition. J Neurosci 24:6144–6151.

Wong CW, Quaranta V, Glenner GG (1985) Neuritic plaques and cerebrovascular amyloid in Alzheimer disease are antigenically related. Proc Natl Acad Sci USA 82:8729–8732.

Chapter 6

Cholesterol and β-Amyloid

Henri J. Huttunen and Dora M. Kovacs*

Neurobiology of Disease Laboratory, Genetics and Aging Research Unit
MassGeneral Institute for Neurodegenerative Disease (MIND)
Massachusetts General Hospital, Harvard Medical School
Charlestown, MA 02129

1.   Cholesterol homeostasis in cells and the brain

Cholesterol can be considered a molecule of life and death: an essential component of cellular membranes and a precursor of many hormones, and yet a risk factor for many life-threatening diseases, such as Niemann Pick type C disease, Smith-Lemli-Opitz syndrome and cardiovascular disease. Because of cholesterol's central role in determining the physical properties of cellular membranes, even small changes in cholesterol homeostasis will affect numerous cellular events. A good example is a mutation in the NPC1 gene leading to the intracellular accumulation of cholesterol due to a failure to transport unesterified cholesterol out of late endosomal/lysosomal compartments (Carstea et al., 1997; Ikonen and Holtta-Vuori, 2004; Loftus et al., 1997). As the basic mechanisms of Aβ generation have been reviewed elsewhere in this book (see Chapters 1–3), we will begin this chapter at the basics of cholesterol biology and then review how cholesterol may be involved in the pathophysiology of Alzheimer's disease (AD).

1.1   Intracellular cholesterol compartmentation

In peripheral cells, cholesterol is derived from two sources: circulating low-density lipoproteins LDL and VLDL that are taken up via receptor-mediated

---
* **Correspondence to:** Neurobiology of Disease Laboratory, Genetics and Aging Research Unit, Massachusetts General Hospital, Harvard Medical School, 114 16th St., Charlestown, MA 02129, Email: Dora_Kovacs@hms.harvard.edu

endocytosis; and, *de novo* synthesis in the endoplasmic reticulum (ER). In systemic cholesterol homeostasis, liver is the central organ, as it synthesizes most of the needed cholesterol, processes absorbed dietary cholesterol and is also partially responsible for cholesterol catabolism. The liver releases low-density lipoproteins into the circulation mainly as lipoprotein-bound cholesteryl-esters. After endocytosis in peripheral cells, esterified cholesterol is released from the lipoprotein particles in the endo/lysosomal compartment and unesterified cholesterol is transported either to the ER or through the Golgi complex to the plasma membrane (reviewed in Soccio and Breslow, 2004; summarized in Figure 1).

The ER is the crucial regulatory compartment in cellular cholesterol homeostasis. First, the entire cholesterol biosynthetic pathway of 19 steps is catalyzed by 9 enzymes that are localized in the ER (in addition, peroxisomes and plasma membrane might play some minor roles in cholesterol synthesis) (Kovacs et al., 2002; Reinhart et al., 1987). The rate-limiting enzyme of the pathway and the target of statins, HMG CoA reductase, is also localized in ER membranes (Reinhart et al., 1987). The ER membranes are low in cholesterol content and serve as a delicate sensor for changes in cellular cholesterol levels. The level of cholesterol in the ER is translated into the expression of genes regulating lipid metabolism through ER-bound transcription factors, called sterol regulatory element binding proteins (SREBP) (Brown and Goldstein, 1997; Brown and Goldstein, 1999). When cholesterol levels are low, active SREBP is released by proteolytic processing after its transport from the ER to the Golgi by SCAP (SREBP-cleavage activating protein). On the other hand, when cholesterol is abundant, SREBP-SCAP complex binds to the ER retention protein Insig, preventing exit of the complex from the ER (reviewed in Anderson, 2003).

The cholesterol:phospholipid ratio is more than ten-fold higher in the plasma membrane as compared to the ER membrane and eight-fold in the Golgi and endosomes (Colbeau et al., 1971; Reinhart, 1990). Cholesterol constitutes ~35–45% of lipid molecules in the plasma membrane. In addition to differences in cholesterol content among intracellular membranes, cholesterol is also compartmentalized within the Golgi and plasma membrane in lipid rafts. A third type of cholesterol compartmentalization is achieved by the budding of neutral lipid vesicles from the ER as a storage form of ER-associated cholesterol. These three lipid compartments are summarized in Figure 1.

Specific intracellular cholesterol trafficking mechanisms create and maintain marked asymmetries in cholesterol contents among intracellular membranes and compartments. To maintain the ER-plasma membrane gradient, cholesterol must be transported from the ER by transport mechanisms. This transport can occur through two general pathways, vesicular and nonvesicular

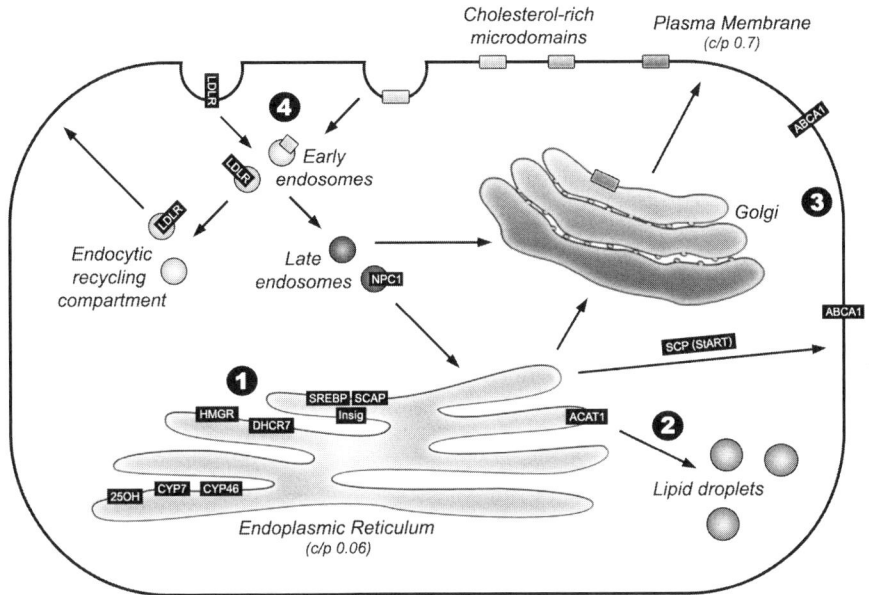

Figure 1. Intracellular trafficking of cholesterol. (1) Endoplasmic reticulum (ER) is the crucial regulatory compartment in cholesterol homeostasis. Synthesis (9 enzymes from HMG CoA reductase (HMGR) to 7-dehydrocholesterol reductase (DHCR7)), esterification (ACAT1) and hydroxylation (e.g. cholesterol 25-hydoxylase (25OH), cholesterol 7α-hydroxylase (CYP7) and cholesterol 24-hydroxylase (CYP46)) of cholesterol occur in the ER, although some hydroxylation may also take place in mitochondria. Furthermore, the cholesterol-sensing mechanism (SREBP, SCAP and Insig) that controls the transcriptional regulation of cholesterogenic enzymes and lipoprotein receptors is located in the ER. Steep concentration gradient of cholesterol is maintained throughout the secretory pathway (cholesterol/phospholipid ratio increases more than ten-fold from ER to plasma membrane). (2) Excess free cholesterol is converted to cholesteryl-esters by ACAT in the ER and stored in neutral, cytoplasmic lipid droplets. (3) Although cholesterol is transported intracellularly in the membranes of vesicles, a nonvesicular, Golgi-independent route appears to be the main ER-to-plasma membrane route for nascent cholesterol. Members of the StAR-related lipid transfer protein (START) family are the main sterol carrier proteins. Reverse cholesterol transport from intracellular pools to extracellular cholesterol acceptors (such as HDL) occurs through ABC-transporters (e.g. ABCA1, ABCG1) located on the plasma membrane. Alternatively, cholesterol/sphingolipid-rich membrane microdomains (caveolae and lipid rafts) are assembled in the Golgi before transport to the plasma membrane. (4) Cholesterol uptake occurs through endocytosis of lipoproteins through lipoprotein receptors such as LDL receptor (LDLR). Released cholesterol is then recycled through the endosomal/lysosomal compartment to the ER, Golgi and plasma membrane. NPC1 is crucial protein in the endocytic sorting plasma membrane-derived cholesterol.

(reviewed in Soccio and Breslow, 2004). Vesicular transport involves the membranes of trafficking vesicles between compartments. Nonvesicular transport between the ER and the plasma membrane is not well defined. Nonvesicular transport, in general, can be mediated by diffusible carrier proteins such as sterol-carrier proteins (SCP) and members of StAR-related lipid transfer protein (START) family that have hydrophobic cavities to bind cholesterol (Soccio and Breslow, 2003). Cholesterol-rich microdomains such as caveolae and lipid rafts form in the Golgi where cholesterol molecules assemble together with sphingomyelin and glycosphingolipids. Cholesterol-rich microdomains are then transported to the plasma membrane to constitute functionally important protein-lipid assemblies (reviewed in Brown and London, 1998; Quest et al., 2004; Simons and Ikonen, 2000). Finally, acyl coenzyme A:cholesterol transferase (ACAT) converts excess unesterified cholesterol to inert cholesterylesters for storage (Buhman et al., 2000; Chang et al., 2001). Cholesteryl-esters are highly hydrophobic and form neutral lipid droplets that pinch off from the ER into the cytoplasm, surrounded by a monolayer of phospholipids and proteins (Martin and Parton, 2005).

Cells can handle excess cellular cholesterol in many ways, via enzymes localized to the ER and also to the mitochondria. As mentioned above, cholesterol can be converted to cholesteryl-esters by the ER-associated ACAT and stored in cytoplasmic lipid storage organelles called lipid droplets or lipid bodies (Martin and Parton, 2005). However, unesterified cholesterol can also be transported out of the cells by plasma membrane ATP-binding cassette transporters (ABC-family proteins; e.g. ABCA1 and ABCA2 in extrahepatic tissues) to become associated with apolipoproteins such as ApoA-I that are ultimately cleared from the plasma by the liver (Santamarina-Fojo et al., 2001). Furthermore, excess unesterified cholesterol can be converted to oxysterols that are excreted from the cells through ABC-family transporters (Knight, 2004). Some cholesterol-metabolizing enzymes such as cholesterol 25-hydroxylase, cholesterol 7α-hydroxylase (CYP7A) and cholesterol 24-hydroxylase (CYP46) (Bjorkhem et al., 1999) are located in the ER (some oxysterol-producing enzymes are located in mitochondria). Oxysterols generated by these enzymes are mostly excreted by the cells but also serve as ligands for liver X receptor (LXR) transcription factors that provide another layer of transcriptional control on cholesterol and fatty acid metabolism (reviewed in Accad and Farese, 1998; Collins, 2004).

Intracellular trafficking of cholesterol is crucial for maintenance of membranes and membrane-dependent cellular functions but also for regulation of cholesterol homeostasis itself. As ER is the central cholesterol sensor of the cell, all pools of cellular cholesterol must have access to the ER in order to regulate cholesterol homeostasis through the SREBP and LXR pathways as well as to be esterified by ACAT for storage in cytoplasmic lipid bodies. Interestingly, changes in ER cholesterol levels not only regulate cholesterol homeostasis but can also affect other cellular processes such as the activation of unfolded protein response (UPR) pathway (Feng et al., 2003).

1.2 Cholesterol metabolism in the brain

Cholesterol is highly enriched in the central nervous system (CNS) compared to other tissues. Although the CNS accounts for only ~2% of whole body mass it contains almost 25% of the total body cholesterol. According to rough estimates, the average concentration of cholesterol in fresh tissues of whole animals is ~2.2 mg/g but in the brain almost 10-fold, in the range of 15–20 mg/g (Dietschy and Turley, 2001; Dietschy and Turley, 2004). Sterols in the CNS are predominantly in the form of unesterified cholesterol with small amounts of cholesterol esters and desmosterol (Dietschy and Turley, 2004). Roughly 70–80% of cholesterol in the adult brain is present in the specialized membranes of myelin, the remainder being in plasma membranes of neurons and glia. As ~90% of the cells in the CNS are glial cells, neurons contain only a small fraction of the total cholesterol pool of the brain (Vance et al., 2005).

Brain cholesterol metabolism is largely independent from systemic cholesterol metabolism. As plasma lipoproteins cannot cross the blood-brain barrier, virtually all cholesterol in the brain is synthesized locally (Dietschy and Turley, 2001). However, it is possible that cerebrovascular endothelial cells could mediate uptake of plasma LDL-cholesterol to the brain to a limited extent (Vance et al., 2005). On the other hand, the blood-brain barrier also restricts the egress of cholesterol from the CNS. For example, in the rat brain cholesterol has a long half-life of 4 to 6 months (Vance et al., 2005). Hydroxylation of cholesterol allows the sterol molecule to be transferred across the blood-brain barrier orders of magnitude faster than cholesterol per se. Thus, the main mechanism by which cholesterol is excreted from the brain is by conversion to 24(S)-hydroxycholesterol via cholesterol 24-hydroxylase (CYP46) (Lund et al., 1999; Lund et al., 2003). However, the mechanism by which 24(S)-hydroxycholesterol crosses the blood–brain barrier is not well understood.

Some of the plasma lipoproteins are also expressed in the CNS, suggesting that these proteins are involved in cholesterol transport among neurons and glia. For example, several members of the LDL receptor family [e.g. LDL receptor (LDLR) and the LDL receptor-related proteins (LRP)] as well as apolipoproteins (Apo) E, A1, D and J, and membrane transporters of the ATP-binding cassette (ABC) family (such as ABCA1, ABCG1 and ABCA4) (Nakamura et al., 2004; Wang et al., 2004), are expressed in the CNS. On the other hand, cerebrospinal fluid has been reported to contain a population of lipoproteins that is distinct from those in plasma (Koch et al., 2001; LaDu et al., 1998; Pitas et al., 1987).

Glia-derived ApoE plays a major role in the cholesterol metabolism in the nervous system. ApoE and ApoJ in the CNS are present in cholesterol-containing lipoprotein particles that are roughly the size and density of plasma HDLs (Koch et al., 2001; LaDu et al., 1998; Pitas et al., 1987). The ApoE-containing lipoproteins in the CNS are proposed to bind to neuronal surface receptors of the LDL receptor family (such as LDLR, LRP and the Apo ER2 receptor) that can mediate uptake of the ApoE-containing lipoproteins. Three common alleles of ApoE are found in humans – ApoE2, E3 and E4; Apo E3 being the most frequently expressed ApoE allele. Importantly, inheritance of the ApoE4 allele is the strongest known risk factor for the development of late-onset Alzheimer's disease (Tanzi and Bertram, 2005; other chapters in this book). Intriguingly, the synthesis of ApoE by glial cells increases dramatically, by up to 150-fold, after injury to the nervous system, suggesting that ApoE is a key molecule for repair processes in the nervous system.

Changes in CNS cholesterol metabolism may be etiologically related to the onset of several neurological diseases. Smith-Lemli-Opitz syndrome (charac-terized by marked abnormalities in brain development and function due to defective cholesterol synthetic pathway) and Niemann-Pick type C disease (characterized by fatal neurodegeneration due to intracellular cholesterol accu-mulation in the endo/lysosomal compartment) represent neurological disorders caused by genetic defects in genes regulating cholesterol homeostasis. Interest-ingly, it has been reported that in patients suffering from Alzheimer's disease the rate of sterol flux across the CNS is elevated and is proportional to the severity of dementia (Lutjohann et al., 2000).

2.   Cholesterol in Aβ biology

Cholesterol homeostasis appears to be intimately linked to various aspects of Aβ biology (summarized in Figure 2). Changes in cholesterol levels can be manifested on cellular physiology on many levels including changes in membrane cholesterol asymmetry, lipid raft structure and function, membrane lipid fluidity and protein conformation. On a mechanistic level, how these changes in cellular cholesterol homeostasis are linked to various aspects of APP and Aβ biology is not well understood. However, statins and ACAT inhibitors appear to be promising therapeutic strategies for AD since both reduce amyloid plaque density in animal models of the disease (discussed later in this chapter).

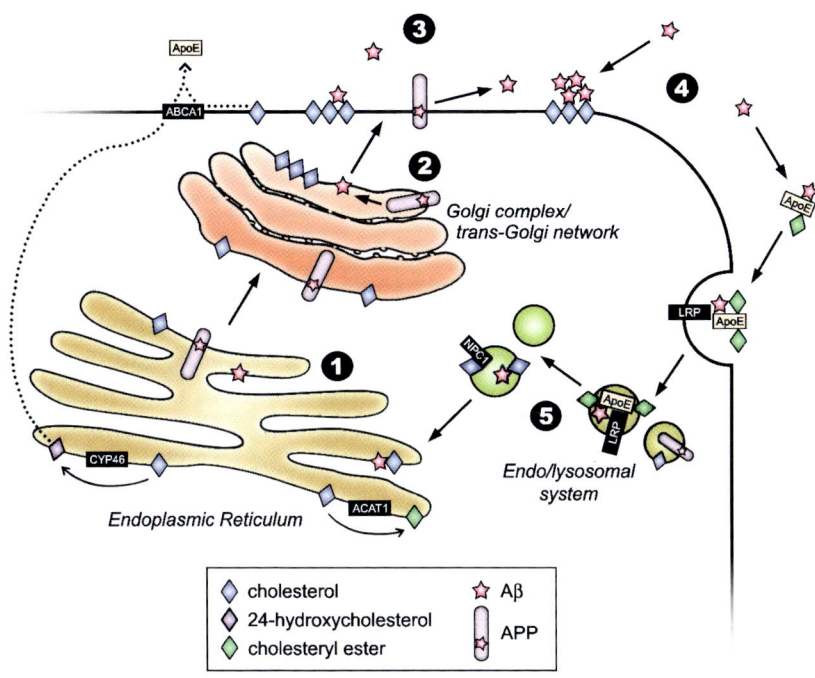

Figure 2. Cholesterol and the cell biology of Aβ. (1) Newly synthesized APP is N-glycosylated in the endoplasmic reticulum (ER). The ER is also the site of cholesterol esterification and hydroxylation, although how these processes mechanistically relate to APP is not well understood. (2) Partially matured APP is transported to the Golgi complex where it completes maturation and is finally transported to the plasma membrane. β- and γ-secretases activities can be found in newly assembled cholesterol/sphingolipid-rich membrane microdomains in the Golgi. APP processing is thought to occur mostly in the trans-Golgi network, the plasma membrane, and endosomes. (3) Unprocessed, mature APP is transported to the plasma membrane. The newly generated Aβ is either released to the extracellular space or remains associated with the plasma membrane and lipid raft structures. Excess unesterified cholesterol in the plasma membrane and ER- and mitochondria-derived oxysterols are loaded to extracellular lipid acceptors such as ApoE by ABCA1. (4) Monomeric Aβ can bind to numerous molecules in the extracellular space. Binding of Aβ to ganglioside GM1 in the cholesterol/sphingolipid-rich membrane microdomains (lipid rafts) strongly favors Aβ aggregation. Binding of ApoE to Aβ regulates aggregation but also cellular uptake of Aβ. (5) ApoE-Aβ complexes are taken up by the cells through receptor-mediated endocytosis mediated by e.g. LRP (LDL receptor-related protein) and LDLR. Lipoprotein-bound cholesteryl esters are hydrolyzed in the endo/lysosomal system and cholesterol is transported through NPC1-containing compartment to the ER and Golgi. Endocytosed Aβ also has access to other subcellular compartments through the vesicular transport system. In addition, Aβ may be generated in the endo/lysosomal system from endocytosed APP.

2.1   Cholesterol and Aβ generation

Of all aspects of Aβ biology linked to cholesterol, Aβ generation is the one supported by strongest experimental evidence. Aβ is generated from amyloid precursor protein (APP) by two successive proteolytic cleavages that are catalyzed by β-secretase (BACE1) and γ-secretase (a heteromultimeric complex containing presenilin, nicastrin, Aph-1 and Pen-2). APP can also be cleaved by α-secretases (ADAM10, ADAM17 or TACE) within the Aβ domain (for details, see Chapters 1–3).

Cholesterol is a major constituent of lipid rafts together with sphingolipids (sphingomyelin and glycosphingolipids) and glycosylphosphatidylinositol (GPI)-linked proteins. As several transmembrane receptors and cytoplasmic signaling molecules have been shown to concentrate to lipid rafts, they are considered as functional hot spots of the plasma membrane (reviewed in Cohen et al., 2004; Quest et al., 2004; Simons and Ikonen, 1997). Lipid rafts are assembled in the Golgi complex (Simons and Ikonen, 1997) and are constantly endocytosed. Amyloidogenic processing of APP depends on cholesterol-rich membrane domains whereas "nonamyloidogenic" α-cleavage of APP occurs mainly in the phospholipid-rich domain of plasma membrane (Ehehalt et al., 2003). Both BACE1 and γ-secretase complex have been shown to localize to lipid rafts (Riddell et al., 2001; Vetrivel et al., 2004). BACE1 appears to be particularly sensitive to membrane cholesterol content and targeting of BACE1 to the lipid rafts by adding a GPI-anchor strongly upregulates its activity (Cordy et al., 2003). Importantly, the enzymes of the α-secretase pathway (e.g. ADAM10, ADAM17 and TACE) preferentially partition to the phospholipid-rich (cholesterol-poor) domain of the plasma membrane (Kojro et al., 2001). A recent study suggests that segregation of β- and γ-secretases to cholesterol-rich membrane domains restricts their access to APP and that aberrant access of APP to these membrane domains results in increased Aβ generation (Abad-Rodriguez et al., 2004). Thus, localization of APP to the subdomains of the plasma membrane could directly regulate whether nonamyloidogenic or amyloidogenic pathway is utilized for APP processing (Abad-Rodriguez et al., 2004). Interestingly, both monomeric and dimeric species of Aβ were shown to be concentrated in lipid rafts both in the transgenic mouse model (Tg2576) of AD as well as human AD brains (Kawarabayashi et al., 2004; Lee et al., 1998). It is possible that these lipid raft-associated dimeric Aβ species were produced from dimerized APP.

Changes in cellular levels of cholesterol and distribution of cholesterol induced by pharmacological agents such as statins and ACAT inhibitors have repeatedly been shown to modulate Aβ generation and will be reviewed below. In animal models, hypercholesterolaemia induced by high cholesterol diets increased deposition of Aβ in brains of rabbits (Sparks et al., 1994) and double-mutant transgcnic mice (PSAPP) that overexpress APP (Tg2567) and presenilin 1 (PS1) (Refolo et al., 2000). However, one study showed that diet-induced hypercholesterolaemia caused a decrease in brain levels of secreted Aβ in transgenic mice expressing APP with Swedish mutation (Howland et al., 1998).

2.2 Cholesterol in Aβ aggregation and clearance

Altered Aβ aggregation and/or clearance have repeatedly been implicated in the pathogenesis of late-onset AD. Ganglioside GM1-bound Aβ has been shown to act as an efficient seed for Aβ aggregation (Choo-Smith et al., 1997; Yanagisawa and Matsuzaki, 2002; Yanagisawa et al., 1995). The binding of Aβ to GM1 is markedly accelerated in cholesterol-rich environments and it has been suggested that soluble Aβ recognizes the GM1 cluster in cholesterol-rich membrane domains as a receptor and binds to it to initiate Aβ aggregation (Kakio et al., 2001). Increases in membrane cholesterol content, which are likely to occur during aging, could favor amyloid fibril formation through this mechanism.

ApoE is involved in many aspects of Aβ metabolism, including Aβ deposition and clearance (DeMattos, 2004; DeMattos et al., 2004; Hone et al., 2003). The cellular mechanisms underlying the effects of ApoE on amyloid deposition and plaque formation are not well understood but are likely to be independent of cholesterol and involve a direct interaction between ApoE and Aβ that alters the metabolism and/or clearance of Aβ (Bales et al., 2002; LaDu et al., 1994). For further details on Aβ aggregation and clearance, see Chapter 10 of this volume.

2.3 Aβ and cholesterol homeostasis

Changes in cholesterol levels alter APP expression and production of Aβ (Gibson Wood et al., 2003). Interestingly, several studies have suggested that Aβ itself may modulate cholesterol homeostasis and properties of cellular membranes. Extracellular Aβ has been shown to stimulate uptake of cholesterol-containing ApoE-lipid complexes into hippocampal neurons and Chinese hamster ovary cells (Beffert et al., 1998; Yang et al., 1999). Oligomeric $Aβ_{1-40}$ induced removal of cholesterol, PC and GM1-ganglioside from cultured rat neurons and astrocytes without an exogenous lipid acceptor (Michikawa et al., 2001). On the other hand, Aβ can promote production of reactive oxygen species (Butterfield et al., 2002) that could promote oxidation of lipoproteins. Oxidized lipoproteins are taken up into the cells faster than their unoxidized counterparts. Whether these contrasting effects of Aβ on cholesterol uptake and release have disturbing effects on cellular cholesterol equilibrium is currently not known.

Several studies have suggested that even low concentrations of $A\beta$ can alter the fluidity of cellular membranes (Muller et al., 2001). $A\beta$ aggregates bind to lipids, preferentially to cholesterol, and it has been suggested that extracellular $A\beta$ aggregates could extract cholesterol directly from membranes (Avdulov et al., 1997). Alternatively, extracellular $A\beta$ could interfere with cholesterol efflux pathways by complexing with plasma membrane proteins involved in cholesterol efflux. Acceptors for ABCA1-mediated cholesterol efflux include apolipoproteins such as ApoA-I that contain amphipathic helices. As $A\beta$ is an amphipathic molecule that binds lipids in an aggregated state (Avdulov et al., 1997) it is possible that $A\beta$ could modulate ABCA1-mediated cholesterol efflux. Finally, it has been suggested that $A\beta$ may inhibit cholesterol esterification and lead to altered distribution of unesterified cholesterol in cells (Liu et al., 1998).

3.   Potential $A\beta$-lowering therapeutics based on cholesterol

Altogether, it appears that the balance between amyloidogenic and non-amyloidogenic APP processing is highly sensitive to changes in intracellular cholesterol distribution. In this section, we will briefly summarize cell-based, animal and clinical studies assessing how cholesterol affects $A\beta$ production as this approach may hold great promise as future therapeutics for Alzheimer's disease.

3.1   Inhibition of HMG CoA reductase (statins)

Statins are a highly successful class of drugs that inhibit HMG CoA reductase, the rate-limiting enzyme in the cholesterol synthetic pathway. These drugs have a good safety profile and are widely used to effectively reduce serum cholesterol. Statins have been shown to attenuate $A\beta$ production in cell-based and animal models of AD and in humans. In cell-based studies, physiological concentrations of lovastatin lower cellular cholesterol levels and specifically inhibit $A\beta$ generation (Buxbaum et al., 2001). Similarly, simvastatin was shown to reduce the production of $A\beta$ by APP-transfected neurons *in vitro* (Fassbender et al., 2001).

Several animal studies, including studies in rabbits, guinea pigs and transgenic mice, have shown a strong connection between plasma cholesterol and $A\beta$ production (Fassbender et al., 2001; Refolo et al., 2000; Sparks et al., 2000; Sparks et al., 1994). Various statins introduced into these animal models of AD have produced highly promising results in terms of reducing cerebral $A\beta$ levels. Simvastatin treatment strongly reduced $A\beta$ levels in both cerebrospinal fluid (CSF) and brain homogenate of guinea pigs (Fassbender et al., 2001). In another study, atorvastatin treatment was shown to reduce $A\beta$ deposition in PSAPP transgenic mouse model of AD (Petanceska et al., 2002). Another cholesterol-lowering drug, BM15.766, strongly reduced Aβ generation and plaque load in Tg2576 transgenic mouse model of AD (Refolo et al., 2001). Statins do not appear to modulate brain cholesterol levels in ApoE knockout mice suggesting that statins regulate brain cholesterol through a mechanism that requires ApoE (Eckert et al., 2001).

Epidemiological studies indicate that the prevalence and incidence of AD in subjects taking statins is remarkably decreased (Jick et al., 2000; Wolozin et al., 2000). In these initial retrospective studies, lovastatin and pravastatin but not simvastatin reduced the prevalence of AD in cardiovascular patients by up to 73%. Treating subjects with doses of statins used in clinical management of hypercholesterolemia has been shown to reduce Aβ in human plasma and in cerebrospinal fluid by almost 40% (Friedhoff et al., 2001).

Prospective studies on statin therapy for AD have produced mixed results. Simvastatin treatment was reported to reduce cognitive decline in a small cohort of 26 mild AD subjects (Simons et al., 2002). However, pravastatin failed to reduce the incidence of AD in the PROSPER study (a large prospective analysis of ~6000 subjects with high cholesterol who were at risk for cardiovascular disease) (Shepherd et al., 2002). The reasons for the failure of pravastatin to prevent AD in this study are unclear. Although statin treatment effectively blocks cholesterol synthesis in the human CNS (Fassbender et al., 2002), it only reduces Aβ levels in the CSF of subjects without or with only mild dementia (Simons et al., 2002). Although it seems unlikely that statins would be beneficial for patients with severe AD, statin treatment seems to delay the cognitive and behavioral decline in mild to moderate AD patients (Sparks et al., 2005). Several clinical trials are ongoing to test the efficacy of various statins as a preventative or therapeutic treatment for AD.

Both hydrophobic statins and hydrophilic statins appear to reduce the levels of 24-hydroxycholesterol (Vega et al., 2003). This is surprising considering that the blood-brain barrier permeability of statins differs greatly depending on lipophilicity. Whether the statin-induced reduction in cerebral Aβ levels depends on the cholesterol-lowering effects of statins or on other mechanisms remains an open question. Cholesterol-independent, indirect anti-inflammatory and antioxidant effects might be important for the efficacy of statins towards AD (Akiyama et al., 2000; Cordle and Landreth, 2005; Kwak et al., 2000; Menge et al., 2005). For example, statins inhibit the production of L-mevalonate, a key intermediate in the cholesterol synthetic pathway. Metabolites of L-mevalonate are involved in post-translational modification of proteins, regulation of gene expression, cell proliferation and differentiation (Menge et al., 2005). Interestingly, recent results suggest that statins modulate APP processing independently of their cholesterol-lowering effect as statin-induced inhibition of the isoprenoid pathway increases α-secretase cleavage of APP in cell-based studies (Pedrini et al., 2005). Statin-induced inhibition of the isoprenoid pathway was also shown to attenuate $A\beta$-induced microglial activation in cells (Cordle and Landreth, 2005). These ambiguous actions of statins on $A\beta$ production and other aspects of AD *in vivo* may partially explain mixed clinical results examining whether statins are beneficial in treating AD.

3.2   Inhibition of ACAT

Although the beneficial effects of statins for AD may be due to their pleiotropic actions, other cholesterol-modifying strategies have been shown to attenuate $A\beta$ production. ACAT is an ER-resident enzyme responsible for conversion of excess free cholesterol to cholesteryl esters (Buhman et al., 2000; Chang et al., 1997; Chang et al., 2001). Inhibition of ACAT function in cells either by genetic or pharmacological means has been shown to efficiently suppress $A\beta$ generation (Puglielli et al., 2001). Remarkably, in a cell line (AC29) derived from Chinese hamster ovary cells that overproduces cholesterol but lacks functional ACAT, both β- and γ-cleavages of APP are almost completely inhibited.

Several small molecule inhibitors of ACAT have been developed for treatment of hyperlipidaemia and atherosclerosis (Giovannoni et al., 2003). To date, none of these have proved superior to statins in clinical studies in terms of efficacy. However, the long continued interest towards ACAT in pharmaceutical industry has provided ACAT inhibitors that are safe for human use and can be used to study the role of ACAT in Alzheimer's disease. The ACAT inhibitor CP-113,818 was recently tested in a mouse model of AD with encouraging results. Two-month treatment with CP-113,818 remarkably reduced amyloid pathology and correlated with improved spatial learning in transgenic mice expressing human APP751 containing the London (V717I) and Swedish (K670M/N671L) mutations (Hutter-Paier et al., 2004). As the study of ACAT inhibitors for AD are still in the preclinical stage, no clinical data is available yet.

3.3   Increased cholesterol efflux and oxysterol mimics

In addition to esterification, cells handle excess unesterified cholesterol through hydroxylation and secretion of resulting oxysterols. Oxysterols stimulate transcription of the cholesterol transporter ABCA1, and ABCA1 mediates the efflux of cholesterol and oxysterols from the cell (Knight, 2004). ABCA1 is partially regulated by the liver X transcription factors (LXR) and synthetic agonists of the LXR receptor have been shown reduce $A\beta$ secretion (Brown et al., 2004; Koldamova et al., 2003; Sun et al., 2003). Interestingly, overexpression of ABCA1 in cells resulted in decreased levels β- and γ-cleavage products of APP independently of cellular lipid efflux (Sun et al., 2003). Furthermore, the natural LXR agonists, oxysterols, also inhibit β-amyloid secretion supporting a role for LXR in regulating β-amyloid secretion (Brown et al., 2004).

Although modulation of ABCA1 expression/activity has not been tested in clinical settings, a synthetic LXR receptor agonist, T0901317, has been shown to lower amyloidogenic processing of APP in a mouse model of AD (Koldamova et al., 2005). Remarkably, both ABC transporters and CYP46, the main oxysterol-producing enzyme in the CNS, have been linked to the pathophysiology of Alzheimer's disease. Polymorphisms in ABCA1, ABCA2 and CYP46 genes have been associated with the prevalence of Alzheimer's disease (Mace et al., 2005; Papassotiropoulos et al., 2003; Wollmer et al., 2003). CYP46 expression was shown to increase around neuritic plaques likely reflecting the need to remove cholesterol from dystrophic neurites (Brown et al., 2004).

ApoE plays a critical role in determining when and where in the brain Aβ will aggregate. ABCA1 is an important regulator of the lipidation state and levels of ApoE in the brain. Although the role of ABCA1 in Aβ biology appears to be complex, it may turn out to be a good therapeutic target for AD. Clearly, further studies are required to fully evaluate how ABCA1, reverse cholesterol transport or mimicking oxysterol action in the brain could be used as a therapeutic approach in Alzheimer's disease.

3.4 Modulation of NPC1 activity

Niemann-Pick type C disease and Alzheimer's disease share several histopathological similarities (Nixon, 2004). Similar patterns of neurofibrillary tangles, tau phosphorylation and endosomal abnormalities are found in the brains of both AD and NPC disease patients (reviewed in Vance et al., 2005). In a mouse model of Niemann-Pick type C disease accumulation of cholesterol in late endosomes/lysosomes is associated with abnormal distribution of presenilin-1 and increased generation of β-CTFs of APP as well as Aβ (Burns et al., 2003). Either the lack of functional NPC1 protein (as in NPC disease) or treatment of cells with the cholesterol transport inhibitor U18666A leads to increased γ-secretase activity, abnormal APP processing and increased Aβ production through accumulation of cholesterol in the endosomal trafficking system (Burns et al., 2003; Runz et al., 2002). These findings suggest that modulation of NPC1 activity could be used to lower Aβ production in cells.

4. Conclusions and future prospects

The presence of cholesterol in biological membranes strongly affects their properties. APP is a transmembrane protein, and the last step in the cascade that produces Aβ occurs through intramembrane proteolysis. Thus, it is not surprising that cholesterol plays a central role in Aβ generation. However, the relationship between cholesterol and Aβ is more complex. In addition to Aβ generation, cholesterol also regulates Aβ aggregation and clearance and might affect the neurotoxic mechanisms of Aβ. Conversely, Aβ has effects on cholesterol homeostasis. Studies with cell-based and animal models as well as studies in clinical settings have provided a wealth of new information on the role of cholesterol in Aβ biology and AD pathophysiology. These advances are expected to provide hope for the development of a potential cholesterol-based therapeutic strategy for this disease. Statins are already widely and effectively used to lower serum cholesterol. The efficacy of statins towards AD still remains an open question and requires further clinical proof. In addition to HMG CoA reductase, other targets such as ACAT, ABC-family transporters and liver X transcription factors have recently emerged as potential new targets for pharmacological intervention and treatment of AD.

References

Abad-Rodriguez J, Ledesma MD, Craessaerts K, Perga S, Medina M, Delacourte A, Dingwall C, De Strooper B, Dotti CG (2004) Neuronal membrane cholesterol loss enhances amyloid peptide generation. J Cell Biol 167:953–960.

Accad M, Farese RV Jr (1998) Cholesterol homeostasis: a role for oxysterols. Curr Biol 8:R601–604.

Akiyama H, Barger S, Barnum S, Bradt B, Bauer J, Cole GM, Cooper NR, Eikelenboom P, Emmerling M, Fiebich BL, et al. (2000) Inflammation and Alzheimer's disease. Neurobiol Aging 21:383–421.

Anderson RG (2003) Joe Goldstein and Mike Brown: from cholesterol homeostasis to new paradigms in membrane biology. Trends Cell Biol 13:534–539.

Avdulov NA, Chochina SV, Igbavboa U, Warden CS, Vassiliev AV, Wood WG (1997) Lipid binding to amyloid beta-peptide aggregates: preferential binding of cholesterol as compared with phosphatidylcholine and fatty acids. J Neurochem 69:1746–1752.

Bales KR, Dodart JC, DeMattos RB, Holtzman DM, Paul SM (2002) Apolipoprotein E, amyloid, and Alzheimer disease. Mol Interv 2:363–375, 339.

Beffert U, Aumont N, Dea D, Lussier-Cacan S, Davignon J, Poirier J (1998) Beta-amyloid peptides increase the binding and internalization of apolipoprotein E to hippocampal neurons. J Neurochem 70:1458–1466.

Bjorkhem I, Diczfalusy U, Lutjohann D (1999) Removal of cholesterol from extrahepatic sources by oxidative mechanisms. Curr Opin Lipidol 10:161–165.

Brown DA, London E (1998) Functions of lipid rafts in biological membranes. Annu Rev Cell Dev Biol 14:111–136.

Brown J, 3rd, Theisler C, Silberman S, Magnuson D, Gottardi-Littell N, Lee JM, Yager D, Crowley J, Sambamurti K, Rahman MM, et al. (2004) Differential expression of cholesterol hydroxylases in Alzheimer's disease. J Biol Chem 279:34674–34681.

Brown MS, Goldstein JL (1997) The SREBP pathway: regulation of cholesterol metabolism by proteolysis of a membrane-bound transcription factor. Cell 89:331–340.

Brown MS, Goldstein JL (1999) A proteolytic pathway that controls the cholesterol content of membranes, cells, and blood. Proc Natl Acad Sci USA 96:11041–11048.

Buhman KF, Accad M, Farese RV (2000) Mammalian acyl-CoA:cholesterol acyltransferases. Biochim Biophys Acta 1529:142–154.

Burns M, Gaynor K, Olm V, Mercken M, LaFrancois J, Wang L, Mathews PM, Noble W, Matsuoka Y, Duff K (2003) Presenilin redistribution associated with aberrant cholesterol transport enhances beta-amyloid production in vivo. J Neurosci 23:5645–5649.

Butterfield DA, Griffin S, Munch G, Pasinetti GM (2002) Amyloid beta-peptide and amyloid pathology are central to the oxidative stress and inflammatory cascades under which Alzheimer's disease brain exists. J Alzheimers Dis 4:193–201.

Buxbaum JD, Geoghagen NS, Friedhoff LT (2001) Cholesterol depletion with physiological concentrations of a statin decreases the formation of the Alzheimer amyloid Abeta peptide. J Alzheimers Dis 3:221–229.

Carstea ED, Morris JA, Coleman KG, Loftus SK, Zhang D, Cummings C, Gu J, Rosenfeld MA, Pavan WJ, Krizman DB, et al. (1997) Niemann-Pick C1 disease gene: homology to mediators of cholesterol homeostasis. Science 277:228–231.

Chang TY, Chang CC, Cheng D (1997) Acyl-coenzyme A:cholesterol acyltransferase. Annu Rev Biochem 66:613–638.

Chang TY, Chang CC, Lin S, Yu C, Li BL, Miyazaki A (2001) Roles of acyl-coenzyme A:cholesterol acyltransferase-1 and -2. Curr Opin Lipidol 12:289–296.

Choo-Smith LP, Garzon-Rodriguez W, Glabe CG, Surewicz WK (1997) Acceleration of amyloid fibril formation by specific binding of Abeta-(1-40) peptide to ganglioside-containing membrane vesicles. J Biol Chem 272:22987–22990.

Cohen AW, Hnasko R, Schubert W, Lisanti MP (2004) Role of caveolae and caveolins in health and disease. Physiol Rev 84:1341–1379.

Colbeau A, Nachbaur J, Vignais PM (1971) Enzymic characterization and lipid composition of rat liver subcellular membranes. Biochim Biophys Acta 249:462–492.

Collins JL (2004) Therapeutic opportunities for liver X receptor modulators. Curr Opin Drug Discov Devel 7:692–702.

Cordle A, Landreth G (2005) 3-hydroxy-3-methylglutaryl-coenzyme A reductase inhibitors attenuate beta-amyloid-induced microglial inflammatory responses. J Neurosci 25:299–307.

Cordy JM, Hussain I, Dingwall C, Hooper NM, Turner AJ (2003) Exclusively targeting beta-secretase to lipid rafts by GPI-anchor addition up-regulates beta-site processing of the amyloid precursor protein. Proc Natl Acad Sci USA 100:11735–11740.

DeMattos RB (2004) Apolipoprotein E dose-dependent modulation of beta-amyloid deposition in a transgenic mouse model of Alzheimer's disease. J Mol Neurosci 23:255–262.

DeMattos RB, Cirrito JR, Parsadanian M, May PC, O'Dell MA, Taylor JW, Harmony JA, Aronow BJ, Bales KR, Paul SM, Holtzman DM (2004) ApoE and Clusterin Cooperatively Suppress Abeta Levels and Deposition. Evidence that ApoE Regulates Extracellular Abeta Metabolism In Vivo. Neuron 41:193–202.

Dietschy JM, Turley SD (2001) Cholesterol metabolism in the brain. Curr Opin Lipidol 12:105–112.

Dietschy JM, Turley SD (2004) Thematic review series: brain Lipids. Cholesterol metabolism in the central nervous system during early development and in the mature animal. J Lipid Res 45:1375–1397.

Eckert GP, Kirsch C, Mueller WE (2001) Differential effects of lovastatin treatment on brain cholesterol levels in normal and apoE-deficient mice. Neuroreport 12:883–887.

Ehehalt R, Keller P, Haass C, Thiele C, Simons K (2003) Amyloidogenic processing of the Alzheimer beta-amyloid precursor protein depends on lipid rafts. J Cell Biol 160:113–123.

Fassbender K, Simons M, Bergmann C, Stroick M, Lutjohann D, Keller P, Runz H, Kuhl S, Bertsch T, von Bergmann K, et al. (2001) Simvastatin strongly reduces levels of Alzheimer's disease beta - amyloid peptides Abeta 42 and Abeta 40 in vitro and in vivo. Proc Natl Acad Sci USA 98:5856–5861.

Fassbender K, Stroick M, Bertsch T, Ragoschke A, Kuehl S, Walter S, Walter J, Brechtel K, Muehlhauser F, Von Bergmann K, and Lutjohann D (2002) Effects of statins on human cerebral cholesterol metabolism and secretion of Alzheimer amyloid peptide. Neurology 59:1257–1258.

Feng B, Yao PM, Li Y, Devlin CM, Zhang D, Harding HP, Sweeney M, Rong JX, Kuriakose G, Fisher EA, et al. (2003) The endoplasmic reticulum is the site of cholesterol-induced cytotoxicity in macrophages. Nat Cell Biol 5:781–792.

Friedhoff LT, Cullen EI, Geoghagen NS, Buxbaum JD (2001) Treatment with controlled-release lovastatin decreases serum concentrations of human beta-amyloid (Abeta) peptide. Int J Neuropsychopharmacol 4:127–130.

Gibson Wood W, Eckert GP, Igbavboa U, Muller WE (2003) Amyloid beta-protein interactions with membranes and cholesterol: causes or casualties of Alzheimer's disease. Biochim Biophys Acta 1610:281–290.

Giovannoni MP, Piaz VD, Vergelli C, Barlocco D (2003) Selective ACAT inhibitors as promising antihyperlipidemic, antiathero-sclerotic and anti-Alzheimer drugs. Mini Rev Med Chem 3:576–584.

Hone E, Martins IJ, Fonte J, Martins RN (2003) Apolipoprotein E influences amyloid-beta clearance from the murine periphery. J Alzheimers Dis 5:1–8.

Howland DS, Trusko SP, Savage MJ, Reaume AG, Lang DM, Hirsch JD, Maeda N, Siman R, Greenberg BD, Scott RW, Flood DG (1998) Modulation of secreted beta-amyloid precursor protein and amyloid beta-peptide in brain by cholesterol. J Biol Chem 273:16576–16582.

Hutter-Paier B, Huttunen HJ, Puglielli L, Eckman CB, Kim DY, Hofmeister A, Moir RD, Domnitz SB, Frosch MP, Windisch M, Kovacs DM (2004) The ACAT Inhibitor CP-113,818 Markedly Reduces Amyloid Pathology in a Mouse Model of Alzheimer's Disease. Neuron 44:227–238.

Ikonen E, Holtta-Vuori M (2004) Cellular pathology of Niemann-Pick type C disease. Semin Cell Dev Biol 15:445–454.

Jick H, Zornberg GL, Jick SS, Seshadri S, Drachman DA (2000) Statins and the risk of dementia. Lancet 356:1627–1631.

Kakio A, Nishimoto SI, Yanagisawa K, Kozutsumi Y, Matsuzaki K (2001) Cholesterol-dependent formation of GM1 ganglioside-bound amyloid beta-protein, an endogenous seed for Alzheimer amyloid. J Biol Chem 276:24985–24990.

Kawarabayashi T, Shoji M, Younkin LH, Wen-Lang L, Dickson DW, Murakami T, Matsubara E, Abe K, Ashe KH, Younkin SG (2004) Dimeric amyloid beta protein rapidly accumulates in lipid rafts followed by apolipoprotein E and phosphorylated tau accumulation in the Tg2576 mouse model of Alzheimer's disease. J Neurosci 24:3801–3809.

Knight BL (2004) ATP-binding cassette transporter A1: regulation of cholesterol efflux. Biochem Soc Trans 32:124–127.

Koch S, Donarski N, Goetze K, Kreckel M, Stuerenburg HJ, Buhmann C, Beisiegel U (2001) Characterization of four lipoprotein classes in human cerebrospinal fluid. J Lipid Res 42:1143–1151.

Kojro E, Gimpl G, Lammich S, Marz W, Fahrenholz F (2001) Low cholesterol stimulates the nonamyloidogenic pathway by its effect on the alpha-secretase ADAM 10. Proc Natl Acad Sci USA 98:5815–5820.

Koldamova RP, Lefterov IM, Ikonomovic MD, Skoko J, Lefterov PI, Isanski BA, DeKosky ST, Lazo JS (2003) 22R-hydroxycholesterol and 9-cis-retinoic acid induce ATP-binding cassette transporter A1 expression and cholesterol efflux in brain cells and decrease amyloid beta secretion. J Biol Chem 278:13244–13256.

Koldamova RP, Lefterov IM, Staufenbiel M, Wolfe D, Huang S, Glorioso JC, Walter M, Roth MG, Lazo JS (2005) The liver X receptor ligand T0901317 decreases amyloid beta production in vitro and in a mouse model of Alzheimer's disease. J Biol Chem 280:4079–4088.

Kovacs WJ, Olivier LM, Krisans SK (2002) Central role of peroxisomes in isoprenoid biosynthesis. Prog Lipid Res 41:369–391.

Kwak B, Mulhaupt F, Myit S, Mach F (2000) Statins as a newly recognized type of immunomodulator. Nat Med 6:1399–1402.

LaDu MJ, Falduto MT, Manelli AM, Reardon CA, Getz GS, Frail DE (1994) Isoform-specific binding of apolipoprotein E to ß-amyloid. J.Biol.Chem. 269:23403–23406.

LaDu MJ, Gilligan SM, Lukens JR, Cabana VG, Reardon CA, Van Eldik LJ, Holtzman DM (1998) Nascent astrocyte particles differ from lipoproteins in CSF. J Neurochem 70:2070–2081.

Lee SJ, Liyanage U, Bickel PE, Xia W, Lansbury PTJr, Kosik KS (1998) A detergent-insoluble membrane compartment contains A beta in vivo. Nat Med 4:730–734.

Liu Y, Peterson DA, Schubert D (1998) Amyloid beta peptide alters intracellular vesicle trafficking and cholesterol homeostasis. Proc Natl Acad Sci USA 95:13266–13271.

Loftus SK, Morris JA, Carstea ED, Gu JZ, Cummings C, Brown A, Ellison J, Ohno K, Rosenfeld MA, Tagle DA, et al. (1997) Murine model of Niemann-Pick C disease: mutation in a cholesterol homeostasis gene. Science 277:232–235.

Lund EG, Guileyardo JM, Russell DW (1999) cDNA cloning of cholesterol 24-hydroxylase, a mediator of cholesterol homeostasis in the brain. Proc Natl Acad Sci USA 96:7238–7243.

Lund EG, Xie C, Kotti T, Turley SD, Dietschy JM, Russell D W (2003) Knockout of the cholesterol 24-hydroxylase gene in mice reveals a brain-specific mechanism of cholesterol turnover. J Biol Chem 278:22980–22988.

Lutjohann D, Papassotiropoulos A, Bjorkhem I, Locatelli S, Bagli M, Oehring RD, Schlegel U, Jessen F, Rao ML, von Bergmann K, Heun R (2000) Plasma 24S-hydroxycholesterol (cerebrosterol) is increased in Alzheimer and vascular demented patients. J Lipid Res 41:195–198.

Mace S, Cousin E, Ricard S, Genin E, Spanakis E, Lafargue-Soubigou C, Genin B, Fournel R, Roche S, Haussy G, et al. (2005) ABCA2 is a strong genetic risk factor for early-onset Alzheimer's disease. Neurobiol Dis 18:119–125.

Martin S, Parton RG (2005) Caveolin, cholesterol, and lipid bodies. Semin Cell Dev Biol 16:163–174.

Menge T, Hartung HP, Stuve O (2005) Statins–a cure-all for the brain? Nat Rev Neurosci 6:325–331.

Michikawa M, Gong JS, Fan QW, Sawamura N, Yanagisawa K (2001) A novel action of alzheimer's amyloid beta-protein (Abeta): oligomeric Abeta promotes lipid release. J Neurosci 21:7226–7235.

Muller WE, Kirsch C, Eckert GP (2001) Membrane-disordering effects of beta-amyloid peptides. Biochem Soc Trans 29:617–623.

Nakamura K, Kennedy MA, Baldan A, Bojanic DD, Lyons K, Edwards PA (2004) Expression and regulation of multiple murine ATP-binding cassette transporter G1 mRNAs/isoforms that stimulate cellular cholesterol efflux to high density lipoprotein. J Biol Chem 279:45980–45989.

Nixon RA (2004) Niemann-Pick Type C disease and Alzheimer's disease: the APP-endosome connection fattens up. Am J Pathol 164:757–761.

Papassotiropoulos A, Streffer JR, Tsolaki M, Schmid S, Thal D, Nicosia F, Iakovidou V, Maddalena A, Lutjohann D, Ghebremedhin E, et al. (2003) Increased brain beta-amyloid load, phosphorylated tau, and risk of Alzheimer disease associated with an intronic CYP46 polymorphism. Arch Neurol 60:29–35.

Pedrini S, Carter TL, Prendergast G, Petanceska S, Ehrlich ME, Gandy S (2005) Modulation of statin-activated shedding of Alzheimer APP ectodomain by ROCK. PLoS Med 2:e18.

Petanceska SS, DeRosa S, Olm V, Diaz N, Sharma A, Thomas-Bryant T, Duff K, Pappolla M, Refolo LM (2002) Statin therapy for Alzheimer's disease: will it work? J Mol Neurosci 19:155–161.

Pitas RE, Boyles JK, Lee SH, Hui D, Weisgraber KH (1987) Lipoproteins and their receptors in the central nervous system. Characterization of the lipoproteins in cerebrospinal fluid and identification of apolipoprotein B,E(LDL) receptors in the brain. J Biol Chem 262:14352–14360.

Puglielli L, Konopka G, Pack-Chung E, Ingano LA, Berezovska O, Hyman BT, Chang TY, Tanzi RE, Kovacs DM (2001) Acyl-coenzyme A: cholesterol acyltransferase modulates the generation of the amyloid beta-peptide. Nat Cell Biol 3:905–912.

Quest, AF, Leyton L, Parraga M (2004) Caveolins, caveolae, and lipid rafts in cellular transport, signaling, and disease. Biochem Cell Biol 82:129–144.

Refolo LM, Malester B, LaFrancois J, Bryant-Thomas T, Wang R, Tint GS, Sambamurti K, Duff K, Pappolla MA (2000) Hypercholesterolemia accelerates the Alzheimer's amyloid pathology in a transgenic mouse model. Neurobiol Dis 7:321–331.

Refolo LM, Pappolla MA, LaFrancois J, Malester B, Schmidt SD, Thomas-Bryant T, Tint GS, Wang R, Mercken M, Petanceska SS, Duff KE (2001) A Cholesterol-Lowering Drug Reduces beta-Amyloid Pathology in a Transgenic Mouse Model of Alzheimer's Disease. Neurobiol Dis 8:890–899.

Reinhart MP (1990) Intracellular sterol trafficking. Experientia 46:599–611.

Reinhart MP, Billheimer JT, Faust JR, Gaylor JL (1987) Subcellular localization of the enzymes of cholesterol biosynthesis and metabolism in rat liver. J Biol Chem 262:9649–9655.

Riddell DR, Christie G, Hussain I, Dingwall C (2001) Compartmentalization of beta-secretase (Asp2) into low-buoyant density, noncaveolar lipid rafts. Curr Biol 11:1288–1293.

Runz H, Rietdorf J, Tomic I, de Bernard M, Beyreuther K, Pepperkok R, Hartmann T (2002) Inhibition of intracellular cholesterol transport alters presenilin localization and amyloid precursor protein processing in neuronal cells. J Neurosci 22:1679–1689.

Santamarina-Fojo S, Remaley AT, Neufeld EB, Brewer HB, Jr (2001) Regulation and intracellular trafficking of the ABCA1 transporter. J Lipid Res 42:1339–1345.

Shepherd J, Blauw GJ, Murphy MB, Bollen EL, Buckley BM, Cobbe SM, Ford I, Gaw A, Hyland M, Jukema JW, et al. (2002) Pravastatin in elderly individuals at risk of vascular disease (PROSPER): a randomised controlled trial. Lancet 360:1623–1630.

Simons K, Ikonen E (1997) Functional rafts in cell membranes. Nature 387:569–572.

Simons K, Ikonen E (2000) How cells handle cholesterol. Science 290:1721–1726.

Simons M, Schwarzler F, Lutjohann D, von Bergmann K, Beyreuther K, Dichgans J, Wormstall H, Hartmann T, Schulz JB (2002) Treatment with simvastatin in normocholesterolemic patients with Alzheimer's disease: A 26-week randomized, placebo-controlled, double-blind trial. Ann Neurol 52:346–350.

Soccio RE, Breslow JL (2003) StAR-related lipid transfer (START) proteins: mediators of intracellular lipid metabolism. J Biol Chem 278:22183–22186.

Soccio RE, Breslow JL (2004) Intracellular cholesterol transport. Arterioscler Thromb Vasc Biol 24:1150–1160.

Sparks DL, Kuo YM, Roher A, Martin T, Lukas RJ (2000) Alterations of Alzheimer's disease in the cholesterol-fed rabbit, including vascular inflammation. Preliminary observations. Ann NY Acad Sci 903:335–344.

Sparks DL, Sabbagh MN, Connor DJ, Lopez J, Launer LJ, Browne P, Wasser D, Johnson-Traver S, Lochhead J, Ziolwolski C (2005) Atorvastatin for the treatment of mild to moderate Alzheimer disease: preliminary results. Arch Neurol 62:753–757.

Sparks DL, Scheff SW, Hunsaker JC, 3rd, Liu H, Landers T, Gross DR (1994) Induction of Alzheimer-like beta-amyloid immunoreactivity in the brains of rabbits with dietary cholesterol. Exp Neurol 126:88–94.

Sun Y, Yao J, Kim TW, Tall AR (2003) Expression of liver X receptor target genes decreases cellular amyloid beta peptide secretion. J Biol Chem 278:27688–27694.

Tanzi RE, Bertram L (2005) Twenty years of the Alzheimer's disease amyloid hypothesis: a genetic perspective. Cell 120:545–555.

Vance JE, Hayashi H, Karten B (2005) Cholesterol homeostasis in neurons and glial cells. Semin Cell Dev Biol 16:193–212.

Vega GL, Weiner MF, Lipton AM, Von Bergmann K, Lutjohann D, Moore C, Svetlik D (2003) Reduction in levels of 24S-hydroxycholesterol by statin treatment in patients with Alzheimer disease. Arch Neurol 60:510–515.

Vetrivel KS, Cheng H, Lin W, Sakurai T, Li T, Nukina N, Wong PC, Xu H, Thinakaran G (2004) Association of gamma-secretase with lipid rafts in post-golgi and endosome membranes. J Biol Chem.

Wang N, Lan D, Chen W, Matsuura F, Tall AR (2004) ATP-binding cassette transporters G1 and G4 mediate cellular cholesterol efflux to high-density lipoproteins. Proc Natl Acad Sci USA 101:9774–9779.

Wollmer MA, Streffer JR, Lutjohann D, Tsolaki M, Iakovidou V, Hegi T, Pasch T, Jung HH, Bergmann K, Nitsch RM, et al. (2003) ABCA1 modulates CSF cholesterol levels and influences the age at onset of Alzheimer's disease. Neurobiol Aging 24:421–426.

Wolozin B, Kellman W, Ruosseau P, Celesia GG, Siegel G (2000) Decreased prevalence of Alzheimer disease associated with 3-hydroxy-3-methyglutaryl coenzyme A reductase inhibitors. Arch Neurol 57:1439–1443.

Yanagisawa K, Matsuzaki K (2002) Cholesterol-dependent aggregation of amyloid beta-protein. Ann NY Acad Sci 977:384–386.

Yanagisawa K, Odaka A, Suzuki N, Ihara Y (1995) GM1 ganglioside-bound amyloid beta-protein (Abeta): a possible form of preamyloid in Alzheimer's disease. Nat Med 1:1062–1066.

Yang DS, Small DH, Seydel U, Smith JD, Hallmayer J, Gandy SE, Martins RN (1999) Apolipoprotein E promotes the binding and uptake of beta-amyloid into Chinese hamster ovary cells in an isoform-specific manner. Neuroscience 90:1217–1226.

Chapter 7

Aβ Structure and Aggregation

Charles Glabe[1] and Ashley I. Bush[2]

[1] Department of Molecular Biology and Biochemistry
University of California, Irvine
Irvine, CA 92697
Email: cglabe@uci.edu

[2] Oxidation Disorders Laboratory
Mental Health Research Institute of Victoria
155 Oak Street, Parkville, Victoria 3052 Australia
Email: abush@mhri.edu.au

[2] Genetics and Aging Research Unit
Massachusetts General Hospital
Blgd 114, 16th Street
Charlestown, MA 02129

1.   Introduction

A large and growing number of degenerative disease, including Alzheimer's, Parkinson's, Huntington's diseases and type II diabetes are characterized by the accumulation of specific misfolded proteins as amyloid fibers. Recent insight into the structure of amyloid fibrils and biochemical analysis of the fibril assembly pathways indicates that many self-aggregating polypeptides share a common structural motif and aggregation pathway regardless of their protein sequences. Indeed, the formation of amyloids is not restricted to disease-related proteins, indicating that the amyloid lattice is a fundamental protein structural motif that has been selected against during evolution. The structural and mechanistic commonalities provide a powerful conceptual framework for understanding pathogenic pathways and the contribution of disease-specific factors to the individual diseases. Pathogenic events that are held in common by multiple diseases may constitute a core pathway leading to degeneration while disease-specific events may be likely to contribute to increasing the concentration of misfolded proteins that initiate the aggregation pathway. To the extent that the diseases share a common core pathway of degeneration, they may be amenable to the same types of therapeutic interventions.

Aβ has been described to form aggregates in two fundamental types of reactions: A. non-metal dependent association ("apo-Aβ"), B. metal-dependent association. Non-metallated aggregates of apo-Aβ form amyloid fibrils and soluble oligomers principally by intermolecular hydrogen bonding and hydrophobic interactions. Metallated Aβ (usually binding Cu^{2+}, Zn^{2+} or Fe^{3+}) forms ionically-bridged aggregates, covalently crosslinked oligomers (when bound to Cu^{2+}), and seeds apo-Aβ fibrillization. There is evidence of both fibrillar Aβ and metal-associated Aβ aggregates in the brain in Alzheimer's disease (AD). Therefore, the characterization of both pathways of aggregation are important for understanding the pathogenesis of AD. Furthermore, metal-dependent aggregation mechanisms for other self-aggregating proteins implicated in disease have become increasingly recognized. Therefore, the structural commonalities of self-aggregating proteins in their apo and metal-associated states may emerge as key to several pathogenic mechanisms.

Here we review the structure of Aβ in its apo- and its metal bound state.

2.   Aβ

Human Aβ1-40   DAEFRHDSGYEVHHQKLVFFAEDVGSNKGAIIGLMVGGVV

Human Aβ1-42   DAEFRHDSGYEVHHQKLVFFAEDVGSNKGAIIGLMVGGVV*IA*

Rat Aβ1-40   DAEF*G*HDSG*F*EV*R*HQKLVFFAEDVGSNKGAIIGLMVGGVV

Figure 1. The primary amino acid sequence of Aβ. The key residues involved in transition metal binding are indicated in red. The 42 residue form is enriched in plaque pathology. The rat sequence is the same as the mouse sequence, with two of the substitutions involving metal coordinates. The sole methionine at residue 35, which plays important roles in membrane insertion, redox activity and toxicity is highlighted.

The sequence of Aβ was first discovered from the insoluble component of amyloid plaques and congophilic angiopathy (Glenner and Wong, 1984). Aβ is a group of 39-43 residue amphipathic polypeptides that are constitutively expressed in all tissues and biological fluids (Haass et al., 1992; Seubert et al., 1992; Shoji et al., 1992). The peptide's generation per se does not cause self-aggregation, since Alzheimer-like pathology does not develop until advanced adult life. In the healthy brain, Aβ is resident in the membrane fraction where it may be normally embedded in a Cu/Zn-bound oligomer (Curtain et al., 2001; Curtain et al., 2003). Soluble forms of Aβ are only detected in AD-affected tissue (McLean et al., 1999) or the brains of APP transgenic mice.

Aβ may be a constitutive Cu/Zn-binding metalloprotein, since the peptide coordinates these metals in plaques (Lovell et al., 1998; Opazo et al., 2002; Dong et al., 2003), and since its expression may directly impact on transition metal homeostasis (Maynard et al., 2002; Bayer et al., 2003; Phinney et al., 2003). Therefore, it is important to consider the structure of Aβ in its metal-bound and it apo-state. Both hypermetallation and undermetallation may lead to the peptide's aggregation.

3.   Aβ aggregation

3.1   Fibril structure

Early studies on the x-ray fiber diffraction of amyloid fibrils indicated that the peptide is organized in a characteristic "cross β" pattern within the fiber lattice, where the extended polypeptide backbone is perpendicular and the hydrogen bonding is parallel to the fiber axis (Eanes and Glenner, 1968; Kirschner et al., 1986). Recent advances in the structural biology of amyloid fibrils have led to the identification of a parallel, exact register motif for the organization of extended polypeptide strand as a common underlying structural organization of disease-related amyloid fibrils. Some fibrils derived from short peptide segments of Aβ are also known to be antiparallel (Petkova et al., 2004). The first evidence for this simple and intriguing structural motif was obtained from solid state NMR analysis of amyloid Aβ fibrils (Benzinger et al., 1998; Antzutkin et al., 2000; Balbach et al., 2002). This was independently confirmed by site-directed spin labeling and electron paramagnetic resonance (SDSL-EPR) spectroscopy for Aβ. Subsequent work employing SDSL-EPR has shown that this parallel, exact register structural motif is conserved in amyloid fibrils from alpha synuclein, islet amyloid polypeptide (IAPP) and tau filaments (Der-Sarkissian et al., 2003; Jayasinghe and Langen, 2004; Margittai and Langen, 2004). Recently, a high resolution x-ray crystal structure was reported for a seven residue segment of the yeast prion sup35, that is also a parallel, exact register sheet structure (Nelson et al., 2005). This structure provides additional details about the sheet stacking interactions between sheets in the fibril. The parallel, exact register structural motif gives rise to a highly ordered linear "zipper" array of stacked amino acid side chains up and down the sheet that has also been observed on a smaller scale in parallel beta helix proteins (Jurnak et al., 1994; Jenkins and Pickersgill, 2001). This stacking of individual amino acid side chains in a linear array gives rise to a molecular "bar code" (Fig. 2) for each sheet in the fibril that varies with the individual amino acid sequence of the amyloid forming protein and may form the basis for observed seeding specificity of amyloid fibrils in nucleating the assembly of proteins of similar or identical sequence (Krebs et al., 2004).

V*H*H*Q*K*L*V*FF

V*H*H*Q*K*L*V*FF

V*H*H*Q*K*L*V*FF

V*H*H*Q*K*L*V*FF

V*H*H*Q*K*L*V*FF

V*H*H*Q*K*L*V*FF

Figure 2. Schematic representation of the "bar code" stacking of amino acid side chains in a parallel exact register sheet. Alternating amino acids, shown in italics, have side chains facing on the opposite side of the sheet. The exact registration of the polypeptide chains gives rise to vertical stacks of side chain "steric zippers" running parallel to the fiber axis. This results in the formation of a series of side chain ridges and grooves that constitute a molecular bar code that is unique for each amyloid sequence. The molecular complementarity of the sheet surfaces forms the basis of the sheet stacking interaction during fibril assembly.

The crystal structure of sup35 heptapeptide reveals that alternating side chains of extended beta strands form two distinct sheet interfaces; one that is non-hydrated or "dry" and another that is hydrated or "wet", even though both faces are composed predominantly of polar amide stacks. The bar code grooves and ridges of the dry face is self complementary and the polar Asn and Gln side chains stacks interdigitate forming van der Walls interactions. The sheet spacing across the dry face is separated by only 8.5 Å, while the wet face is completely hydrated and the distance across this interface longer. It is not yet clear whether this alternating dry, wet sheet interface is a characteristic of other amyloid fibers as the wet interface has the characteristics of a crystal lattice contact and may not exist as amyloid fibrils (Nelson et al., 2005). Each polypeptide chain forms 11 hydrogen bonds. The 5 main chain hydrogen bonds are intermolecular and four additional hydrogen bonds are contributed by side chain amide "stacks" that force the polypeptides to align in parallel and in register (Nelson et al., 2005). Although only Asn, Gln and Tyr side chain stacks were observed in the Sup35 heptapeptide crystal structure, it is reasonable to envision that other amino acid side chains also form these type of linear stacks in more complex amyloid sequences.

3.2   Fibril polymorphisms

The extended parallel, in register generic amyloid motif is sufficiently broad to accommodate a wide range of potential variation in the length of beta strands; the number of strands and the location of turns or hairpin bends still satisfy the critical features of this common motif. It is also consistent with the hollow helical nanotube structure proposed for polyglutamine peptides on the basis of x-ray fiber diffraction (Perutz et al., 2002). Small differences in the location of turns, the length of sheets or the stacking of the sheets may give rise to self-propagating variants that can explain prion strains. It should also be noted that some fibrils might not be parallel, in-register structures. Small peptides can also form cross beta fibrils in an antiparallel fashion (Petkova et al., 2004). Others appear to involve minimal reorganization of the native structure, such as the fibrils that accumulate in the "serpinopathies". These appear to polymerize by a domain swapping mechanism (Lomas and Carrell, 2002).

Aβ appears to be organized as a hairpin structure containing two parallel, in register strands separated by a bend or turn. There is general agreement that amino terminal 10 residues are disordered, residues 11-20 constitute the first strand, followed by a turn from residues 21-27 and then residues 28-38 make up the second strand although the exact boundaries of these features is not entirely clear and there are some discrepancies that need to be resolved among the different approaches. The last few residues also appear to be disordered or more mobile than the strand regions. This hairpin could conceivably have two possible orientations within the fiber where the strands are oriented "cross beta" or perpendicular to the fiber axis and the hydrogen bonding is parallel to the fiber axis. One is with the plane of the hairpin perpendicular to the fiber axis. In this orientation, the all of the strands are parallel, the main chain hydrogen within the sheet is strictly intermolecular and the amino acid side chains in the sheet are all stacked the entire length of the sheet. The other possible orientation is where the plane of the hairpin is parallel to the fiber axis. In this orientation, the main chain hydrogen bonding alternates between parallel intermolecular and antiparallel intramolecular. Since in this latter orientation the hairpin is the fundamental building block of a single sheet, instead of a pair of sheets separated by a turn in the first model, only pairs of identical amino acid side would stack, rather than linear stretches that run the entire length of the sheet. The phenomenon of spin exchange has been observed in tau fibrils (Margittai and Langen, 2004). This phenomenon depends on extended linear arrays of spin labeled side chains that would allow free diffusion of electrons in the spin label analogous to the case in a crystal lattice, indicating that the main chain hydrogen bonding is strictly parallel and in register throughout as observed in the crystal structure of the sup35 heptapeptide.

The details of the intersheet side chain packing also remains to be established for Aβ fibrils. Some progress is beginning to be made using disulfide cross-linking of cysteine substituted Aβ polypeptides (Shivaprasad and Wetzel, 2004). These studies indicated that residues 17 and 34 could be efficiently intramolecularly crosslinked by a disulfide bond within preformed fibrils, indicating that they are located proximally inside the hairpin turn. Residues 17/35 and 17/36 were not efficiently crosslinked in fibrils and no intermolecular crosslinking was observed. These results are inconsistent with the hairpin model based the electrostatic interaction of the side chains of Glu 22 and Lys 28 (Tycko, 2003). However, there appears to be considerable flexibility regarding the registration of side chains on the inside of the hairpin with respect to the ability to form fibrils. Intramolecular disulfides also form in Aβ monomers in the 17/35 and 17/36 double mutants and these purified, disulfide bonded monomers are able to assemble into amyloid fibrils that resemble the wild type fibrils by several criteria (Shivaprasad and Wetzel, 2004). This is particularly surprising for 17/35, because this species predicts that the registration of residues on the C-terminal strand would be flipped 180 degrees with respect to the inside and outside of the hairpin, suggesting that there is also considerable structural flexibility of the sheet stacking interactions on the inside and outside faces of this sheet. This observation of structural variability is consistent with the finding that Aβ can assemble into at least two slightly different forms, reminiscent of the strain phenomenon in prions (Harper et al., 1997; Goldsbury et al., 2005; Petkova et al., 2005). While this seems to add another layer of complexity to the problem of understanding the fibril structure it is also possible that small differences in the fibril structure may not be pathologically significant.

3.3   Oligomer structure

In contrast to the fibril structure, relatively little is known about the structure of amyloid oligomers. Structural characterization of oligomers is complicated by the difficulty in preparing homogeneous populations of oligomers and the fact that oligomers are transient and not as stable as fibrils (Kayed et al., 2003). The observation that they can be stabilized by detergents may help to alleviate this later problem (LeVine, 2004). Aβ oligomers are extended or beta structures as analyzed by circular dichroism and infrared spectroscopy (Walsh et al., 1999; Kayed et al., 2003). Hydrogen-deuterium exchange analysis of oligomers indicates that 40% of the total backbone hydrogen bonds are resistant to exchange in the oligomeric conformation, indicating they have a stable core that is consistent with substantial beta sheet character (Kheterpal et al., 2003). In contrast, 50% of the backbone hydrogen bonds are resistant to exchange in the mature amyloid fibril, indicating that there is a small increase in main chain hydrogen bonding that accompanies the transition to the fibrillar conformation (Kheterpal et al., 2000).

The size distribution of Aβ oligomers is heterogeneous. There is broad consensus for the preferential accumulation of a soluble high molecular weight species of approximately 100–200 kDa under relatively physiological conditions in vitro (Soreghan et al., 1994; Walsh et al., 1997; Goldsbury et al., 2000; Nichols et al., 2002; Kayed et al., 2003; Lashuel et al., 2003; LeVine, 2004). The oligomeric forms of Aβ secreted in the culture medium of transfected cells appear to be somewhat smaller by gel filtration although their distribution overlaps with the elution of soluble APP, which is approximately 70 kDa (Cleary et al., 2005).

3.4 Aggregation pathways

The relationship between oligomers and fibrils remains to be established. There seems to be some structural relationship because both appear to be extended or beta structures and they display similar amounts of main chain hydrogen bonding that is resistant to exchange. On the other hand, amyloid oligomers and fibrils appear to contain mutually exclusive and non-overlapping conformations in terms of their reactivities to antibodies that recognize generic epitopes that are common to amyloids of different sequence (O'Nuallain and Wetzel, 2002; Kayed et al., 2003). It also remains to be established whether oligomers represent a fundamental unit of amyloid structure that assembles en block into fibrils or whether it is an off pathway structure that serves to buffer the concentration of monomers. Oligomers are a kinetic intermediate, waxing at early times and waning as fibrils evolve (Harper et al., 1997; Kayed et al., 2003), but this does not necessarily imply that they are on pathway intermediates. Moreover, at early times oligomers appear as spherical aggregates and at later times appear to elongate by coalescence of spherical subunits with a "string of beads" appearance leading to the view that these "protofibrils" are precursors on the pathway to mature fibrils (Harper et al., 1997; Walsh et al., 1997; Blackley et al., 2000). However, other studies suggest that the spherical oligomers may be off pathway intermediates that serve to buffer Aβ monomer concentration (Goldsbury et al., 2005) or that both on pathway and off pathway fates are possible under different conditions (Nichols et al., 2002).

4. Metal-Aβ aggregation

4.1 Cu, Zn and Fe interaction with Aβ

In vitro studies have shown that Zn(II), Cu(II) and Fe(III) are able to induce protease resistant aggregation and precipitation of Aβ (Bush et al., 1994a; Bush et al., 1994b; Huang et al., 1997; Atwood et al., 1998; Atwood et al., 2000). Up to 3.5 equivalents of Zn^{2+} or Cu^{2+} can bind synthetic Aβ (Atwood et al., 2000), which is likely to be accomplished by the peptide coordinating multiple metal ions by forming oligomers. The measured saturable affinities of Aβ1-40 for Zn^{2+} are ≈ 5μM and ≈ 100 nM using [65]Zn as a tracer with immobilized peptide (Bush et al., 1994a; Bush et al., 1994b), while for Cu^{2+} the lowest apparent affinity is ≈ 1 nM (for both Aβ1-40 and Aβ1-42, using chelators of known affinity as competitors and directly measuring metal bound to either soluble or precipitated peptide). The highest apparent affinity, using competition against chelators for Cu^{2+}, is much greater for $A\beta1$-42 ($\approx 10^{-18}$) than for $A\beta1$-40 ($\approx 10^{-12}$) (Atwood et al., 2000), which may be due to positive cooperativity in forming a hexameric assembly (Curtain et al., 2001) or may be due to a perturbed equilibrium caused by $A\beta1$-42 more readily precipitating from solution. Assessment of transition metal affinity with $A\beta1$-40 and $A\beta1$-42 in solution using tyrosine fluorescence produced results that are at variance with the earlier estimations (Garzon-Rodriguez et al., 1999). Specifically, affinities were about 3 orders of magnitude lower than with the other reported methods using unlabeled peptide. The discordance may be explained possibly by inability of tyrosine fluorescence spectroscopy to detect the high affinity interactions. The study by Garzon-Rodriguez et al. (1999) also described evidence of positive cooperativity between Cu^{2+} and Zn^{2+} binding as well as Fe^{2+} binding affinity of a similar affinity to that of Cu^{2+}.

The ionic assembly of $A\beta$ aggregates by Cu^{2+} and Zn^{2+} is reversible with chelation *in vitro* (Huang et al., 1997; Atwood et al., 1998; Atwood et al., 2000) and in post-mortem AD brain tissue (Cherny et al., 1999), which may be the basis for a therapeutic approach (Cherny et al., 2001; Ritchie et al., 2003). The interaction of synthetic peptide with transition metals is so sensitive that the apparent self-aggregation of $A\beta$ (into amyloid precipitates and soluble SDS-resistant oligomers) in solution has been shown to be initiated by high nanomolar levels of metals that common contaminate laboratory buffers (Huang et al., 2004). This does not mean that $A\beta$ will not self-aggregate in the absence of metal ions, but that the presence of metal ions even at trace concentrations will markedly accelerate aggregation and precipitation possibly by initiating nucleation. Apart from aggregating $A\beta$, Cu^{2+} also reacts with $A\beta$ yielding covalent crosslinking of the peptide (Atwood et al., 1998; Atwood et al., 2000). Dityrosine crosslinking is much more efficient when $A\beta$ is aggregated in the fibrillar state, perhaps as a result of the proximal positioning of adjacent tyrosine residues in the parallel, in register fibril structure (Yoburn et al., 2003).

While these metals play important roles in normal physiology, they are found in relatively high concentrations in the neocortical regions of the brain most susceptible to AD neurodegeneration. During neurotransmission high concentrations of Zn (300 μM) (Frederickson, 1989) and Cu (30 μM) (Schlief et al., 2005) are released, which may explain why $A\beta$ precipitation into amyloid commences in the synapse (Terry et al., 1991). The amyloid plaques could be described as metallic sinks as high concentrations of Cu (400 μM), Zn (1 mM) and Fe (1 mM) (3 to 4-fold background brain tissue concentrations) have been found in amyloid deposits in AD-affected brains (Lovell et al., 1998). Zn(II) released by synaptic transmission has been demonstrated to play a major role in Aβ plaque formation. Genetic ablation of the zinc transporter 3, that transports zinc into synaptic vesicles, markedly inhibits cortical amyloid plaque deposition (Lee et al., 2002; Lee et al., 2004) and congophilic angiopathy (Friedlich et al., 2004) in the brains of the Tg2576 transgenic mouse model for AD.

Nature also provides an interesting "control peptide" in the rat/mouse homologue of Aβ, that differs from human Aβ by three substitutions, with Arg5, Tyr10 and His13 of human Aβ becoming Gly5, Phe10 and Arg13 (Shivers et al., 1988). The behavior of this peptide is important since rats and mice do not form brain amyloid deposits with age (Vaughan and Peters, 1981), even in mutant presenilin 1 transgenic mice (Duff et al., 1996), while other mammalian species expressing the human Aβ sequence do deposit brain amyloid (Johnstone et al., 1991). The substitutions in the rat/mouse Aβ attenuates metal affinity, preventing the precipitation by Zn^{2+} or Cu^{2+} (Bush et al., 1994b; Atwood et al., 1998), and abolishing the peptide's electrochemical reduction of Cu^{2+} and Fe^{3+} and its associated toxicity (Huang et al., 1999a; Huang et al., 1999b).

The histidine residues of Aβ at positions 6, 13 and 14 in its N-terminus (Figure 1), are a structural element that enables Aβ to coordinate up to 3.5 moles of Cu or Zn with a range of affinities (Atwood et al., 2000). In aqueous solution Aβ undergoes rapid conformational exchanges making it difficult to determine the precise nature of the metal binding sites. The question has been approached using various spectroscopic techniques, including Raman spectroscopy, circular dichroism and magnetic resonance. Raman spectroscopy has been used to study the binding modes of Zn^{2+} and Cu^{2+} to Aβ in solution and insoluble aggregates. Two different modes of metal-Aβ binding were reported, one characterized by metal binding to the imidazole N_τ atom of histidine, producing insoluble aggregates, the other involving metal binding to the N_π, but not the N_τ atom of histidine as well as to main-chain amide nitrogens, giving soluble complexes. Zn^{2+} binds to Aβ only via the N_τ regardless of pH, while the Cu^{2+} binding mode is pH dependent. At mildly acidic pH, Cu^{2+} binds to Aβ in the former mode, whereas the latter mode is predominant at neutral pH. From this data it was proposed that the transition from one binding mode to the other explained the strong pH dependence of Cu^{2+}-induced Aβ aggregation (Atwood et al., 1998).

The interactions of synthetic Aβ with copper, zinc and iron have been probed using Nuclear Magnetic Resonance (NMR) and Electron paramagnetic resonance (EPR) spectroscopy. Huang et al. (Huang et al., 1999b) used multi-frequency EPR (S-, X- and Q-band) to show that copper coordinates tightly to Aβ1-40 and that an approximately equimolar mixture of peptide and $CuCl_2$ produced a single Cu^{2+}-peptide complex. Computer simulation of the S band spectrum with an axially symmetrical spin Hamiltonian and the *g* and *A* matrices ($g_{||}$, 2.295; $g_\square$, 2.073; $A_{||}$163.60, $A_\perp$10.0 X 10^{-4} cm^{-1}) suggested a tetragonally distorted geometry, which is commonly found in type 2 copper proteins. Expansion of the $M_I = -1/2$ resonance revealed nitrogen ligand hyperfine coupling. Computer simulation of these resonances indicated the presence of at least three nitrogen atoms. This, and the magnitude of the $g_{||}$ and $A_{||}$ values, together with Peisach and Blumberg plots, are consistent with a fourth equatorial ligand binding to copper via an oxygen. Therefore, the coordination sphere for the copper-peptide complex was considered to be 3N1O.These authors also used EPR spectroscopy to measure residual Cu^{2+} remaining after incubating stoichiometric ratios of CuCl$_2$ with Aβ1-40. There was a 76% loss of the Cu^{2+} signal, compatible with peptide-mediated reduction of Cu^{2+} to diamagnetic Cu$^+$, which is undetectable by EPR, in agreement with the corresponding concentration of Cu$^+$ measured by bioassay.

A variety of spectroscopic studies have confirmed that the histidine residues constitute the principal site of metal coordination, and both in aqueous solution and membrane-mimetic environments, coordination of the metal ion is consistent, establishing that the three histidine residues are all involved in the coordination of the first metal ion (Curtain et al., 2001; Curtain et al., 2003). At Cu(II)/peptide molar ratios > 0.3, exchange broadening was observed in the EPR spectra, indicating that the peptide was coordinating a second Cu(II) atom in a highly cooperative manner (Curtain et al., 2001; Curtain et al., 2003). This is consistent with previous studies that demonstrated the cooperative binding of Cu (Garzon-Rodriguez et al., 1999). Similar broadening of Cu EPR spectra have been observed in mimetics of superoxide dismutase (SOD) which contain two copper atoms bridged by imidazole (Ohtsu et al., 2000). Methylation of the N$^{\varepsilon 2}$ atoms in the imidazole sidechains of Aβ abolished this second binding site (Tickler et al., 2005). Therefore, the second binding site might result in a histidine residue bridge between metal ions (distance between metals ~6 Å); the resulting coordination sphere would be reminiscent of that observed in the active site of SOD1. A Raman spectroscopy study showed that all three histidine residues are involved in the coordination of Cu and Zn by Aβ, the exact nature of the coordination sphere being pH dependent (Miura et al., 2000). While the data from these EPR studies which have been performed in phosphate buffered saline (PBS) are consistent with the formation of SOD-like structures, a number of other EPR studies have given a range of different results (Antzutkin, 2004; Karr et al., 2004; Syme et al., 2004). While, these studies agree that the histidine residues of Aβ are critical for Cu coordination, Cu induced broadening in EPR was not observed. Of critical importance is that different solution conditions (i.e. different buffers) were used in each of these studies and that Aβ is a pleiomorphic molecule, its structure is dependent on the local conditions; for instance the Zn induced aggregation of Aβ is dependent on NaCl being present in the buffer (Huang et al., 1997).

These different solution conditions are likely to give rise to different structural propensities for Aβ and may explain the variations in the reported results. Moreover, Tyr10 was identified as a ligand for Cu but not Zn. A Raman study (Dong et al., 2003) of the core of senile plaques confirmed that Cu and Zn but not Fe were coordinated to the histidine residues of the deposited Aβ. Zn^{2+} was coordinated to the histidine N_τ and the Cu^{2+} to the N_π, confirming that the metal binding mode was the same in both the synthetic peptide and its aggregates and the naturally occurring plaques. Fe in the plaque vicinity is contained within ferritin found inside the neuritic processes that penetrate the plaque (Grundke-Iqbal et al., 1990).

The bridging histidine may be responsible for the reversible metal induced aggregation of Aβ. The bridging histidine residues may also explain the multiple metal binding sites observed for each peptide and the high degree of cooperativity evident for subsequent metal binding. With three histidines bound to the metal center a large scope exists for metal-mediated cross-linking of the peptides leading to aggregation, which will be reversible when the metal is removed by chelation. It is quite possible that metal-induced precipitation of Aβ is quite different from that induced prolonged incubation of monomeric peptide in the absence of metal.

Rat Aβ forms a metal complex via two histidine residues and two oxygen ligands rather than three histidine residues and one oxygen ligand, compared with human Aβ where the side-chain of His13 of human Aβ is ligated to the metal ion. This was borne out by the EPR spectrum, which was typical of a square planar 2N2O Cu^{2+} coordination.

4.2 Aβ redox activity

Oxidative stress markers characterize the neuropathology both of Alzheimer's disease and of amyloid-bearing transgenic mice (Smith et al., 1996; Smith et al., 1998). The neurotoxicity of Aβ has been linked to hydrogen peroxide generation in cell cultures by a mechanism that is still being fully described but is likely to be dependent on Aβ coordinating redox active metal ions (Barnham et al., 2004). Huang et al. showed that human Aβ directly produces H_2O_2 by a mechanism that involves the reduction of metal ions, Fe^{3+} or Cu^{2+} (Huang et al., 1999a; Huang et al., 1999b). Spectrophotometry was used to show that the Aβ peptide reduced Fe^{3+} and Cu^{2+} to Fe^{2+} and Cu^+ and that molecular oxygen was then trapped by Aβ and reduced to H_2O_2 in a reaction that is driven by substoichiometric amounts of Fe^{2+} or Cu^+. In the presence of Cu^{2+} or Fe^{3+}, Aβ produced a positive thiobarbituric-reactive substance, compatible with the generation of the hydroxyl radical (OH*). Tabner et al. used spin-trapping to identify the radical produced by Aβ in the presence of Fe^{2+}, concluding that it was OH* (Tabner et al., 2002). However they also found OH* was produced in the presence of Fe^{2+} by $A\beta25$-35, which does not contain a strong metal binding site.

Incubation with Cu^{2+} causes SDS-resistant oligomerization of $A\beta$ (Atwood et al., 1998; Atwood et al., 2000), which is also found in the neurotoxic soluble $A\beta$ extracted from the AD brain (Roher et al., 1996). Atwood et al. (Atwood et al., 2004) found that Cu^{2+} induced SDS-resistant oligomers of $A\beta$ gave a fluorescence signal characteristic of the cross-linking of the peptide's tyrosine 10. This finding was confirmed by directly identifying the dityrosine by electrospray ionization mass spectrometry and by the use of a specific dityrosine antibody. The addition of H_2O_2 strongly promoted Cu^{2+}-induced dityrosine cross-linking of $A\beta1$-28, $A\beta1$-40, and $A\beta1$-42, and Atwood et al. (Atwood et al., 2004) suggested that the oxidative coupling was initiated by interaction of H_2O_2 with a Cu^{2+} tyrosinate. The dityrosine modification is significant since it is highly resistant to proteolysis and would be important in increasing the structural strength of the plaques.

Density functional theory calculations elucidated the chemical mechanisms underlying the catalytic production of H_2O_2 by $A\beta/Cu$ and the production of dityrosine (Haeffner et al., In press). Tyrosine10 (Y10) was identified as the critical residue. This finding accords with the growing awareness that the O_2 activation ability of many cupro-enzymes is also coupled to the redox properties of tyrosine and the relative stability of tyrosyl radicals. The latter play important catalytic roles in photosystem II, ribonucleotide reductase, COX-2, DNA photolyase, galactose oxidase, and cytochrome-c oxidase.

The Cu/tyrosinate hypothesis was tested using an $A\beta1$-42 peptide with tyrosine 10 substituted with alanine (Y10A) (Barnham et al., 2004). Both peptides gave rise to similar ^{65}Cu EPR spectra with the strong single $g_\perp$ resonance characteristic of an axially symmetric square planar complex, although there was a significant increase in the $g_{||}$ value of Y10A. The increase was probably due to some distortion of the coordination sphere because the oxygen ligand, which was possibly from Y10, was now derived from another oxygen donor (e.g. H_2O, phosphate, or carboxylate from the peptide). While wild-type $A\beta42$ rapidly reduces Cu^{2+} to Cu^+ in aqueous solution, with near-complete reduction taking 80 min. the mutation of Y10 to alanine markedly decreased the ability of Ab to reduce Cu^{2+}. Further, spin-trapping studies also confirmed the DFT observation that Y10 acts as a gate that facilitates the electron transfer needed to reduce Cu^{2+} to Cu^+. When the spin trap 2-methyl-2-nitrosopropane was added to the reaction mixture $w.t.A\beta1$-42/Cu^{2+}/ascorbate a broad line triplet characteristic of a trapped carbon-centered radical bound to a peptide appeared in the EPR spectra. However, if Y10A peptide were substituted for the wild-type peptide, formation of this triplet was inhibited.

4.3 The effect of metal binding on the interaction of $A\beta$ with membranes

Numerous reports have described the effects of $A\beta$ on membranes and lipid systems and their possible roles in its neurotoxicity. Structural studies in different membrane-mimetic systems have demonstrated considerable variation in peptide conformation. There is much experimental evidence, from far-ultraviolet circular dichroism spectroscopy and FT-IR spectroscopy that the $A\beta$ peptides can be membrane associated in the $\beta^\sim$ configuration, although there are reports of membrane-associated α-helices being found in the presence of gangliosides, cholesterol and Cu^{2+} or Zn^{2+}.

EPR spectroscopy, using spin labelled lipid chains or protein segments, has been used extensively to study translational and rotational dynamics in biological membranes. Lipids at the hydrophobic interface between lipid and transmembrane protein segments and peptides in their monomeric and oligomeric states have their rotational motion restricted. This population of lipids can be resolved in the EPR spectrum as a motionally restricted component distinct from the fluid bilayer lipids which can be quantified to give both the stoichiometry and selectivity of the first shell of lipids interacting directly with membrane-penetrant peptides. The stoichiometric data can give an estimate of the number of subunits in a membrane-associated oligomeric structure. Using this approach it was shown that $A\beta1$-40 and $A\beta1$-42 bound to Cu^{2+} or Zn^{2+} bound to bilayers of negatively charged, but not zwitterionic lipid, giving rise to such a partly immobilised component in the spectrum (Curtain et al., 2001; Curtain et al., 2003). When the peptide:lipid was increased, the relationship between the mole fraction of peptide and proportion of slow component was linear. Even at a fraction of 15%, all of the peptide was associated with the lipid, suggesting that the structure associating with the lipid membrane was well defined. The lipid:peptide ratio is approximately 4:1 (Curtain et al., 2003). This stoichiometry can be satisfied by 6 helices arranged in association, surrounded by 24 boundary lipids. In the presence of Zn^{2+}, $A\beta1$-40 and $A\beta1$-42 both attached to the bilayer over the pH range 5.5–7.5, as did $A\beta1$-42 in the presence of Cu^{2+}. However, only $A\beta40$ associated with the lipid bilayer in the presence of Cu^{2+} at pH 5.5–6.5; at higher pH, there was a change in the Cu^{2+} coordination sphere that inhibited membrane association. Increasing the cholesterol component of the lipid to 0.2 mole fraction of the total lipid inhibited insertion of $A\beta1$-40 and $A\beta1$-42 (Curtain et al., 2003). This is important, because elevated cholesterol is a risk factor AD, appears in amyloid plaques and exaggerates transgenic mouse neuropathology (Refolo et al., 2000; Mori et al., 2001). Therefore, elevated membrane cholesterol may cause Cu^{2+}-bound $A\beta$ species to emerge outside of cells, where the toxic redox activity of $A\beta$ is inititiated by the availability of Met35 (Varadarajan et al.,

1999; Curtain et al., 2001). Some soluble forms of this released Aβ may drift into the synapse and cerebral blood vessel walls where very high concentrations of Zn^{2+} cause the peptide to precipitate (Lee et al., 2002; Bush, 2003; Friedlich et al., 2004; Lee et al., 2004).

CD spectroscopy revealed that the Aβ peptides had a high α-helix content when membrane penetrant, but were predominantly β-strand when not. It was also possible to gain an estimate of the secondary structure of the peptide from the degree of immobilisation of the lipid in the shell; the more immobilised the more likely the peptide is present as a β structure. Simulation of the spectra and calculation of the on-off rates suggested that the peptide was most likely penetrating as an α-helix (Curtain et al., 2001; Curtain et al., 2003). In membrane-mimetic environments, coordination of the metal ion is the same as in aqueous solution, with the three-histidine residues, at sequence positions 6, 13 and 14, all involved in the coordination, along with an oxygen ligand. As had been observed at Cu^{2+}/peptide molar ratios > 0.3 in aqueous solution, line broadening was detectable in the EPR spectra, indicating that the peptide was coordinating a second Cu^{2+} atom in a highly cooperative manner at a site less than 6 Å from the initial binding site. So, there appear to be two switches, metal ions (Zn^{2+} and Cu^{2+}) and negatively charged lipids, needed to change the conformation of the peptide from β-strand non-penetrant to α-helix penetrant.

5.   Conclusions

The structure of Aβ aggregates and the pathways of aggregation remains a challenging problem, although considerable progress has been made in recent years. Exact register parallelism is emerging as a common structural motif for amyloid fibrils and may also represent a key organizing principle for amyloid oligomers. Also, characterization of the interactions of Aβ with transition metals has opened up a new vista of potential pathogenic interactions and structural consequences that also may impact upon the disease state. These new insights into the structures and pathways of aggregation may help to understand the mechanisms of amyloid pathogenesis in degenerative diseases and ultimately lead to new therapeutic strategies that prevent the formation of toxic aggregates or interfere with the mechanism of toxicity.

References

Antzutkin ON (2004) Magn Reson Chem 42:231–246.
Antzutkin ON, Balbach JJ, Leapman RD, Rizzo NW, Reed J, Tycko R (2000) Multiple quantum solid-state NMR indicates a parallel, not antiparallel, organization of beta-sheets in Alzheimer's beta-amyloid fibrils. Proc Natl Acad Sci USA 97:13045–13050.

Atwood CS, Scarpa RC, Huang X, Moir RD, Jones WD, Fairlie DP, Tanzi RE, Bush AI (2000) Characterization of copper interactions with Alzheimer Aβ peptides-identification of an attomolar affinity copper binding site on Aβ1-42. J Neurochem 75:1219–1233.

Atwood CS, Moir RD, Huang X, Bacarra NME, Scarpa RC, Romano DM, Hartshorn MA, Tanzi RE, Bush AI (1998) Dramatic aggregation of Alzheimer Aβ by Cu(II) is induced by conditions representing physiological acidosis. J Biol Chem 273:12817–12826.

Atwood CS, Perry G, Zeng H, Kato Y, Jones WD, Ling KQ, Huang X, Moir RD, Wang D, Sayre LM, Smith MA, Chen SG, Bush AI (2004) Copper Mediates Dityrosine Cross-Linking of Alzheimer's Amyloid-beta. Biochemistry 43:560–568.

Balbach JJ, Petkova AT, Oyler NA, Antzutkin ON, Gordon DJ, Meredith SC, Tycko R (2002) Supramolecular Structure in Full-Length Alzheimer's beta-Amyloid Fibrils: Evidence for a Parallel beta-Sheet Organization from Solid- State Nuclear Magnetic Resonance. Biophys J 83:1205–1216.

Barnham KJ, Haeffner F, Ciccotosto GD, Curtain CC, Tew D, Mavros C, Beyreuther K, Carrington D, Masters CL, Cherny RA, Cappai R, Bush AI (2004) Tyrosine gated electron transfer is key to the toxic mechanism of Alzheimer's disease beta-amyloid. Faseb J 18:1427–1429.

Bayer TA, Schafer S, Simons A, Kemmling A, Kamer T, Tepests R, Eckert A, Schussel K, Eikenberg O, Sturchler-Pierrat C, Abramowski D, Staufenbiel M, Multhaup G (2003) Dietary Cu stabilizes brain superoxide dismutase 1 activity and reduces amyloid Aβ production in APP23 transgenic mice. Proc Natl Acad Sci USA 100:14187–14192.

Benzinger TL, Gregory DM, Burkoth TS, Miller-Auer H, Lynn DG, Botto RE, Meredith SC (1998) Propagating structure of Alzheimer's beta-amyloid(10-35) is parallel beta-sheet with residues in exact register. Proc Natl Acad Sci USA 95:13407–13412.

Blackley HK, Sanders GH, Davies MC, Roberts CJ, Tendler SJ, Wilkinson MJ (2000) In-situ atomic force microscopy study of beta-amyloid fibrillization. J Mol Biol 298:833–840.

Bush AI (2003) The Metallobiology of Alzheimer's disease. TINS 26:207–214.

Bush AI, Pettingell WH, Jr., Paradis MD, Tanzi RE (1994a) Modulation of Aβ adhesiveness and secretase site cleavage by zinc. J Biol Chem 269:12152–12158.

Bush AI, Pettingell WH, Multhaup G, Paradis Md, Vonsattel JP, Gusella JF, Beyreuther K, Masters CL, Tanzi RE (1994b) Rapid induction of Alzheimer Aβ amyloid formation by zinc. Science 265:1464–1467.

Cherny RA, Legg JT, McLean CA, Fairlie D, Huang X, Atwood CS, Beyreuther K, Tanzi RE, Masters CL, Bush AI (1999) Aqueous dissolution of Alzheimer's disease Aβ amyloid deposits by biometal depletion. J Biol Chem 274:23223–23228.

Cherny RA, Atwood CS, Xilinas ME, Gray DN, Jones WD, McLean CA, Barnham KJ, Volitakis I, Fraser FW, Kim Y-S, Huang X, Goldstein LE, Moir RD, Lim JT, Beyreuther K, Zheng H, Tanzi RE, Masters CL, Bush AI (2001) Treatment with a copper-zinc chelator markedly and rapidly inhibits β-amyloid accumulation in Alzheimer's disease transgenic mice. Neuron 30:665–676.

Cleary JP, Walsh DM, Hofmeister JJ, Shankar GM, Kuskowski MA, Selkoe DJ, Ashe KH (2005) Natural oligomers of the amyloid-beta protein specifically disrupt cognitive function. Nat Neurosci 8:79–84.

Curtain C, Ali F, Volitakis I, Cherny R, Norton R, Beyreuther K, Barrow C, Masters C, Bush A, Barnham K (2001) Alzheimer's disease amyloid- binds Cu and Zn to generate an allosterically-ordered membrane-penetrating structure containing SOD-like subunits. J Biol Chem 276:20466–20473.

Curtain CC, Ali FE, Smith DG, Bush AI, Masters CL, Barnham KJ (2003) Metal ions, pH and cholesterol regulate the interactions of Alzheimer's disease amyloid-β peptide with membrane lipid. J Biol Chem 278.

Der-Sarkissian A, Jao CC, Chen J, Langen R (2003) Structural organization of alpha-synuclein fibrils studied by site-directed spin labeling. J Biol Chem 278:37530–37535.

Dong J, Atwood CS, Anderson VE, Siedlak SL, Smith MA, Perry G, Carey PR (2003) Metal binding and oxidation of amyloid-beta within isolated senile plaque cores: Raman microscopic evidence. Biochemistry 42:2768–2773.

Duff K, Eckman C, Zehr C, Yu X, Prada CM, Perez-tur J, Hutton M, Buee L, Harigaya Y, Yager D, Morgan D, Gordon MN, Holcomb L, Refolo L, Zenk B, Hardy J, Younkin S (1996) Increased amyloid-beta42(43) in brains of mice expressing mutant presenilin 1. Nature 383:710–713.

Eanes ED, Glenner GG (1968) X-ray diffraction studies on amyloid filaments. J Histochem Cytochem 16:673–677.

Frederickson CJ (1989) Neurobiology of zinc and zinc-containing neurons. Int Rev Neurobiol 31:145–328.

Friedlich AL, Lee JY, van Groen T, Cherny RA, Volitakis I, Cole TB, Palmiter RD, Koh JY, Bush AI (2004) Neuronal zinc exchange with the blood vessel wall promotes cerebral amyloid angiopathy in an animal model of Alzheimer's disease. J Neurosci 24:3453–3459.

Garzon-Rodriguez W, Yatsimirsky AK, Glabe CG (1999) Binding of Zn(II), Cu(II), and Fe(II) ions to Alzheimer's A beta peptide studied by fluorescence. Bioorg Med Chem Lett 9:2243–2248.

Glenner GG, Wong CW (1984) Alzheimer's disease: initial report of the purification and characterization of a novel cerebrovascular amyloid protein. Biochem Biophys Res Commun 120:885–890.

Goldsbury C, Frey P, Olivieri V, Aebi U, Muller SA (2005) Multiple Assembly Pathways Underlie Amyloid-beta Fibril Polymorphisms. J Mol Biol.

Goldsbury CS, Wirtz S, Muller SA, Sunderji S, Wicki P, Aebi U, Frey P (2000) Studies on the in vitro assembly of $A\beta$1-40: Implications for the search for $A\beta$ fibril formation inhibitors. J Str Biol 130:217–231.

Grundke-Iqbal I, Fleming J, Tung YC, Lassmann H, Iqbal K, Joshi JG (1990) Ferritin is a component of the neuritic (senile) plaque in Alzheimer dementia. Acta Neuropathol (Berl) 81:105–110.

Haass C, Schlossmacher MG, Hung AY, Vigo-Pelfrey C, Mellon A, Ostaszewski BL, Lieberburg I, Koo EH, Schenk D, Teplow DB, et a (1992) Amyloid β-peptide is produced by cultured cells during normal metabolism. Nature 359:322–325.

Haeffner F, Smith D, Barnham K, Bush A (In press) Model Studies of Cholesterol and Ascorbate Oxidation by Copper Complexes: Relevance to Alzheimer's Disease b–amyloid Metallochemistry. J Inorg Biochem.

Harper JD, Wong SS, Lieber CM, Lansbury PT (1997) Observation of metastable Abeta amyloid protofibrils by atomic force microscopy. Chem Biol 4:119–125.

Huang X, Atwood CS, Moir RD, Hartshorn MA, Tanzi RE, Bush AI (2004) Trace metal contamination initiates the apparent auto-aggregation, amyloidosis, and oligomerization of Alzheimer's Abeta peptides. J Biol Inorg Chem 9:954–960.

Huang X, Atwood CS, Moir RD, Hartshorn MA, Vonsattel J-P, Tanzi RE, Bush AI (1997) Zinc-induced Alzheimer's $A\beta$1-40 aggregation is mediated by conformational factors. J Biol Chem 272:26464–26470.

Huang X, Atwood CS, Hartshorn MA, Multhaup G, Goldstein LE, Scarpa RC, Cuajungco MP, Gray DN, Lim J, Moir RD, Tanzi RE, Bush AI (1999a) The $A\beta$ peptide of Alzheimer's Disease directly produces hydrogen peroxide through metal ion reduction. Biochemistry 38:7609–7616.

Huang X, Cuajungco MP, Atwood CS, Hartshorn MA, Tyndall J, Hanson GR, Stokes KC, Leopold M, Multhaup G, Goldstein LE, Scarpa RC, Saunders AJ, Lim J, Moir RD, Glabe C, Bowden EF, Masters CL, Fairlie DP, Tanzi RE, Bush AI (1999b) Cu(II) potentiation of Alzheimer Aβ neurotoxicity: correlation with cell-free hydrogen peroxide production and metal reduction. J Biol Chem 274:37111–37116.

Jayasinghe SA, Langen R (2004) Identifying structural features of fibrillar islet amyloid polypeptide using site-directed spin labeling. J Biol Chem.

Jenkins J, Pickersgill R (2001) The architecture of parallel beta-helices and related folds. Prog Biophys Mol Biol 77:111–175.

Johnstone EM, Chaney MO, Norris FH, Pascual R, Little SP (1991) Conservation of the sequence of the Alzheimer's disease amyloid peptide in dog, polar bear and five other mammals by cross-species polymerase chain reaction analysis. Mol Brain Res 10:299–305.

Jurnak F, Yoder MD, Pickersgill R, Jenkins J (1994) Parallel beta-domains: a new fold in protein structures. Curr Opin Struct Biol 4:802–806.

Karr J, Kaupp L, Szalai V (2004) J Am Chem Soc 126:13534–13538.

Kayed R, Head E, Thompson JL, McIntire TM, Milton SC, Cotman CW, Glabe CG (2003) Common structure of soluble amyloid oligomers implies common mechanism of pathogenesis. Science 300:486–489.

Kheterpal I, Zhou S, Cook KD, Wetzel R (2000) Abeta amyloid fibrils possess a core structure highly resistant to hydrogen exchange. Proc Natl Acad Sci USA 97:13597–13601.

Kheterpal I, Lashuel HA, Hartley DM, Walz T, Lansbury PT, Jr., Wetzel R (2003) Abeta protofibrils possess a stable core structure resistant to hydrogen exchange. Biochemistry 42:14092–14098.

Kirschner DA, Abraham C, Selkoe DJ (1986) X-ray diffraction from intraneuronal paired helical filaments and extraneuronal amyloid fibers in Alzheimer disease indicates cross-beta conformation. Proc Natl Acad Sci USA 83:503–507.

Krebs MR, Morozova-Roche LA, Daniel K, Robinson CV, Dobson CM (2004) Observation of sequence specificity in the seeding of protein amyloid fibrils. Protein Sci 13:1933–1938.

Lashuel HA, Hartley DM, Petre BM, Wall JS, Simon MN, Walz T, Lansbury PT, Jr (2003) Mixtures of wild-type and a pathogenic (E22G) form of Abeta40 in vitro accumulate protofibrils, including amyloid pores. J Mol Biol 332:795–808.

Lee J-Y, Cole TB, Palmiter RD, Suh SW, Koh J-Y (2002) Contribution by synaptic zinc to the gender-disparate plaque formation in human Swedish mutant APP transgenic mice. Proc Natl Acad Sci USA 99:7705–7710.

Lee J-Y, Kim J-H, Hong SH, Lee JY, Cherny RA, Bush AI, Palmiter RD, Koh J-Y (2004) Estrogen decreases zinc transporter 3 expression and synaptic vesicle zinc levels in mouse brain. J Biol Chem 279:8602–8607.

LeVine H, 3rd (2004) Alzheimer's beta-peptide oligomer formation at physiologic concentrations. Anal Biochem 335:81–90.

Lomas DA, Carrell RW (2002) Serpinopathies and the conformational dementias. Nat Rev Genet 3:759–768.

Lovell MA, Robertson JD, Teesdale WJ, Campbell JL, Markesbery WR (1998) Copper, iron and zinc in Alzheimer's disease senile plaques. J Neurol Sci 158:47–52.

Margittai M, Langen R (2004) Template-assisted filament growth by parallel stacking of tau. Proc Natl Acad Sci USA 101:10278–10283.

Maynard CJ, Cappai R, Volitakis I, Cherny RA, White AR, Beyreuther K, Masters CL, Bush AI, Li Q-X (2002) Overexpression of Alzheimer's disease β-amyloid opposes the age-dependent elevations of brain copper and iron levels. J Biol Chem 277:44670–44676.

McLean C, Cherny R, Fraser F, Fuller S, Smith M, Beyreuther K, Bush A, Masters C (1999) Soluble pool of $A\beta$ amyloid as a determinant of severity of neurodegeneration in Alzheimer's Disease. Ann Neurol 46:860–866.

Miura T, Suzuki K, Kohata N, Takeuchi H (2000) Metal binding modes of Alzheimer's amyloid β-peptide in insoluble aggregates and soluble complexes. Biochemistry 39:7024–7031.

Mori T, Paris D, Town T, Rojiani AM, Sparks DL, Delledonne A, Crawford F, Abdullah LI, Humphrey JA, Dickson DW, Mullan MJ (2001) Cholesterol accumulates in senile plaques of Alzheimer disease patients and in transgenic APP(SW) mice. J Neuropathol Exp Neurol 60:778–785.

Nelson R, Sawaya MR, Balbirnie M, Madsen AO, Riekel C, Grothe R, Eisenberg D (2005) Structure of the cross-beta spine of amyloid-like fibrils. Nature 435:773–778.

Nichols MR, Moss MA, Reed DK, Lin WL, Mukhopadhyay R, Hoh JH, Rosenberry TL (2002) Growth of beta-amyloid(1-40) protofibrils by monomer elongation and lateral association. Characterization of distinct products by light scattering and atomic force microscopy. Biochemistry 41:6115–6127.

Ohtsu H, Shimazaki Y, Odani A, Yamauchi O, Mori W, Itoh S, Fukuzumi S (2000) J Am Chem Soc 122:5733–5741.

O'Nuallain B, Wetzel R (2002) Conformational Abs recognizing a generic amyloid fibril epitope. Proc Natl Acad Sci USA 99:1485–1490.

Opazo C, Huang X, Cherny R, Moir R, Roher A, White A, Cappai R, Masters C, Tanzi R, Inestrosa N, Bush A (2002) Metalloenzyme-like activity of Alzheimer's disease β-amyloid: Cu-dependent catalytic conversion of dopamine, cholesterol and biological reducing agents to neurotoxic H_2O_2. J Biol Chem 277:40302–40308.

Perutz MF, Finch JT, Berriman J, Lesk A (2002) Amyloid fibers are water-filled nanotubes. Proc Natl Acad Sci USA 99:5591–5595.

Petkova AT, Buntkowsky G, Dyda F, Leapman RD, Yau WM, Tycko R (2004) Solid State NMR Reveals a pH-dependent Antiparallel beta-Sheet Registry in Fibrils Formed by a beta-Amyloid Peptide. J Mol Biol 335:247–260.

Petkova AT, Leapman RD, Guo Z, Yau WM, Mattson MP, Tycko R (2005) Self-propagating, molecular-level polymorphism in Alzheimer's beta-amyloid fibrils. Science 307:262–265.

Phinney AL, Drisaldi B, Schmidt SD, Lugowski S, Coronado V, Liang Y, Horne P, Yang J, Sekoulidis J, Coomaraswamy J, Chishti MA, Cox DW, Mathews PM, Nixon RA, Carlson GA, George-Hyslop PS, Westaway D (2003) In vivo reduction of amyloid-{beta} by a mutant copper transporter. Proc Natl Acad Sci USA 100:14193–14198.

Refolo LM, Malester B, LaFrancois J, Bryant-Thomas T, Wang R, Tint GS, Sambamurti K, Duff K, Pappolla MA (2000) Hypercholesterolemia accelerates the Alzheimer's amyloid pathology in a transgenic mouse model. Neurobiol Dis 7:321–331.

Ritchie CW, Bush AI, Mackinnon A, Macfarlane S, Mastwyk M, MacGregor L, Kiers L, Cherny RA, Li QX, Tammer A, Carrington D, Mavros C, Volitakis I, Xilinas M, Ames D, Davis S, Beyreuther K, Tanzi RE, Masters CL (2003) Metal-protein attenuation with iodochlorhydroxyquin (clioquinol) targeting $A\beta$ amyloid deposition and toxicity in Alzheimer's disease: a pilot phase 2 clinical trial. Arch Neurol 60:1685–1691.

Roher AE, Chaney MO, Kuo YM, Webster SD, Stine WB, Haverkamp LJ, Woods AS, Cotter RJ, Tuohy JM, Krafft GA, Bonnell BS, Emmerling MR (1996) Morphology and toxicity of $A\beta$-(1-42) dimer derived from neuritic and vascular amyloid deposits of Alzheimer's disease. J Biol Chem 271:20631–20635.

Schlief ML, Craig AM, Gitlin JD (2005) NMDA receptor activation mediates copper homeostasis in hippocampal neurons. J Neurosci 25:239–246.

Seubert P, Vigo-Pelfrey C, Esch F, Lee M, Dovey H, Davis D, Sinha S, Schlossmacher M, Whaley J, Swindlehurst C, et al. (1992) Isolation and quantification of soluble Alzheimer's β-peptide from biological fluids. Nature 359:325–327.

Shivaprasad S, Wetzel R (2004) An intersheet packing interaction in A beta fibrils mapped by disulfide cross-linking. Biochemistry 43:15310–15317.

Shivers BD, Hilbich C, Multhaup G, Salbaum M, Beyreuther K, Seeburg PH (1988) Alzheimer's disease amyloidogenic glycoprotein: expression pattern in rat brain suggests a role in cell contact. EMBO J 7:1365–1370.

Shoji M, Golde TE, Ghiso J, Cheung TT, Estus S, Shaffer LM, Cai X-D, McKay DM, Tintner R, Frangione B, Younkin SG (1992) Production of the Alzheimer amyloid β protein by normal proteolytic processing. Science 258:126–129.

Smith MA, Perry G, Richey PL, Sayre LM, Anderson VE, Beal MF, Kowall N (1996) Oxidative damage in Alzheimer's. Nature 382:120–121.

Smith MA, Hirai K, Hsiao K, Pappolla MA, Harris P, Siedlak S, Tabaton M, Perry G (1998) Amyloid-beta deposition in Alzheimer transgenic mice is associated with oxidative stress. J Neurochem 70:2212–2215.

Soreghan B, Kosmoski J, Glabe C (1994) Surfactant properties of Alzheimer's A beta peptides and the mechanism of amyloid aggregation. J Biol Chem 269:28551–28554.

Syme C, Nadal R, Rigby S, Viles J (2004) J Biol Chem 279:18169–18177.

Tabner BJ, Turnbull S, El-Agnaf OM, Allsop D (2002) Formation of hydrogen peroxide and hydroxyl radicals from A(beta) and alpha-synuclein as a possible mechanism of cell death in Alzheimer's disease and Parkinson's disease. Free Radic Biol Med 32:1076–1083.

Terry RD, Masliah E, Salmon DP, Butters N, DeTeresa R, Hill R, Hansen LA, Katzman R (1991) Physical basis of cognitive alterations in Alzheimer's disease: synapse loss is the major correlate of cognitive impairment. Ann Neurol 30:572–580.

Tickler AK, Smith DG, Ciccotosto GD, Tew DJ, Curtain CC, Carrington D, Masters CL, Bush AI, Cherny RA, Cappai R, Wade JD, Barnham KJ (2005) Methylation of the imidazole side chains of the Alzheimer disease amyloid-beta peptide results in abolition of superoxide dismutase-like structures and inhibition of neurotoxicity. J Biol Chem 280:13355–13363.

Tycko R (2003) Insights into the amyloid folding problem from solid-state NMR. Biochemistry 42:3151–3159.

Varadarajan S, Yatin S, Kanski J, Jahanshahi F, Butterfield DA (1999) Methionine residue 35 is important in amyloid-beta-peptide associated free radical oxidative stress. Brain Res Bull 50:133–141.

Vaughan DW, Peters A (1981) The structure of neuritic plaques in the cerebral cortex of aged rats. J Neuropathol Exp Neurol 40:472–487.

Walsh DM, Lomakin A, Benedek GB, Condron MM, Teplow DB (1997) Amyloid beta-protein fibrillogenesis. Detection of a protofibrillar intermediate. J Biol Chem 272:22364–22372.

Walsh DM, Hartley DM, Kusumoto Y, Fezoui Y, Condron MM, Lomakin A, Benedek GB, Selkoe DJ, Teplow DB (1999) Amyloid beta-protein fibrillogenesis. Structure and biological activity of protofibrillar intermediates. J Biol Chem 274:25945–25952.

Yoburn JC, Tian W, Brower JO, Nowick JS, Glabe CG, Van Vranken DL (2003) Dityrosine cross-linked Abeta peptides: fibrillar beta-structure in Abeta(1-40) is conducive to formation of dityrosine cross-links but a dityrosine cross-link in Abeta(8-14) does not induce beta-structure. Chem Res Toxicol 16:531–535.

Chapter 8

Signaling Mechanisms that Mediate Aβ Induced Neuronal Dysfunction

Carl W. Cotman[1], Ph.D. and Jorge Busciglio, Ph.D.[2]

[1]*Institute for Brain Aging and Dementia*
University of California, Irvine
1113 GNRF ZOT 4540
Irvine CA 92697-4540
Email: cwcotman@uci.edu

[2]*Department of Neurobiology and Behavior*
School of Biological Sciences
University of California, Irvine
2146 McGaugh Hall Mail Code: 4500
Irvine, CA 92697
Email: jbuscigl@uci.edu

1.   Introduction

After the original proposal of the "amyloid hypothesis", genetic data and transgenic models have provided strong support for the notion that Aβ plays a central role in the disease process (Selkoe 1991). While it is known that Aβ is neurotoxic *in vivo* and *in vitro* (Cotman and Pike 1994; Yankner 1996), cognitive impairment appears to precede high levels of Aβ accumulation in these models (Morgan, Diamond et al. 2000). Further, while Aβ accumulates in these models, neuronal death is uncommon. In AD, cognitive loss also appears to precede the deposition of Aβ plaques and pronounced neuronal degeneration in the brain (Morris, Storandt et al. 1996; Hsia, Masliah et al. 1999; Lue, Kuo et al. 1999; Naslund, Haroutunian et al. 2000). Thus if Aβ is contributing to brain dysfunction, it must have an ability to compromise neuronal function by mechanisms in addition to neuronal death.

An emerging view is that Aβ adopts different aggregated conformations in vivo (e.g., oligomers, protofibrils and fibrils; see Chapter 6), and that each of these species alters neuronal homeostasis in distinct, yet not completely specified ways. Recent results indicate that, in addition to Aβ fibrils, Aβ oligomers are also present in the AD brain (Kayed et al. 2003; Lacor et al. 2004), supporting the idea that multiple Aβ conformations contribute to AD etiopathology by different toxic mechanisms. Furthermore, the characterization of intracellular Aβ accumulation as a relevant pathological event significantly increases the complexity of the neurotoxic puzzle. While it is well known Aβ fibrils and oligomers can cause neuronal death *in vitro* and *in vivo*, only recently have mechanisms studies demonstrated that Aβ can cause neuronal dysfunction in the absence of cell death. (Cotman and Anderson 1995; Yankner 1996; Mattson and Chan 2003). It is possible that the presence of Aβ may promote changes on neuronal plasticity and/or synaptic connectivity, leading to the early cognitive deficits characteristic of the disease process.

In this Chapter, we will present new mechanisms by which Aβ may compromise neuronal function in the absence of cell death. The first section summarizes experiments directed to understand the effect of Aβ on NMDA and BDNF/TrkB signal transduction. NMDA and BDNF/TrkB signal transduction are critical for neuronal communication and brain plasticity such as learning and memory. The second section presents a mechanism by which chronic exposure to Aβ mediates the development of neuronal dystrophy. The mechanism involves aberrant activation of focal adhesion signals. Thus Aβ can act through normal signaling systems to cause aberrant signaling and neuronal dysfunction in the absence of cell death.

2.   Aβ impairs NMDA and BDNF/TrkB signal transduction

Because of the protracted and progressive nature of the disease, Aβ may be present in the brain at sublethal concentrations for extended periods. Although at these levels Aβ does not compromise neuron survival, Aβ may affect critical signal transduction processes that mediate plastic neuronal changes involved in learning and memory. Though learning and memory utilizes many mechanisms, there are two key signaling pathways: 1. NMDA receptor activation and the regulation of downstream signal transduction pathways and 2. BDNF signaling, Trk B activation and the regulation of downstream pathways. Recently, as discussed below, it has been demonstrated that Aβ at nontoxic levels can impair neuronal plasticity processes via interference with NMDA and BDNF/TrkB signal transduction.

3. NMDA receptor activation and learning and memory

NMDA receptor activation is essential for certain forms of learning and memory and for long term potentiation. For example, NMDA antagonists block the induction of LTP and defects in NMDA receptor activation compromise learning and memory (Balazs 2005). NMDA regulates several gene transcriptional products critical for learning and memory such as CREB.

4. BDNF signaling is critical for learning and memory

BDNF is critical for neuronal survival, protection from injury, and activity-dependent plasticity including learning and memory mechanisms, encoding LTP, neurogenesis, and synaptic growth/sprouting (Bramham and Messaoudi 2005). Mice deficient in either BDNF or its receptor TrkB exhibit impaired dendritic and axonal arborization, synaptic activity and neuronal plasticity, including impairment in long-term potentiation (LTP) and learning and memory processes (Korte et al. 1995; Patterson et al. 1996; Causing et al. 1997; Martinez et al. 1998; Minichiello et al. 1999). BDNF can induce its own synthesis and stimulate selective increases in several genes and proteins involved in synaptic plasticity, e.g., synaptotagamin, synapsin, NR2B, PSD95, and proteins involved in receptor trafficking. BDNF can regulate expression in the cell body but also locally at the synapse. Recently, BDNF has been shown to regulate the local translation of a set of proteins (Steward and Schuman 2001), including CaMKII and Homer-2, in the dendritic compartment (Schratt et al. 2004). This mechanism may provide a local regulatory mechanism for mediating synaptic plasticity. Thus, impaired BDNF signaling will compromise many aspects of brain function.

5. Can β-amyloid impair signaling at the level of NMDA or BDNF activation?

If Aβ compromises NMDA receptor and BDNF signaling, neuronal function would be impaired prior to neuronal degeneration and might be a new mechanism to account for early cognitive decline prior to accumulation of extensive pathology and degeneration. If so, then Aβ serves to suppress key plasticity mechanism and may compromise learning and memory in the absence of neuronal degeneration.

6.   Is NMDA receptor function impaired in the presence of sub-lethal levels of AB?

Do $A\beta_{1-42}$ levels which do not compromise the survival of cortical neurons interfere with NMDA signaling functions critical for neuronal plasticity such as CREB activation? The transcription factor CREB, which regulates expression of cAMP response element (CRE)-containing genes, plays an essential role in learning and memory processes in a variety of species ranging from Drosophila to mammals (Bourtchuladze, Frenguelli et al. 1994; Yin et al. 1994; Tully 1997; Abel and Kandel 1998).

NMDA evoked a marked increase in P-CREB levels, and this effect was suppressed by sublethal concentrations of $A\beta_{1-42}$, to approximately one-half. The depolarization-induced elevation of P-CREB was also suppressed via high K+ to approximately one-half, $40\pm2.0\%$ of untreated controls after treatment with $A\beta_{1-42}$ without affecting total CREB levels (Tong et al. 2001).

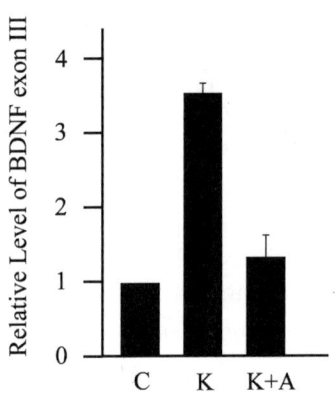

Figure 1. Exon III BDNF mRNA levels using RT-PCR increased in the high K^+-exposed cells: pretreatment with 5 μM $A\beta_{1-42}$ attenuated the high K^+-induced elevation.

$A\beta$ forms small diffusible oligomers (also referred to as ADDLs, for $A\beta$-derived diffusible ligands), which are highly neurotoxic in the low micromolar to high nanomolar range (Oda et al. 1995; Lambert et al. 1998). At the sublethal concentrations of 100 nM, $A\beta$ oligomers elicited a marked suppression of the high K^+-induced increase in P-CREB content to $62\pm13.6\%$ of the levels in the untreated controls. The effect was specific to $A\beta_{1-42}$. $A\beta_{1-42}$ as a random sequence [$A\beta_{1-42}(R)$] failed to influence the high K^+ induced increase of P-CREB levels. Importunately, a derivative of $A\beta_{1-42}$, $A\beta_{25-35}$ which causes neuron degeneration and is assumed to have a similar action did not interfere with CREB phosphorylation at the sublethal concentrations showing the mechanism for suppression is distinct from degeneration.

Sublethal concentrations of $A\beta_{1-42}$ decreases CRE transcriptional activity and the induction of BDNF Exon III to 33% of controls (Fig. 1).

Taken together these data support the hypothesis that can $A\beta_{1-42}$ will interfere with critical signal transduction and transcriptional activity of CREB regulated gene products. Recently the effect of $A\beta_{1-42}$ on CREB activation has been confirmed and the mechanism further identified (Snyder, et al. 2005). $A\beta$ promotes the endocytosis of surface NMDA receptors. The effect of $A\beta_{1-42}$

appeared to be mediated via nicotinic $\alpha7$ receptors which in turn activates downstream phosphatases that regulate NMDA receptor internalization.

Overall, these data suggest that $A\beta_{1-42}$ may compromise NMDA receptor signaling as $A\beta$ accumulates in the aging and AD brain. Thus, nondegenerative mechanisms may participate in the cascade of events causing brain dysfunction and cognitive decline. Neuronal function may be further impaired by BDNF signal suppression.

7. $A\beta_{1-42}$ decreases the ability of BDNF to regulate the transcription factors, CREB and Elk-1, and compromises the ability of BDNF to protect neurons

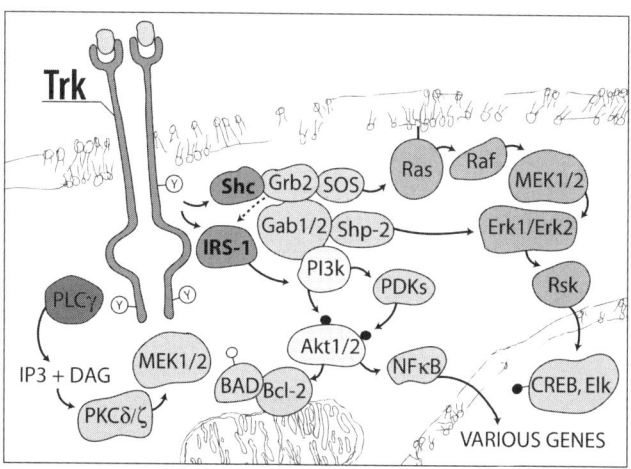

Figure 2. Stimulation of the receptor tyrosine kinase TrkB activates three major transduction pathways: phosphatidylinositol-3-kinase (PI3-K)/Akt, Ras/MAPK, and PLCγ/PKC pathways.

BDNF signals cells via a receptor Tyrosine kinase (RTK), TrkB. BDNF binding results in the dimerization and autophosphorylation of TrkB receptors that trigger directly or through binding of adapter proteins the activation of signal transduction pathways, such as the Ras/MAPK, Akt/PI 3-K and phospholipase C (PLCγ) pathways, thereby accounting for the pleiotropic effects of BDNF. Figure 2 illustrates the central signaling pathways and molecules involved in regulating Trk receptor function.

Pretreatment with $A\beta_{1-42}$ in the subtoxic range results in a concentration-dependent decrease in the effect of BDNF on P-CREB, without affecting the amount of T-CREB. Pretreatment with $A\beta_{1-42}$ reduces P-CREB levels to 66 $\pm7\%$. $A\beta_{1-42}$ also causes a significant decrease in the amount of P-Elk-1 in the BDNF-treated cultures. BDNF increases the levels of phosphorylated Elk-1 (P-Elk-1) and pretreatment with $A\beta_{1-42}$ also causes a significant decrease in P-Elk-1. Pretreatment with $A\beta_{1-42}$ reduces BDNF-induced CRE transcription activity to $55\pm12\%$ and SRE activity to $56\pm12\%$.

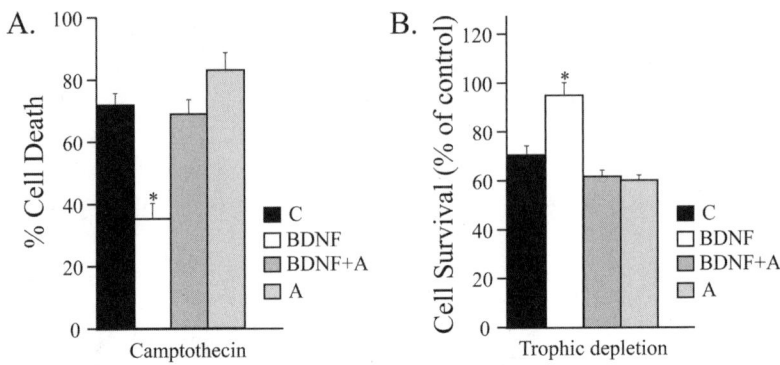

Figure 3. Aβ_{1-42} abrogated BDNF protection of cortical neurons from apoptosis induced either by exposure to camptothecin (A) or by deprivation from trophic support (B).

An enhancement of neuronal survival is one of important functions of BDNF. It is possible that the Aβ mediated impairment of cell signaling could compromise the capacity of BDNF to protect neurons. Removal of trophic support results in significant cell loss. The protective effect of BDNF is severely compromised by sublethal concentrations of Aβ_{1-42} (Fig. 3).

BDNF can also protect cortical neurons from apoptosis induced by camptothecin an inhibitor of DNA topoisomerase-1 that causes DNA damage (Morris et al. 1996). Camptothecin treatment causes massive cell degeneration that is markedly reduced by BDNF treatment. Exposure to 5 μM Aβ_{1-42} completely abolishes BDNF protection. The survival promoting capacity of BDNF with reduced trophic support depends on the PI3-K pathway while that in the presence of camptothecin depends on the MAPK pathway suggesting Aβ may have a broad range of action.

8.   Aβ_{1-42} suppresses signaling of the PI3-K and MAPK pathways

What is the mechanism by which Aβ_{1-42} reduces the ability of neurons to withstand these insults and reduce transcriptional activity? Brief exposure to BDNF (10 min) causes a robust increase in the amount of both the phosphorylated p42 and p44 isoforms of MAPK. Pretreatment with sublethal concentrations of Aβ_{1-42} for 2 h has no significant effect on the basal level of phosphorylated MAPK, but results in a decrease in the BDNF-elicited activation of both MAPK isoforms (58$\pm$12% for p44, 45$\pm$18% for p42) as well as P-Raf and P-MEK1/2. Aβ oligomers at non-toxic levels (ADDLs) are also effective in suppressing the BDNF-induced activation of the MAPK, 68.45% $\pm$6.03%.

Similarly, PI3-K signaling pathway are also activated by BDNF (Huang and Reichardt 2001). Exposure of cortical neurons to BDNF results in a marked increase in the level of activated Akt which is markedly reduced after $A\beta_{1-42}$ pretreatment ($62\pm12\%$).

9. **$A\beta_{1-42}$ does not interfere with the BDNF-induced phosphorylation of Trk B receptors nor PLCγ phosphorylation but rather acts at the level of the docking proteins**

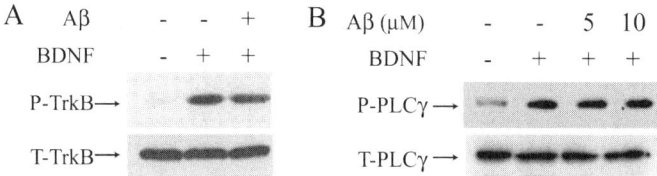

Figure 4. $A\beta_{1-42}$ treatment does not interfere with the activation of TrkB (A) and PLCγ (B). IRS-1 and Shc. The Shc isoforms (kDa: 66,52 and 46 respectively) are indicated as a, b and c.

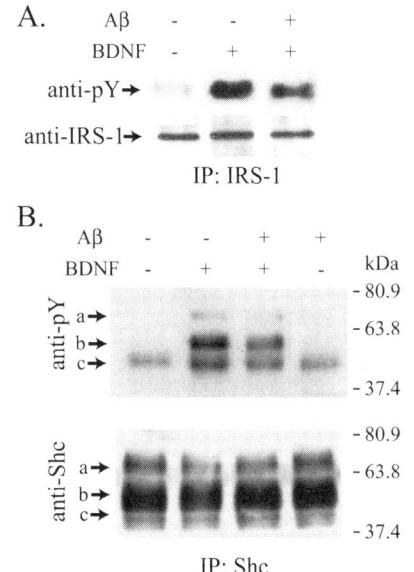

Figure 5. $A\beta_{1-42}$ causes a reduction in the level of the activated IRS-1 and suppresses the Tyr phosphorylation of the different Shc isoforms, $57\pm8\%$ for p66, $75\pm7\%$ for p52 and $63\pm4\%$ for p46.

Since the BDNF-activation of two of the major signaling pathways were suppressed by sublethal levels of $A\beta_{1-42}$, it is possible that the peptide interferes with the activation of the BDNF receptor TrkB. BDNF induced a pronounced Tyr-autophosphorylation of the TrkB receptor and this is not compromised by pretreatment with $A\beta_{1-42}$ (Fig. 4A). Thus the peptide interferes with BDNF-induced signaling at steps downstream of the BDNF receptor and upstream of MAPK and PI3-K.

The activation of the PI3-K and Ras-MAPK/ERK that were affected by $A\beta_{1-42}$, pathways, requires interaction of the activated BDNF receptor with docking proteins that after phosphorylation mediate the TrkB signaling to the protein kinase cascades. In contrast, the activated TrkB receptor directly binds and phosphorylates PLCγ, and Aβ_{1-42} does not affect this process (Fig. 4B). It is possible, therefore, that Aβ_{1-42} interferes with signal transduction at the level of the docking proteins. The state of Tyr phosphorylation of IRS-1 that activates PI3-K and the phosphorylation of Shc that mediates signaling via Grb2/SOS to Ras (Fig. 5).

This data suggests that low levels of Aβ_{1-42} impair BDNF-TrkB transcriptional activity and neuron survival probably by a mechanism involving regulation of the docking proteins. The regulation of docking proteins has also been shown to participate in insulin resistance.

10.   The neurotrophic factor resistance model of brain aging and neurodegenerative diseases

Thus, Aβ can interfere with NMDA receptor function and BDNF/TrkB signaling, two mechanisms involved in learning and memory and brain plasticity. These two mechanisms alone or in combination could interfere with brain function in the absence of degeneration. The fact that Aβ oligomers as well as fibrillar Aβ can suppress cell signaling suggests that the peptide may be a factor in the early decline seen even prior to plaque formation and place the brain at risk for further functional decline. Interestingly, the mechanism may share features in common to insulin resistance which predisposes an individual for type II diabetes.

Our data suggests that, similar to insulin resistance, Aβ generates a state of neurotrophin resistance. These two mechanisms – neurotrophin resistance and insulin resistance – share a similar regulatory mechanism involving the docking proteins that convey receptor activation to the PI3-K/Akt and Ras/Erk pathways (Zick 2003). If this is the case, the model provides a new mechanism for understanding brain dysfunction and developing new means to improve function before irreversible neuronal degeneration evolves. In the case of insulin resistance, a condition evolves that if left unchecked causes pancreatic β cell degeneration, but if managed the risk for pancreatic dysfunction and degeneration is reduced. Similar opportunities may exist for AD.

11.   Fibrillar Aβ induces neuronal dystrophy and synaptic loss

Early studies on Aβ neurotoxicity reported the induction of neuronal dystrophy by Aβ (Busciglio, Lorenzo et al. 1992; Pike et al. 1992). Neuronal dystrophy is specifically associated with AD and is not present in other neurodegenerative conditions that lack amyloid plaques (Benzing et al.

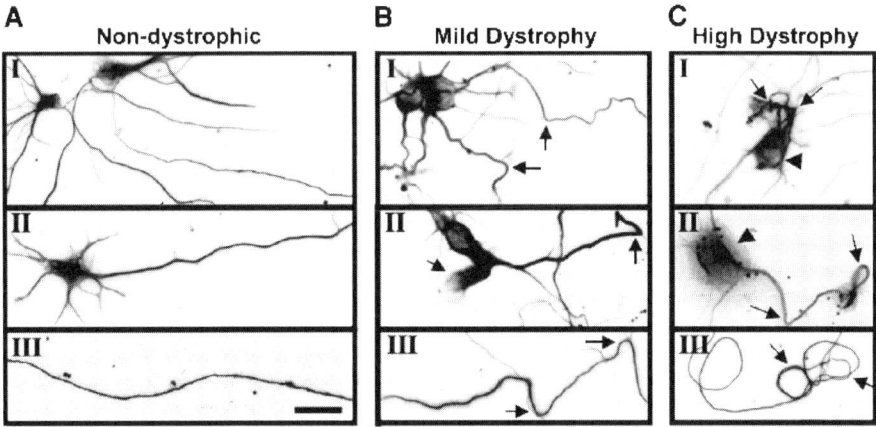

Figure 6. Morphological assessment of neuronal dystrophy induced by fibrillar Aβ. Neurons were characterized as non-dystrophic (A), mildly dystrophic (B), and highly dystrophic (C). Note the smooth appearance of neuronal processes in non-dystrophic neurons (A I, II, III). Mildly dystrophic neurons exhibit sharp angles in neuronal processes and lamellipodia-like structures growing from cell bodies (arrows, B I, II, III). Highly dystrophic neurons present neuritic retraction and severe tortuosity including sharp angles and loops (arrows, C I, II, III). Cortical cultures were treated with 20 μM fibrillar Aβ for 4 days, fixed and immunostained with anti-tubulin class III. Fluorescent images were converted to gray scale and inverted to enhance morphological detail. Scale bar = 15 μm.

1993). In AD, the number of dystrophic neurites correlates with the degree of dementia (McKee et al. 1991), and synaptic loss is observed in areas of aberrant neuronal sprouting (Geddes et al. 1986; Masliah et al. 1991). Thus, a pathological relationship between Aβ deposition, neuronal dystrophy and synaptic loss may lead to cognitive impairment in AD. We quantified the dystrophic effect of fibrillar amyloid β (Aβ) and its relationship with neurotoxicity and synaptic loss in cultured neurons. Treatment with fibrillar Aβ led to the development of neuritic dystrophy in the majority of neurons present in the culture (Fig. 6) (Grace et al. 2002). Morphometric analysis and viability assays showed that neuronal dystrophy appeared significantly earlier and at lower Aβ concentrations than neurotoxicity, suggesting that both effects are generated independently. A dramatic reduction in the density of synaptophysin immunoreactivity was closely associated with dystrophic changes in viable neurons suggesting that: (1) Aβ fibrils scattered in the brain parenchyma may induce dystrophic changes but not neuronal death; and (2) as the aberrant changes in neuronal morphology proceed, they lead to a concurrent loss of synaptic connections and, at later stages of the disease process, to neuronal degeneration (Grace et al. 2002).

Novel in vivo imaging techniques have provided direct evidence of the dystrophic effect of fibrillar Aβ in the brain. Two recent studies utilizing transcranial multi-photon live imaging in AD transgenic models show dramatic dystrophic and synaptotoxic effects of fibrillar amyloid deposits in living animals (Tsai et al. 2004; Spires et al. 2005). In the first study, Tsai et al. describe spine loss, dendritic shaft atrophy, and large axonal varicosities associated with neurite breakage and large-scale, permanent disruption of neuronal connections in close apposition with Aβ deposits in APP/PS1 double transgenic mice. In the second study, Spires et al. observed disrupted neurite trajectories, marked reductions in spine density, and loss of pre-and post-synaptic markers in association with Aβ deposits in Tg2576 APP transgenic mice. These results underscore the relevance of fibrillar Aβ as a neurotoxic agent, and its critical role in the development of neuritic and synaptic pathology in AD.

12.   Molecular signals that mediate fibrillar Aβ-induced neuronal dystrophy

Extracellular signals producing alterations in the cytoskeleton are often transduced through adhesion proteins (Mueller et al. 1989). Focal adhesion (FA) sites are integrin-based structures that mediate cell-substrate adhesion and the bidirectional exchange of information between extracellular molecules and the cytoplasm. Previous studies showed that Aβ binds to integrins (Kowalska and Badellino 1994; Sabo et al. 1995; Goodwin et al. 1997), and activates the FA proteins paxillin and focal adhesion kinase (FAK), which are downstream of integrin receptors, (Zhang et al. 1994; Zhang et al. 1996; Berg et al. 1997; Williamson et al. 2002). During the assembly of FA structures, paxillin is tyrosine-phosphorylated, and translocates to the cytoskeleton where it acts as a scaffold that binds integrin receptors, kinases such as FAK, phosphatases such as the protein tyrosine phosphatase PEST (PTP-PEST), adapter proteins and actin-binding proteins (Fig. 8) (Schaller et al. 1995; Turner 2000). This FA complex mediates spatial and temporal interactions that are initiated upon integrin receptor activation. The interplay between kinases and phosphatases leads to the initiation of multiple signal transduction cascades that promote local changes in the cytoskeleton that result in modifications in cell morphology (Aplin et al. 1998; Turner 2000). We addressed the role of FA signaling in Aβ-induced neuronal dystrophy, by analyzing the expression and activity of FA proteins in cultured neurons and in the AD brain. In culture, fibrillar Aβ treatment induced integrin receptor clustering, paxillin tyrosine phospho-rylation and translocation to the cytoskeleton, and promoted the formation of aberrant FA-like structures, suggesting the activation of focal adhesion

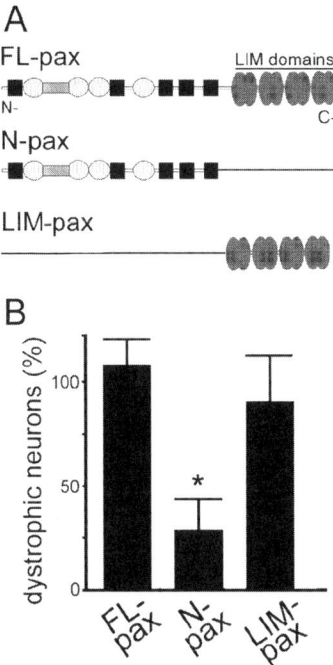

Figure 7. Paxillin LIM domains are required for Aβ-induced neuronal dystrophy. **(A)** Cortical neurons were transfected with GFP or GFP-paxillin expression vectors containing full length paxillin (FL-pax), or a deletion of the LIM domains (N-pax), or the LIM domains (LIM-pax) respectively. **(B)** Cortical neurons were transfected at day 5 and treated with fibrillar Aβ. The number of dystrophic neurons was quantified 24 hr later and expressed as a percent of the number of Aβ-induced dystrophic neurons expressing GFP (100%). Note the significant inhibition of neuronal dystrophy in neurons expressing N-pax (28.6±15.1%). FL-pax and LIM-pax were not significantly different than GFP (FL-pax: 108±12%, LIM-pax 90.4±21.9%). Values are the mean ± SEM; n=4 independent experiments; 250 neurons were scored per experiment. *p < 0.05 relative to GFP-transfected cells by Anova.

signaling cascades (Grace and Busciglio 2003). Deletion experiments indicated that the LIM domains in the C-terminus of paxillin mediate Aβ-induced neuronal dystrophy (Fig. 7 and 8). LIM domains are zinc-binding, cysteine rich motifs consisting of two zinc fingers repeated in tandem which mediate protein-protein interactions. LIM domains are critical to bring together several components into the focal adhesion complex.

Deletion and mutational studies showed that the recruitment of PTP-PEST to the FA complex by binding to paxillin LIM domains is a critical step for the development of Aβ-induced neuronal dystrophy (Fig. 7B). PTP-PEST is required for the turnover of FA contacts (Angers-Loustau et al. 1999). In the absence of PTP-PEST, FA contacts may become stabilized, preventing further

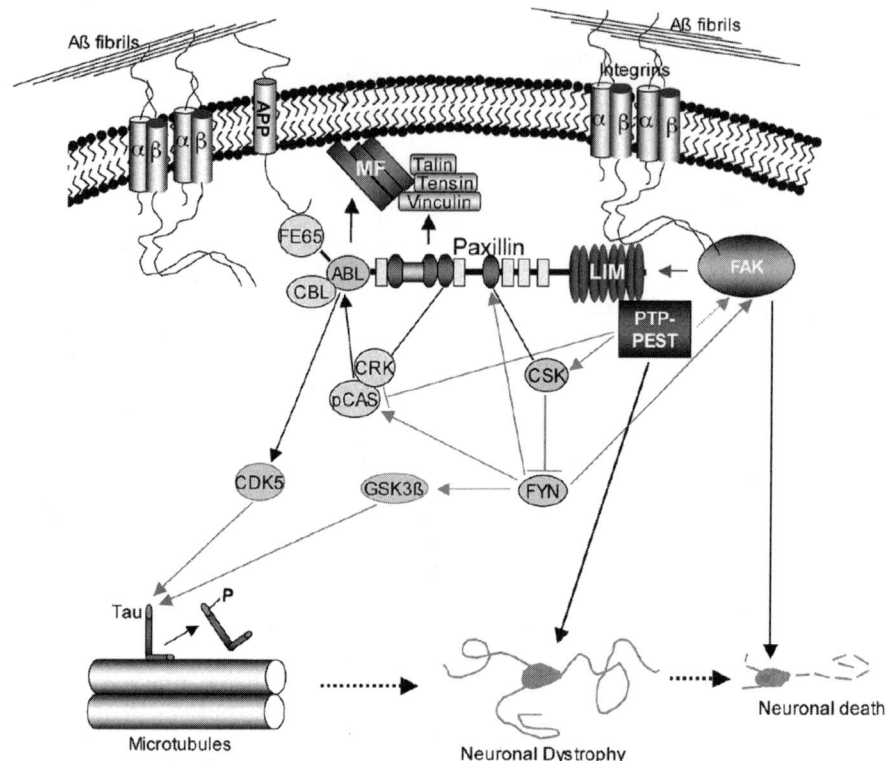

Figure 8. FA pathways involved in Aβ-induced neuronal dystrophy. Fibrillar Aβ binds to and induces the clustering of integrin receptors, leading to the activation of paxillin and FAK and their translocation to the nascent FA complex. Paxillin binds to vinculin, which promote microfilament stabilization at the FA site. PTP-PEST binds to paxillin leading to dephosphorylation of several FA proteins, which prevents the stabilization of the FA contact, allowing the neuron to continuously respond to fibrillar Aβ stimuli. Alternatively, APP binds to Aβ fibrils, bringing them in contact with integrins and/or activating FA signaling through FE65, which binds to the C-terminus of APP and associates with c-abl, which in turn binds and phosphorylates paxillin. A pathway involving fyn, which is downstream of PTP-PEST promotes GSK3β activity, while the interaction of cbl with c-abl increases cdk5 activity (Zukerberg, Patrick et al. 2000). Both Cdk5 and GSK3β hyperphosphorylate tau, leading to microtubular destabilization and neuronal dystrophy. An alternative pathway involving FAK activity leads to neuronal cell death but not neuronal dystrophy. Paxillin contains four SH2 binding domains, five leucine-rich LD domains, 1 proline-rich SH3 binding domain and 4 LIM domains (adapted from Grace & Busciglio, 2003).

response to extracellular stimuli. FRNK, a dominant negative form of FAK, did not alter Aβ-induced neuronal dystrophy, suggesting that although FAK is activated by fibrillar Aβ (Zhang et al. 1994; Zhang et al. 1996; Williamson et al. 2002), it is not involved in the dystrophic response.

Fibrillar Aβ has been shown to bind to cell-surface APP (Lorenzo et al. 2000), which co-localizes with integrins in neurons (Storey et al. 1996; Yamazaki et al. 1997) and participates in FA signaling (Sabo et al. 2001). APP levels were upregulated by Aβ treatment and APP was significantly enriched in aberrant FA-like structures (Heredia et al. 2004). APP may bring Aβ fibrils into physical contact with integrin receptors. It may also activate paxillin through the adapter protein FE65 (Sabo et al. 2001), which associates with c-abl, a non-receptor tyrosine kinase that phosphorylates paxillin (Salgia et al. 1995; Sabo et al. 2001; Fig. 3). Interestingly, a polymorphism in the FE65 gene that reduces its binding to APP appears to confer resistance to late-onset AD (Hu et al. 1998; Lambert et al. 2000; Hu et al. 2002). Ongoing experiments are directed to establish the role of APP in the transduction of fibrillar Aβ-induced neuronal dystrophy.

13. Abnormal activation of focal adhesion proteins in the Alzheimer's brain

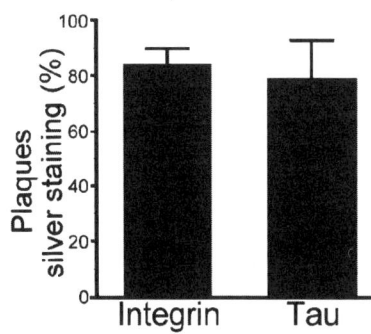

Figure 9. Quantification of integrin- and hyperphosphorylated tau-positive plaques shows that 84±6% of silver stained plaques were positive for integrin immunoreactivity and 79±14% were positive for hyperphosphorylated tau. Ten to twenty microscopic fields were analyzed in adjacent sections of temporal cortex of four AD brain cases. At least 50 plaques were scored per silver stained section. Scale bars A-F: 50μm; G and H: 250μm.

Previous studies indicate that integrins are abnormally expressed in the AD brain (Akiyama et al. 1991; Frohman et al. 1991; Eikelenboom et al. 1994), suggesting that Aβ-induced neuronal dystrophy may be mediated by integrin signaling. Integrin β1, the subunit that directly binds to and recruits paxillin to the FA complex (Fig. 8; Schaller et al. 1995), was enriched in dystrophic neurites and cell bodies surrounding amyloid cores. Most importantly, we detected increased levels of both tyrosine-phosphorylated activated paxillin and FAK in dystrophic neurites surrounding plaque cores, consistent with the activation of FA signaling cascades in neuronal processes in contact with fibrillar Aβ (Grace and Busciglio 2003). In fact, most senile plaques exhibited strong integrin immunoreactivity, indicating that FA protein activation may be a common feature in the AD brain (Fig. 9).

In summary, the aberrant activation of FA pathways appears to be critically involved in fibrillar $A\beta$-induced neuronal dystrophy. The ability of neurons to respond dynamically to extracellular cues is reminiscent of plasticity mechanisms. Brain regions with the highest plasticity are the most vulnerable in AD (Small 1998). In this regard, maladaptive neuronal plasticity has been proposed to play a major role in AD (Cotman et al. 1998; Mesulam 1999). Alterations in the composition of the extracellular environment in the AD brain, e.g. accumulation of $A\beta$ fibrils, may stimulate cellular responses consistent with misregulation of FA signaling. Thus, under pathological conditions, increased neuronal plasticity may result in loss of synaptic contacts and neuronal dysfunction.

14.   Summary

In this chapter, we present a set of mechanisms by which $A\beta$ can compromise neuronal function in the absence of degeneration. $A\beta$ has an ability to impair NMDA and BDNF/TrkB signal transduction, two mechanisms critical for neuronal communication and brain plasticity such as learning and memory. Chronic exposure to $A\beta$ also can cause the development of neuronal dystrophy. The mechanism involves aberrant activation of focal adhesion signals.

$A\beta_{1-42}$ impairs BDNF-TrkB transcriptional activity and neuron survival probably by a mechanism involving regulation of the TrkB docking proteins. The regulation of docking proteins has also been shown to participate in insulin resistance and may induce a similar state of neurotrophin resistance in brain. $A\beta$ can also impair NMDA transcriptional activity of genes such as CREB. The mechanism appears to involve $A\beta$ activated internalization of NMDA receptor mediated via α-7 nicotinic receptors.

$A\beta$ can also cause neuronal dystrophy and the loss of synaptic connectivity via focal adhesion signaling pathways. Alterations in the composition of the extracellular environment in the AD brain, e.g. accumulation of $A\beta$ fibrils, may stimulate cellular responses consistent with misregulation of FA signaling. Thus $A\beta$ accesses normal signaling pathways to cause aberrant signaling which can compromise neuronal function dysfunction in the absence of cell death.

Taken together, these normal signaling pathways – compromised NMDA, TrkB and focal adhesion signaling – may contribute to early functional losses prior to the manifestation of classical AD pathology. In addition, these same $A\beta$-mediated mechanisms may also exacerbate brain dysfunction by similar processes once pathology develops. In a sense, $A\beta$ literally high jacks key plasticity mechanisms to impair neuronal function.

References

Abel T, Kandel E (1998) Positive and negative regulatory mechanisms that mediate long-term memory storage. Brain Res Brain Res Rev 26(2-3):360–378.

Akiyama H, Kawamata T, et al. (1991) Immunohistochemical localization of vitronectin, its receptor and beta-3 integrin in Alzheimer brain tissue. J Neuroimmunol 32(1):19–28.

Angers-Loustau A, Cote JF, et al. (1999) Protein tyrosine phosphatase-PEST regulates focal adhesion disassembly, migration, and cytokinesis in fibroblasts. J Cell Biol 144(5):1019–1031.

Aplin AE, Howe A, et al. (1998) Signal transduction and signal modulation by cell adhesion receptors: the role of integrins, cadherins, immunoglobulin-cell adhesion molecules, and selectins. Pharmacol Rev 50(2):197–263.

Balazs RB, Cotman RCW (2005). Excitatory Amino Acid Transmission Book. Oxford Publishers Chapter 11:1–3.

Benzing WC, Mufson EJ, et al. (1993) Alzheimer's disease-like dystrophic neurites characteristically associated with senile plaques are not found within other neurodegenerative diseases unless amyloid beta-protein deposition is present. Brain Res 606(1):10–18.

Berg MM, Krafft GA, et al. (1997) Rapid impact of beta-amyloid on paxillin in a neural cell line. J Neurosci Res 50(6):979–989.

Bourtchuladze R, Frenguelli B, et al. (1994) Deficient long-term memory in mice with a targeted mutation of the cAMP-responsive element-binding protein. Cell 79(1):59–68.

Bramham CR, Messaoudi E (2005) BDNF function in adult synaptic plasticity: The synaptic consolidation hypothesis. Prog Neurobiol 76(2):99–125.

Busciglio J, Lorenzo A, et al. (1992) Methodological variables in the assessment of beta amyloid neurotoxicity. Neurobiol Aging 13(5):609–612.

Causing CG, Gloster A, et al. (1997) Synaptic innervation density is regulated by neuron-derived BDNF. Neuron 18(2):257–267.

Cotman C, Pike C (1994) Beta-amyloid and its contributions to nerodegeneration in Alzheimer' disease. Alzheimer's Disease 17:305–315.

Cotman CW, Anderson AJ (1995) A potential role for apoptosis in neurodegeneration and Alzheimer's disease. Mol Neurobiol 10(1):19–45.

Cotman CW, Hailer NP, et al. (1998) Cell adhesion molecules in neural plasticity and pathology: similar mechanisms, distinct organizations? Prog Neurobiol 55(6):659–669.

Eikelenboom P, Zhan SS, et al. (1994) Cellular and substrate adhesion molecules (integrins) and their ligands in cerebral amyloid plaques in Alzheimer's disease. Virchows Arch 424(4):421–427.

Frohman EM, Frohman TC, et al. (1991) Expression of intercellular adhesion molecule 1 (ICAM-1) in Alzheimer's disease. J Neurol Sci 106(1):105–111.

Geddes JW, Anderson KJ, et al. (1986) Senile plaques as aberrant sprout-stimulating structures. Exp Neurol 94(3):767–776.

Goodwin JL, Kehrli ME, Jr, et al. (1997) Integrin Mac-1 and beta-amyloid in microglial release of nitric oxide. Brain Res 768(1–2):279–286.

Grace EA, Busciglio J (2003) Aberrant activation of focal adhesion proteins mediates fibrillar amyloid beta-induced neuronal dystrophy. J Neurosci 23(2):493–502.

Grace EA, Rabiner CA, et al. (2002) Characterization of neuronal dystrophy induced by fibrillar amyloid beta: implications for Alzheimer's disease. Neuroscience 114(1):265–273.

Heredia L, Lin R, et al. (2004) Deposition of amyloid fibrils promotes cell-surface accumulation of amyloid beta precursor protein. Neurobiol Dis 16(3):617–629.

Hsia AY, Masliah E, et al. (1999) Plaque-independent disruption of neural circuits in Alzheimer's disease mouse models. Proc Natl Acad Sci USA 96(6):3228–3233.

Hu Q, Cool BH, et al. (2002) A candidate molecular mechanism for the association of an intronic polymorphism of FE65 with resistance to very late onset dementia of the Alzheimer type. Hum Mol Genet 11(4):465–475.

Hu Q, Kukull WA, et al. (1998) The human FE65 gene: genomic structure and an intronic biallelic polymorphism associated with sporadic dementia of the Alzheimer type. Hum Genet 103(3):295–303.

Huang EJ, Reichardt LF (2001) Neurotrophins: roles in neuronal development and function. Annu Rev Neurosci 24:677–736.

Kayed R, Head E, et al. (2003) Common structure of soluble amyloid oligomers implies common mechanism of pathogenesis. Science 300(5618):486–489.

Korte M, Carroll P, et al. (1995) Hippocampal long-term potentiation is impaired in mice lacking brain- derived neurotrophic factor. Proc Natl Acad Sci USA 92(19):8856–8860.

Kowalska MA, Badellino K (1994) beta-Amyloid protein induces platelet aggregation and supports platelet adhesion. Biochem Biophys Res Commun 205(3):1829–1835.

Lacor PN, Buniel MC, et al. (2004) Synaptic targeting by Alzheimer's-related amyloid beta oligomers. J Neurosci 24(45):10191–10200.

Lambert JC, Mann D, et al. (2000) A FE65 polymorphism associated with risk of developing sporadic late-onset alzheimer's disease but not with Abeta loading in brains. Neurosci Lett 293(1):29–32.

Lambert MP, Barlow AK, et al. (1998) Diffusible, nonfibrillar ligands derived from Abeta1-42 are potent central nervous system neurotoxins. Proc Natl Acad Sci USA 95(11):6448–6453.

Lorenzo A, Yuan M, et al. (2000) Amyloid beta interacts with the amyloid precursor protein: a potential toxic mechanism in Alzheimer's disease. Nat Neurosci 3(5):460–464.

Lue LF, Kuo YM, et al. (1999) Soluble amyloid beta peptide concentration as a predictor of synaptic change in Alzheimer's disease. Am J Pathol 155(3):853–862.

Martinez A, Alcantara S, et al. (1998) TrkB and TrkC signaling are required for maturation and synaptogenesis of hippocampal connections. J Neurosci 18(18):7336–7350.

Masliah E, Mallory M, et al. (1991) Patterns of aberrant sprouting in Alzheimer's disease. Neuron 6(5):729–739.

Mattson MP, Chan SL (2003) Neuronal and glial calcium signaling in Alzheimer's disease. Cell Calcium 34(4–5):385–397.

McKee AC, Kosik KS, et al. (1991) Neuritic pathology and dementia in Alzheimer's disease. Ann Neurol 30(2):156–165.

Mesulam MM (1999) Neuroplasticity failure in Alzheimer's disease: bridging the gap between plaques and tangles. Neuron 24(3):521–529.

Minichiello L, Korte M, et al. (1999) Essential role for TrkB receptors in hippocampus-mediated learning. Neuron 24(2):401–414.

Morgan D, Diamond DM, et al. (2000) A beta peptide vaccination prevents memory loss in an animal model of Alzheimer's disease. Nature 408(6815):982–985.

Morris JC, Storandt M, et al. (1996) Cerebral amyloid deposition and diffuse plaques in nor-mal aging: Evidence for presymptomatic and very mild Alzheimer's disease. Neurology 46(3):707–719.

Mueller SC, Kelly T, et al. (1989) Dynamic cytoskeleton-integrin associations induced by cell binding to immobilized fibronectin. J Cell Biol 109(6 Pt 2):3455–3464.

Naslund J, Haroutunian V, et al. (2000) Correlation between elevated levels of amyloid beta-peptide in the brain and cognitive decline [see comments]. Jama 283(12):1571–1577.

Oda T, Wals P, et al. (1995) Clusterin (apoJ) alters the aggregation of amyloid beta-peptide (A beta 1-42) and forms slowly sedimenting A beta complexes that cause oxidative stress. Exp Neurol 136(1):22–31.

Patterson SL, Abel T, et al. (1996) Recombinant BDNF rescues deficits in basal synaptic transmission and hippocampal LTP in BDNF knockout mice. Neuron 16(6):1137–1145.

Pike CJ, Cummings BJ, et al. (1992) beta-Amyloid induces neuritic dystrophy in vitro: similarities with Alzheimer pathology. Neuroreport 3(9):769–772.

Sabo S, Lambert MP, et al. (1995) Interaction of beta-amyloid peptides with integrins in a human nerve cell line. Neurosci Lett 184(1):25–28.

Sabo SL, Ikin AF, et al. (2001) The Alzheimer amyloid precursor protein (APP) and FE65, an APP-binding protein, regulate cell movement. J Cell Biol 153(7):1403–1414.

Salgia R, Li JL, et al. (1995) Molecular cloning of human paxillin, a focal adhesion protein phosphorylated by P210BCR/ABL. J Biol Chem 270(10):5039–5047.

Schaller MD, Otey CA, et al. (1995) Focal adhesion kinase and paxillin bind to peptides mimicking beta integrin cytoplasmic domains. J Cell Biol 130(5):1181–1187.

Schratt GM, Nigh EA, et al. (2004) BDNF regulates the translation of a select group of mRNAs by a mammalian target of rapamycin-phosphatidylinositol 3-kinase-dependent pathway during neuronal development. J Neurosci 24(33):9366–9377.

Selkoe DJ (1991) The molecular pathology of Alzheimer's disease. Neuron 6(4):487–498.

Small DH (1998) The role of the amyloid protein precursor (APP) in Alzheimer's disease: does the normal function of APP explain the topography of neurodegeneration? Neurochem Res 23(5):795–806.

Spires TL, Meyer-Luehmann M, et al. (2005) Dendritic spine abnormalities in amyloid precursor protein transgenic mice demonstrated by gene transfer and intravital multiphoton microscopy. J Neurosci 25(31):7278–7287.

Steward O, Schuman EM (2001) Protein synthesis at synaptic sites on dendrites. Annu Rev Neurosci 24:299–325.

Storey E, Beyreuther K, et al. (1996) Alzheimer's disease amyloid precursor protein on the surface of cortical neurons in primary culture co-localizes with adhesion patch components. Brain Res 735(2):217–231.

Tong L, Thornton PL, et al. (2001) Beta -amyloid-(1-42) impairs activity-dependent cAMP-response element-binding protein signaling in neurons at concentrations in which cell survival Is not compromised. J Biol Chem 276(20):17301–17306.

Tsai J, Grutzendler J, et al. (2004) Fibrillar amyloid deposition leads to local synaptic abnormalities and breakage of neuronal branches. Nat Neurosci 7(11):1181–1183.

Tully T (1997) Regulation of gene expression and its role in long-term memory and synaptic plasticity. Proc Natl Acad Sci USA 94(9):4239–4241.

Turner C E (2000) Paxillin interactions. J Cell Sci 113 Pt 23:4139–4140.

Williamson R, Scales T, et al. (2002) Rapid tyrosine phosphorylation of neuronal proteins including tau and focal adhesion kinase in response to amyloid-beta peptide exposure: involvement of Src family protein kinases. J Neurosci 22(1):10–20.

Yamazaki T, Koo EH, et al. (1997) Cell surface amyloid beta-protein precursor colocalizes with beta 1 integrins at substrate contact sites in neural cells. J Neurosci 17(3):1004–1010.

Yankner, B. A. (1996) Mechanisms of neuronal degeneration in Alzheimer's disease. Neuron 16(5):921–932.

Yin JC, Wallach JS, et al. (1994) Induction of a dominant negative CREB transgene specifically blocks long-term memory in Drosophila. Cell 79(1):49–58.

Zhang C, Lambert MP, et al. (1994) Focal adhesion kinase expressed by nerve cell lines shows increased tyrosine phosphorylation in response to Alzheimer's A beta peptide. J Biol Chem 269(41):25247–25250.

Zhang C, Qiu HE, et al. (1996) A beta peptide enhances focal adhesion kinase/Fyn association in a rat CNS nerve cell line. Neurosci Lett 211(3):187–190.

Zick Y (2003) Role of Ser/Thr kinases in the uncoupling of insulin signaling. Int J Obes Relat Metab Disord 27 Suppl 3:S56–60.

Chapter 9

Beta Amyloid and Excitatory Synapses

Roberto Malinow[1,2,3], Helen Hsieh[3] and Wei Wei[2]

[1] Cold Spring Harbor Laboratory
1 Bungtown Road
Cold Spring Harbor, NY 11724
516 367 8416
email: malinow@cshl.org

[2] State University of New York and Stonybrook
Dept. Neurobiology

[3] Watson School of Biological Sciences
Cold Spring Harbor Laboratory

1.   Introduction

In this short review we will focus on the effect of APP and $A\beta$ on excitatory synaptic function, as there is considerable evidence that synapses are a key target site of Alzheimer's Disease (Terry, Masliah et al. 1991). Two aspects of synaptic function affected by $A\beta$ can be considered: 'basal' synaptic transmission and synaptic plasticity. Of course, if synaptic plasticity is altered, and if one believes that the plasticity examined in experimental conditions occurs normally in the animal, one would expect 'basal' transmission also to be altered. It is surprising how little this concept is considered in studies that examine the effect of any genetic perturbation on synaptic plasticity.

The effect of $A\beta$ on synaptic transmission and plasticity can be examined using several experimental protocols, and it is not clear if results from different protocols should be directly compared. Nevertheless, here we will consider results from such different protocols. Essentially there are three methods one can use to determine the effect of $A\beta$ on synaptic function: 1) direct application of synthesized or purified $A\beta$ on neural tissue; 2) examination of transgenic animals which over-express mutant forms of APP that generate increased amounts of Aβ; or 3) transient over-expression of APP in neurons which drives increased amounts of Aβ. Each method has advantages and disadvantages. With the first method, the advantage is that the identity (namely Aβ) and concentration of active compound is known. However, the oligemerizaton state of Aβ must be carefully controlled. Furthermore, with injection of Aβ into the brain it is difficult to control the concentration of Aβ relative to the synapses being examined. With the second method, using transgenic mice, concentrations of Aβ produced can be fairly well determined and of course the effects on behavior can be assessed. However, APP is chronically over-expressed and thus there may be compensatory mechanisms that obscure determining the direct effects of Aβ on synaptic function. Furthermore, different genetic strains may produce different amounts of Aβ, and in different genetic backgrounds, which may complicate interpretations of results. With the third method, transient overexpression, the effects of various mutant forms of APP or Aβ can be compared, and the effects are relatively acute. However, levels of Aβ reaching synapses may not be easy to assess.

2.   Effect of Aβ on basal excitatory transmission

Early studies on the anatomy and immunohistochemistry of brains from AD patients indicated a dramatic loss of synapses and synaptic markers that correlated well with disease severity (Terry, Masliah et al. 1991). Such markers were better correlated than brain plaque content to disease severity. The effect of Aβ on synaptic function was more carefully examined in the PDAPP(717V$\rightarrow$F) transgenic mouse model (Hsia et al., 1999). These animals, showed considerable decrease in basal levels of excitatory synaptic function before the appearance of plaques supporting the view that Aβ has effects on synaptic transmission independent of plaque formation. Similar findings using these mice were reported by Larson and colleagues (Larson et al. 1999). However, a decrease in basal transmission has not been observed in all transgenic mouse strains. Studies employing mice derived from the Tg2576 APP(Sw) line have reported normal basal transmission in a slice preparation (Chapman et al. 1999) as well as in an *in vivo* preparation (Stern et al. 2004). It is not clear if the differences in these results are due to different levels or types of Aβ (e.g. 1-40 vs 1-42) produced by these transgenic animals.

The effect on synapses following direct application of Aβ on neural tissue has been mixed. In some studies enhancement of NMDA component of transmission was found (Wu et al. 1995), other studies showed no effect on basal transmission (Wang et al. 2004), while more recent results have measured a decrease in surface NMDA receptors (Snyder et al. 2005). Part of this complexity may arise from the fact that increased activation of synaptic

NMDA-Rs can lead to their removal (Morishita et al. 2005). It is not clear if application in slice preparations has the same effect as on dissociated cultured neurons. It is also not clear if the levels applied to tissue (even if they correspond to levels measured in extracellular space of AD patients) are those sensed by synapses during the disease process. It is possible that synaptic regions close to focal sources of Aβ will be exposed to much higher levels of Aβ than those levels measured in the extracellular space. Why might there be locally high sources of Aβ during the disease process? Senile plaques, which contain large amounts of Aβ, may release their contents to the nearby extracellular space. Thus, sites close to plaques may be exposed to high levels of Aβ.

An alternative means to determine the effect of APP and its proteolytic products on synapses is to express transiently APP in neurons by acute gene transfer. This can be achieved by viral or biolistic delivery techniques. One advantage of this technique is that different APP-related constructs can be delivered and compared. Furthermore, other constructs can be co-expressed along with APP, allowing one to dissect mechanisms by which APP and its proteolytic products act. We have used acute gene delivery techniques to express APP and related constructs in individual neurons of organotypic hippocampal slices (Kamenetz et al. 2003). Transfected neurons can be identified by the co-expressed GFP. Simultaneous whole-cell patch clamp recordings are obtained form transfected and nearby non-transfected neurons. By evoking synaptic transmission with a stimulus electrode, transmission onto these two neurons can be compared, and thus the effect of expression of a given construct on synaptic transmission can be determined.

We found that expression of APP for 1 day produced a depression of excitatory synaptic transmission with no effect on inhibitory transmission, neuronal input resistance or resting potential. This indicates that the effect on excitatory transmission was not due to a general toxic effect on the neuron. The effect could be narrowed down to formation of Aβ since a) expression of APP with a point mutation that prevents cleavage by beta secretase (APP$_{MV}$) does not produce synaptic depression, b) expression of the beta secretase product of APP (C99) is sufficient to produce synaptic depression and c) expression of C99 in the presence of γ-secretase inhibitors produces no synaptic depression. More recent results indicate that the synaptic depression produced by Aβ is due to synaptic removal of AMPA-Rs (Hsieh & Malinow, unpublished results). For instance, coexpression of an AMPA-R subunit with a mutation that reduces its endocytosis blocks the synaptic depressing effect of APP. Interestingly, expression of this AMPA-R subunit also prevents the depression of NMDA-R mediated transmission (Hsieh and Malinow, unpublished results), suggesting that Aβ-driven endocytosis of NMDA-R (Snyder et al. 2005) requires endocytosis of AMPA-Rs.

The synaptic depressing effects of $A\beta$ are interesting when coupled to findings indicating that increased neural activity can drive the processing of APP to $A\beta$ (Kamenetz et al. 2003). These two findings suggest a negative feedback system: high levels of neural activity drive formation of $A\beta$; formation of $A\beta$ depresses synaptic transmission and thereby reduces high levels of neural activity. We have hypothesized that the $A\beta$-mediated negative feedback system is a normal process which may become dysregulated (e.g. by producing unregulated amounts of $A\beta$) and thereby contribute to AD.

3.   Effect of $A\beta$ on LTP

Long-term potentiation (LTP) is a persistent enhancement in synaptic transmission that follows a brief period of repeated pre- and postsynaptic activity (Bliss and Collingridge 1993). LTP is observed in many CNS synapses, is long-lasting, demonstrates associative properties and thus has served as a powerful model of synaptic plasticity that may underlie associative forms of learning and memory. It is therefore clear why many groups have tested the effects of AD-related molecules and model systems on LTP.

Most studies testing the effects of $A\beta$ on LTP have found that this form of plasticity is reduced (Cullen et al. 1997; Lambert et al. 1998; Chapman et al. 1999; Moechars et al. 1999; Chen et al. 2002; Walsh et al. 2002; Kamenetz et al. 2003). It is notable, however, that a group with much expertise in LTP studies failed to see a deficit of LTP in a transgenic mouse overexpressing APP, despite clear reduction in synaptic transmission (Hsia et al. 1999).

It cannot be over-emphasized that testing the effects of $A\beta$ on LTP is not as trivial as it may seem. When synthetic $A\beta$ is put in solution it can aggregate which will generate different products with potentially different and variable effects. As a small hydrophobic peptide, it can adhere to perfusion tubing which will affect the concentration in solution. When $A\beta$ is injected *in vivo*, it is very difficult to control its concentration, as sites close to the injection will have high concentration, while more distant sites will have lower concentration; the volume over which the injected solution spreads is virtually impossible to measure. In one remarkable study (Walsh et al. 2002), injection of a few picograms of $A\beta$ oligomers into brain ventricles was sufficient to block LTP. Rough calculations suggest that a few thousand molecules per neuron may be sufficient to block LTP.

Transgenic models have provided insight into the effects of $A\beta$ on plasticity but these studies also have limitations. The levels of expression of APP and proteolytic products differ in different strains. Furthermore, genetic backgrounds may modify effects of $A\beta$. It is also important to note that effects of APP or proteolytic products can have effects on transmission that will manifest themselves as effects on LTP. For instance, if increased expression of APP produces a decrease in excitatory transmission, then one may see decreased LTP merely because there is not sufficient depolarization during a tetanic stimulus. In such a case the effect is not directly on the biochemical pathways underlying LTP.

Lastly, it is important to emphasize that physiology experiments should be done, as much as possible, in a blind manner. That is, the experimenter should be blind as to whether a test or control compound (genotype) is being tested. This is because there are many hidden biases that can creep into physiology experiments. For instance, it is not uncommon to remove from analysis a particular experiment because the recording was not stable. The 'stability of a recording' however, is a somewhat subjective judgment. Thus, if one is blind during the experiment and during the analysis, these sorts of hidden biases can be avoided.

4.   Conclusion

In summary, there is ample evidence that APP and its proteolytic products have an impact on synaptic transmission. However, there are still many issues that have not been fully clarified:

1) Where is APP localized in brain neurons (pre-synaptic or postsynaptic or both)?
2) Where is APP processed? Which organelles and in what part of the neuron? Where are the processing enzymes?
3) What neural signals control the processing of APP?
4) Is there a function for APP or proteolytic products in synaptic transmission?
5) Are there toxic effects of acute or chronic presence of APP or proteolytic products on synaptic transmission?
6) What are the downstream signals at synapses produced by APP or proteolytic products?
7) What is the effect of APP or proteolytic products on LTP? Why are there variable results?
8) Which experimental systems that permit analysis of synapses are most informative with respect to the human condition?

These are but just a few of the questions that would help in understanding the role of APP and proteolytic products in synaptic transmission and plasticity.

References

Bliss TV, Collingridge GL (1993) A synaptic model of memory: long-term potentiation in the hippocampus. Nature 361(6407):31–39.

Chapman PF, White GL, et al. (1999) Impaired synaptic plasticity and learning in aged amyloid precursor protein transgenic mice. Nat Neurosci 2(3):271–276.

Chen QS, Wei WZ, et al. (2002) Alzheimer amyloid beta-peptide inhibits the late phase of long-term potentiation through calcineurin-dependent mechanisms in the hippocampal dentate gyrus. Neurobiol Learn Mem 77(3):354–371.

Cullen WK, Suh YH, et al. (1997) Block of LTP in rat hippocampus in vivo by beta-amyloid precursor protein fragments. Neuroreport 8(15):3213–3217.

Hsia AY, Masliah E, et al. (1999) Plaque-independent disruption of neural circuits in Alzheimer's disease mouse models. Proc Natl Acad Sci USA 96(6):3228–3233.

Kamenetz F, Tomita T, et al. (2003) APP processing and synaptic function. Neuron 37(6):925–937.

Lambert MP, Barlow AK, et al. (1998) Diffusible, nonfibrillar ligands derived from Abeta1-42 are potent central nervous system neurotoxins. Proc Natl Acad Sci USA 95(11):6448–6453.

Larson J, Lynch G, et al. (1999) Alterations in synaptic transmission and long-term potentiation in hippocampal slices from young and aged PDAPP mice. Brain Res 840(1–2):23–35.

Moechars D, Dewachter I, et al. (1999) Early phenotypic changes in transgenic mice that over-express different mutants of amyloid precursor protein in brain. J Biol Chem 274(10):6483–6492.

Morishita W, Marie H, et al. (2005) Distinct triggering and expression mechanisms underlie LTD of AMPA and NMDA synaptic responses. Nat Neurosci 8(8):1043–1050.

Selkoe DJ (2001) Alzheimer's disease:genes, proteins, and therapy. Physiol Rev 81(2):741–766.

Snyder EM, Nong Y, et al. (2005) Regulation of NMDA receptor trafficking by amyloid-beta. Nat Neurosci 8(8):1051–1058.

Stern EA, Bacskai BJ, et al. (2004) Cortical synaptic integration in vivo is disrupted by amyloid-beta plaques. J Neurosci 24(19):4535–4540.

Terry RD, Masliah E, et al. (1991) Physical basis of cognitive alterations in Alzheimer's disease:synapse loss is the major correlate of cognitive impairment. Ann Neurol 30(4):572–580.

Walsh DM, Klyubin I, et al. (2002) Naturally secreted oligomers of amyloid beta protein potently inhibit hippocampal long-term potentiation in vivo. Nature 416(6880):535–539.

Wang Q, Rowan MJ, et al. (2004) Beta-amyloid-mediated inhibition of NMDA receptor-dependent long-term potentiation induction involves activation of microglia and stimulation of inducible nitric oxide synthase and superoxide. J Neurosci 24(27):6049–6056.

Wu J, Anwyl R, et al. (1995) beta-Amyloid selectively augments NMDA receptor-mediated synaptic transmission in rat hippocampus. Neuroreport 6(17):2409–2413.

Chapter 10

Aβ Degradation

Malcolm Leissring, Ph.D.[1] and Takaomi C. Saido, Ph.D.[2]

[1]*Brigham & Women's Hospital*
Ctr Neurolog Diseases
77 Ave Louis Pasteur
Harvard Inst of Med, Rm. 740
Boston MA 02115
Email: mleissring@rics.bwh.harvard.edu

[2] *RIKEN Brain Sci Inst*
Lab Proteolytic Neurosci
2-1 Hirosawa Wako-shi
Saitama 351-0198 Japan
Email: saido@brain.riken.go.jp

1.   Introduction

Alzheimer's disease (AD) is characterized by abnormal accumulation of the amyloid β-protein (Aβ) in brain regions important for memory and cognition. Aβ is a normal product of cellular metabolism (Haass et al. 1992) derived from the amyloid precursor protein (APP) by the successive action of the β- and γ-secretases (see Chapters 4 and 3, respectively). The production of Aβ is normally counterbalanced by its elimination *via* any of several processes, including (i) proteolytic degradation, (ii) cell-mediated clearance (which may itself involve proteolytic degradation), (iii) active transport out of the brain (Chapter 10), and (iv) deposition into insoluble aggregates (Chapter 6). While the relative importance of these different pathways remains to be established, a growing body of evidence suggests that proteolytic degradation is a particularly important determinant of cerebral Aβ levels and, by extension, of Aβ-associated pathology.

Characterization of the molecular and pathological phenotypes of familial AD-causing gene mutations identified largely in 1990's played a central role in establishing that Aβ is indeed most like to play a central role in AD pathogenesis and that the increased generation of Aβ, particularly Aβ$_{42}$, is the cause of Aβ accumulation in the brains of familial AD patients harboring mutations in APP or presenilin-1 or -2 (Chapter 1). Nonetheless, there is no clear evidence that increased Aβ production precedes Aβ deposition in most cases of sporadic AD, which account for more than 90% of all AD cases. Moreover, pathological and biochemical analysis of sporadic AD indicates that a selective increase in Aβ$_{42}$ production, as observed in familial AD, is not likely to be a cause of most sporadic AD cases (Lemere et al. 1996; Scheuner et al. 1996). These facts led to the proposal that decreased catabolism rather than increased anabolism of Aβ may be a primary cause for sporadic AD (Saido 1998), for which aging is by far the strongest risk factor, as down-regulation rather than up-regulation of metabolism is a general feature of aging (Lu et al. 2004).

Nonetheless, widespread interest in the proteolytic degradation of Aβ did not take hold until the turn of the 21st Century. This contrasts starkly with studies of Aβ production, which enjoyed intense scrutiny over this same period. While several factors may have contributed to this state of affairs, it can be argued that Aβ degradation is a more challenging topic to address experimentally because it relies upon studies in the living animal to an extent that studies on Aβ production do not. Notably, whereas all of the components involved in Aβ production (APP, β- and γ-secretases) can be recapitulated faithfully and economically within cell culture, in contrast cell culture experiments are of limited value for investigating Aβ degradation, particularly in determining the relative importance of different candidate proteases in regulating the overall economy of brain Aβ. As described in greater detail below, a key turning point in the field came with the first study that was explicitly designed to examine Aβ degradation in the living animal (Iwata et al. 2000). In addition to identifying neprilysin (NEP) as one of the principal Aβ-degrading proteases, this study also served to highlight the significance of Aβ degradation to AD pathogenesis generally, thereby igniting interest in a previously underappreciated aspect of Aβ metabolism. It now appears quite likely that the aging-dependent reduction of Aβ degradation may be a primary cause of Aβ accumulation in aged brains and thus of sporadic AD development.

The rise in interest in Aβ degradation spawned by this and other discoveries is impressive: whereas only a few papers addressing this topic were published in the entire 20th Century, subsequent growth in this field in just a few years has been so great that it is now impossible to cite all the papers in this field in a chapter of this length. Accordingly, this chapter will focus principally on biomedical evidence pertaining to the *in vivo* relevance of Aβ-degrading proteases and their therapeutic potential, and will rely on Chapter 1 to summarize the large body of evidence examining possible genetic linkage between Aβ-degrading proteases and AD. Individual Aβ-degrading proteases will be considered in turn, followed by a general discussion of future research directions.

2.   Neprilysin

NEP (a.k.a. neutral endopeptidase; enkephalinase; CALLA; EC 3.4.24.11) is a ~93-kDa zinc-metalloendopeptidase implicated in the degradation of a wide array of bioactive peptides (Turner and Tanzawa 1997). Several features of NEP appear to make it ideally suited to the degradation of secreted peptides such as Aβ. First, NEP is a type-2, membrane-bound glycoprotein with its active site oriented located in the intralumenal/extracellular space (Turner et al. 2001) where Aβ peptides are normally secreted. Second, NEP is localized to subcellular compartments that overlap with well-established sites of Aβ production (Takahashi et al. 2002), being present for example at the cell surface, and in the early Golgi and endoplasmic reticulum (Turner et al. 2001). Third, within neurons – the principal source of Aβ production – NEP overlaps with known regions of Aβ secretion, particularly presynaptic terminals (Fukami et al. 2002; Iwata et al. 2002; Lazarov et al. 2002; Iwata et al. 2004), where it appears to be remarkably well positioned to play a role in the catabolism of endogenously produced Aβ. Indeed, this idea was tested directly by expressing chimeric variants of NEP engineered to be retained selectively in different subcellular compartments (Hama et al. 2004); it was concluded that the normal localization of NEP, with its localization in multiple compartments, was the most efficient at degrading Aβ.

Aβ was first demonstrated to be proteolyzed by NEP using synthetic Aβ and purified NEP by Howell and colleagues (1995). Despite this early report, the pathophysiological relevance of NEP to the regulation of Aβ was not widely appreciated until the first study showing a role for NEP in the living animal (Iwata et al. 2000). In this study, radiolabeled $Aβ_{1-42}$ was superfused into rat hippocampus in the presence or absence of different protease inhibitors, and cleavage of the labeled Aβ was monitored by high-performance liquid chromatography. Among the different protease inhibitors studied, Aβ degradation was most strongly inhibited by thiorphan, a potent and selective inhibitor of NEP (Iwata et al. 2000). Subsequent work by this same team showed that endogenous murine Aβ levels were significantly elevated (~100%) in mice with targeted disruption of the *NEP* gene (Iwata et al. 2001) (see Table 1). Importantly, heterozygous knockout mice, in which brain neprilysin activity is half that of wild-type mice, show a ~50% increase in brain Aβ levels, indicating that even partial reduction of neprilysin expression/activity, which seems to be a natural process upon aging as described below, will result in elevation of steady-state Aβ levels. It is of high interest that an unbiased genetic screen in *D. melanogaster* independently identified NEP as a key regulator of Aβ-mediated toxicity (Finelli et al. 2004).

The identification of NEP as a physiologically relevant Aβ-degrading protease triggered a large number of studies that support a tight link between NEP expression and the modulation of AD pathology. Reduced levels of NEP mRNA and protein have been observed in brain regions that are susceptible to Aβ deposition (Carpentier et al. 2002; Iwata et al. 2002; Iwata et al. 2004; Mohajeri et al. 2004; Caccamo et al. 2005), as well as in AD patients relative to controls (Yasojima et al. 2001; Wang et al. 2005). NEP levels have also consistently been shown to decrease with aging (Yasojima et al. 2001; Carpentier et al. 2002; Iwata et al. 2002; Caccamo et al. 2005), an effect that has recently been attributed, at least in part, to age-dependent decreases in somatostatin expression (Hama and Saido 2005; Saito et al. 2005). It is notable that the aging-associated reduction of neprilysin expression takes place particularly in the molecular layers of the dentate gyrus, into which neurons of the entorhinal cortex project (Iwata et al. 2002), a finding that may serve to explain why these neurons are particularly vulnerable to amyloid deposition in AD. In addition to changes in overall expression levels, the NEP enzyme may itself be vulnerable to oxidative damage, with a few studies reporting increased levels of NEP (and/or IDE) containing the oxidative adduct hydroxynonenol both in AD brain and in normal aging (Wang et al. 2003; Caccamo et al. 2005), thereby providing a compelling mechanistic link between oxidative damage and AD pathogenesis. More recently, Maruyama and colleagues (Maruyama et al. 2005) demonstrated that CSF neprilysin activities, derived from neurons, are significantly reduced in prodromal AD patients. Taken together, these experimental and clinical observations support the notion that an aging-induced reduction of neprilysin activity may play a causal role in the development of idiopathic AD.

Consideration of the role of Aβ degradation may also force a re-examination of the pathogenic mechanisms underlying certain familial forms of AD. In this regard, it has been reported that synthetic Aβ peptides harboring any of several intra-Aβ mutations are degraded more slowly by NEP (Tsubuki et al. 2003). Thus, whereas most of these mutations had previously been associated with an increased tendency to aggregate (Murakami et al. 2002), decreased clearance via proteolytic degradation may also play a role in the pathogenesis of certain familial forms of AD.

On the therapeutic side, several studies have demonstrated that NEP over-expression can effect dramatic changes in steady-state Aβ levels and in the development of Aβ plaques and associated pathology. Viral overexpression of NEP has been shown to induce local reductions in steady-state Aβ levels and to retard plaque deposition (Marr et al. 2003; Iwata et al. 2004). In addition, transgenic overexpression of NEP restricted to neurons completely prevented plaque formation and its associated pathology, and reversed the premature lethality in APP transgenic mice (Leissring et al. 2003; see Figure 1). These and several related studies (Lazarov et al. 2002; Dolev and Michaelson 2004) have helped to reveal that Aβ deposition is a more reversible process than had previously been appreciated. Nonetheless, earlier interventions have generally proved more effective than later ones in attenuating Aβ plaque formation, suggesting that amyloid deposits ultimately enter into a protease-resistant phase (Marr et al. 2003; Iwata et al. 2004). Although mature plaques may be impossible to remove by proteolysis alone, it is notable that neprilysin can degrade some early aggregates of Aβ, including naturally secreted synapto-toxic Aβ oligomers both *in vitro* and *in vivo* (Walsh et al. 2002; Kanemitsu et al. 2003; Iwata and Saido, unpublished data). This contrasts with insulin-degrading enzyme, which when overexpressed in APP transfected cells, can reduce the production of Aβ oligomers, but cannot degrade them after they are formed (Walsh et al. 2002; Leissring, Walsh and Selkoe, unpublished observations). The latter observations suggest that the pathway leading from Aβ production to the formation of cytotoxic aggregates includes a protease-sensitive, presumably monomeric phase that might be targeted therapeutically.

With respect to the utility of NEP as a therapeutic target in AD, it is notable that NEP activity seems to be unexpectedly selective to Aβ in the brain, because neprilysin deficiency does not seem to alter the levels of putative "neprilysin substrate" neuropeptides such as enkephalin, cholecystokinin, neuropeptide Y, substance P, and somatostatin in the brain (Iwata & Saido, unpublished data). Peptidomic analyses of neprilysin knockout brains did not reveal any differences in these substrates relative to wild-type mice (Tsubuki and Saido, unpublished observations). Moreover, the K_m value of neprilysin for Aβ is lower than those for other substrates (Iwata et al. 2005), and Aβ levels in AD brain are about 10,000 times higher than the normal levels (Funato et al. 1998), indicating that neprilysin is likely to selectively act on Aβ in AD brain.

Several distinct NEP transcripts exist that are each expressed in a tissue-specific manner, presumably regulated by the activity of separate promoter elements within the regulatory region of the NEP gene (D'Adamio et al. 1989). Levels of expression of NEP vary considerably both ontogenetically, and also among tissues. For example, neurons express the type-1 transcript whereas kidney expresses the type-2 transcript, the quantity of which is approximately 1000-fold greater as compared to brain (Barnes et al. 1992). These observations indicate that NEP expression levels are highly regulatable, a feature that might be exploited in the development of therapies based on upregulation of Aβ degradation. The range of factors that has been shown to modulate NEP levels is remarkably diverse, and includes somatostatin (Saito et al. 2005), substance P (Bae et al. 2002), Aβ (Mohajeri et al. 2002; Mohajeri et al. 2004), the intracellular domain of APP (Pardossi-Piquard et al. 2005), estrogen (Huang et al. 2004), green tea extract (Melzig and Janka 2003), and even environmental enrichment (Lazarov et al. 2005). While these results are encouraging for their therapeutic implications, in light of the labile nature of NEP expression and its powerful influence on Aβ homeostasis, factors that down-regulate NEP levels should also be explored as potential risk factors for AD. For instance, somatostatin and estrogen are well known for their aging-dependent reduction of expression in humans (Huang et al. 2004; Hama et al. 2005), providing a possible mechanistic explanation for the aging-dependent reduction of neprilysin activity. Among these, somatostatin receptors are most interesting targets; there are five receptor subtypes with distinct distributions (Saito et al., 2003). Low molecular weight agonists specific for receptor subtypes responsible for neprilysin activation would be medication candidates.

Several human genetic studies link AD development to variability in the *NEP* gene, including new single-nucleotide polymorphisms (Shi et al. 2005), which are summarized elsewhere (Iwata et al. 2005; see also Chapter 1). While specific disease-causing mutations analogous to those in APP or the presenilins have not yet been identified for NEP or any other Aβ-degrading protease, it cannot be emphasized too much that, for sporadic AD, aging is a far stronger risk factor than any genetic factor(s) including apolipoprotein E genotype, the only universally confirmed genetic risk determinant for sporadic AD (Chapter 1).

3.   Endothelin-converting enzymes

Endothelin-converting enzymes-1 and -2 (ECE-1, ECE-2; EC 3.4.24.71) are membrane-bound zinc-metalloproteinases belonging to the same family as NEP (M13). While other members of the M13 family have been shown to be capable of degrading Aβ in cell culture or *in vitro* (Takaki et al. 2000; Shirotani et al. 2001), ECEs stand alone as the only members besides NEP demonstrated to play a role *in vivo*. The involvement of ECEs in the degradation of Aβ was discovered by Eckman and colleagues (Eckman et al. 2001), who noted that levels of Aβ secreted into the medium of cultured cells were significantly elevated by addition of phosphoramidon, a known inhibitor of ECEs. Overexpression of ECE in cultured cells stably expressing APP led to a >90% reduction in levels of secreted Aβ, an effect that was reversed by treatment with phosphoramidon (Eckman et al. 2001). Since ECEs are most active under acidic pH conditions and degrade Aβ inside cells (Eckman et al. 2001), ECEs may be involved in Aβ metabolism in intracellular acidic compartments. Aβ levels were found to be substantially elevated in a gene dosage-dependent manner in knockout mice lacking ECE-2 and also in mice lacking one allele of ECE-1 (Eckman et al. 2003; homozygous null ECE-1 mice were non viable; Table 1). Inhibitors of the ECEs are in use clinically, making it critically important to determine the potential relevance of these proteases to AD pathogenesis.

4. Insulin-degrading enzyme

Insulin-degrading enzyme (IDE; EC3.4.24.56; a.k.a. insulysin, insulinase, insulin protease) is a ~110-kDa zinc-metalloendopeptidase that degrades a broad range of substrates, including several involved in glucose homeostasis and/or diabetes pathogenesis – e.g., insulin, glucagon, and amylin – along with a range of other bioactive peptides (Duckworth et al. 1998). IDE also degrades intracellular substrates, including – notably – the intracellular domain of APP (Edbauer et al. 2002). Interestingly, AICD has recently been implicated in transcriptional regulation of APP (von Rotz et al. 2004) and neprilysin (Pardossi-Piquard et al. 2005) expression, raising the possibility that IDE might indirectly affect Aβ levels via its effects on AICD levels; nonetheless, despite large (several-fold) increases in AICD in IDE knockout mice, no alteration of APP or NEP levels was observed in these animals (Farris et al. 2003; Miller et al. 2003), casting some doubt on the *in vivo* relevance of the latter findings or indicating altered metabolism of APP or NEP.

IDE was actually the first protease shown to degrade Aβ *in vitro* within crude brain homogenates (Kurochkin and Goto 1994), and was later identified independently by Selkoe and colleagues as the major Aβ-degrading activity secreted into the medium of a range of cultured cells (Qiu et al. 1997; Qiu et al. 1998). Subsequent work demonstrated that primary neurons cultured from IDE knockout exhibit a >90% reduction in the degradation of extracellular Aβ (Farris et al. 2003). These observations, together with studies linking genetic variations in the regions of the IDE gene to late-onset-AD (see Chapter 1), suggest that IDE is a particularly strong functional and positional candidate for the regulation of cerebral Aβ levels.

IDE belongs to a superfamily of zinc-metalloendopeptidases (M16) characterized by a zinc-binding motif (HxxEH) that is inverted with respect to the motif (HExxH) present NEP, ECEs and most other zinc-metalloendopeptidases (Becker and Roth 1992). Although IDE's crystal structure is not yet available, its active site shares a high degree of sequence homology with related proteases whose structures have been determined, such as yeast mitochondrial processing peptidase (MPP; Taylor et al. 2001). Interestingly, despite inversion of the zinc-binding motif at the primary sequence level, the active site of MPP was shown to be unexpectedly similar 3-dimensionally to that of the canonical zinc-metalloprotease thermolysin (Makarova and Grishin 1999), suggesting that the catalytic mechanisms utilized by members of the two superfamilies may be similar, representing an interesting case of convergent evolution.

Despite the aforementioned similarities, the M16 superfamily exhibits a unique set of enzymological properties that can be attributed to distinguishing structural features. First, the cleavage sites within known substrates of IDE seem to share little primary sequence homology; instead, secondary or tertiary structural features of substrates seem to strongly influence both substrate and cleavage-site specificity (Duckworth et al. 1998). In this regard, it has been noted that many substrates of IDE share the tendency to form fibrils (e.g., Aβ, insulin, amylin), leading to the conjecture that IDE may recognize structural features in substrates that confer amyloidogenicity (Kurochkin 1998). A second distinguishing enzymological feature of IDE and its homologues is their ability to process relatively large peptide substrates such as insulin and insulin-like growth factors, which exist as highly structured, disulfide-bonded molecules (Duckworth et al. 1998). These enzymological peculiarities were explained by the crystal structures of M16 family members (Iwata et al. 1998; Taylor et al. 2001), which revealed the presence of a large internal cavity involved in substrate binding. Thus, the range of substrates that can be hydrolyzed by IDE is limited to those that can bind within this large central cavity, explaining for example IDE's preference for larger peptides (Duckworth et al. 1998). Once inside the cavity, constraints on the positioning of the substrate strongly influence which peptide bonds gain access to the active site, thus accounting for the lack of cleavage site-specificity exhibited by IDE, and the strong influence of secondary and tertiary structure.

IDE also differs from NEP and ECE-1 and -2 in terms of its subcellular localization. The vast majority of IDE exists within the cytosol (Duckworth et al. 1998), a finding that has served to foster doubts about the relevance of IDE to the degradation of extracellular substrates such as Aβ and insulin. However, as shown by numerous independent groups in work spanning several decades, IDE is also localized to several cellular compartments with clear relevance to Aβ degradation, including the cell surface and the extracellular space (e.g., Qiu et al. 1997; Seta and Roth 1997; Qiu et al. 1998; Vekrellis et al. 2000), as well as to several other subcellular compartments, including endosomes and peroxisomes (Duckworth et al. 1998). The peroxisomal localization of IDE is explained by the presence of a type-1 peroxisomal targeting sequence at the extreme C-terminus (Duckworth et al. 1998), but the mechanism of translocation to the extracellular space or endosomes remains elusive. Recently, a mitochondrial isoform of IDE was identified that is generated by alternative translation initiation, which results in the addition of a mitochondrial targeting sequence at the N-terminus (Leissring et al. 2004). While the significance of IDE's localization to these intracellular compartments is uncertain, a growing body of evidence suggests that low levels of Aβ peptides may reach intracellular compartments such as the cytosol and mitochondria, where their accumulation may be disproportionately toxic by comparison to their extracellular counterparts (LaFerla et al. 1995; Zhang et al. 2002). In this regard, a recent study demonstrated that Aβ generated within the endoplasmic reticulum is transported passively into the cytosol, where non-proteasomal degradation is mediated by IDE (Schmitz et al. 2004). While the pathological significance of these observations remains to be clarified, they raise the interesting possibility that different Aβ-degrading proteases may work in concert to confer protection from different pools of Aβ.

Despite its distinctive subcellular localization, mice with targeted deletion of the IDE gene show significant increases in endogenous brain Aβ levels (Farris et al. 2003; Miller et al. 2003) that are comparable in magnitude to those in NEP and ECE-1 and -2 knockout mice (Table 1). Indeed, in the only side-by-side comparison yet reported (Farris et al. 2003), IDE and NEP knockout mice were found to have a virtually identical percent increase in cerebral Aβ levels when Aβ was extracted in parallel and quantified on the same ELISA plate (Table 1). Notably, as is true for NEP and ECE-2 (Iwata et al. 2001; Eckman et al. 2003), Aβ levels vary in a gene dosage-dependent manner (Miller et al. 2003), suggesting that IDE is rate-limiting in the determination of endogenous Aβ levels. Like NEP, IDE protein levels have been shown to decrease with aging in normal mice (Caccamo et al. 2005) and protein and mRNA levels are notably higher in brain regions that do not normally accrue Aβ deposits, such as the cerebellum (Caccamo et al. 2005; Farris et al. 2005). Significant reductions in hippocampal IDE levels have also been reported in AD patients harboring one or more ApoE4 alleles (Cook et al. 2003).

True to its name, IDE is a major insulin protease, with crude liver homogenates from IDE knockout mice showing a $\sim 98\%$ reduction in insulin degradation (Farris et al. 2003). Several lines of evidence suggest that IDE dysfunction may contribute to the development of type-2 diabetes mellitus (DM2). The GK rat, a well-established model of DM2 used for decades in diabetes research, was found to harbor missense mutations in IDE that contribute to the diabetic phenotype (Fakhrai-Rad et al. 2000), and these mutations were shown to result in a partial loss of function in IDE's ability to degrade both insulin and Aβ (Fakhrai-Rad et al. 2000; Farris et al. 2004). IDE knockout mice were also found to exhibit a diabetic phenotype and to have significant elevations in fasting insulin levels (Farris et al. 2003). Emerging evidence from epidemiological studies suggests there is a high degree of co-morbidity between DM2 and AD (Ott et al. 1999), and a growing body of genetic studies has linked variations near the *IDE* gene to both DM2 (e.g., Karamohamed et al. 2003) and late-onset AD (reviewed in Chapter 1), suggesting that IDE hypofunction may represent a common feature in these two seemingly disparate diseases. In this regard, compelling evidence from human studies suggests that Aβ and insulin may compete for limiting quantities of IDE on a rapid time-scale, as administration of insulin to normal human subjects maintained on a euglycemic clamp was found to result in a rapid rise in Aβ levels in the cerebrospinal fluid (Watson et al. 2003). Conversely, diabetic status may indirectly affect IDE expression levels, as diet-induced insulin resistance induced in APP transgenic mice was found to augment amyloid plaque formation, and was associated with decreased levels of IDE (Ho et al. 2004). It is also possible that IDE may have indirect effects on Aβ metabolism via alteration of insulin levels, given reports that insulin signaling regulates APP processing and trafficking (Solano et al. 2000; Gasparini et al. 2001). This possibility may serve to explain why co-administration of insulin with radiolabeled Aβ in rat hippocampus had no effect on Aβ degradation in a superfusion paradigm (Iwata et al. 2000); on the other hand, it should be noted that insulin – α2-chain, disulfide-bonded peptide with potent biological properties, including effects on Aβ secretion – is clearly a less than ideal inhibitor of IDE, providing an alternative explanation for its poor efficacy in this experimental paradigm.

While IDE is unquestionably a strong functional and positional candidate for the modulation of AD pathogenesis, only a limited amount of work has focused on its *in vivo* therapeutic potential. Transgenic overexpression of IDE in neurons in an APP transgenic mouse model (Mucke et al. 2000) lowered steady-state levels of Aβ and attenuated plaque formation and its associated cytopathology (Leissring et al. 2003). While the latter study also examined the consequences of NEP overexpression in the same animal model, quantitative comparisons were not possible because very different amounts of overexpression of the two proteases were obtained in the different transgenic mouse lines. Specifically, a $\sim 100\%$ increase in IDE levels led to a $\sim 50\%$ reduction in Aβ levels and plaque burden (not shown); whereas a $\sim 700\%$ increase in NEP levels led to a $\sim 90\%$ decrease in Aβ levels and a virtual elimination of Aβ deposition (see Figure 1).

It is notable that a lower percent increase in IDE activity was achieved despite a greater absolute abundance of the transgenic IDE transcript relative to the transgenic NEP transcript (as measured by quantitative PCR with identical primers); this may be explained by the fact that basal levels of IDE are greater by far than NEP in the brains of wild-type mice, making it more difficult to increase IDE levels in overexpression paradigms. Despite these quantitative differences, the effects of both proteases were seemingly very similar qualitatively (see Figure 1). For example, overexpression of either IDE or NEP led to significant decreases in both Aβ_{40} and Aβ_{42}, including both Tris-soluble and Tris-insoluble, guanidinium-extractible pools (Leissring et al. 2003), suggesting that both proteases may target overlapping pools of Aβ. In addition, overexpression of either IDE or NEP also reversed the premature

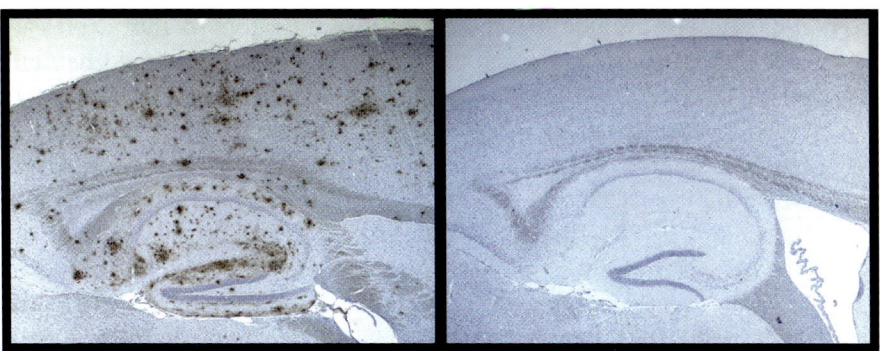

Figure 1. Effects of overexpression of NEP (by 700%) on amyloid pathology in APP transgenic mice. Left panel: APP transgenic mouse, alone; right panel: APP transgenic mouse with NEP overexpression. Panels show representative sections from 14-month-old mice in the indicated transgenic backgrounds.

lethality that occurs in this particularly aggressive APP transgenic model, suggesting that both proteases target neurotoxic species of Aβ. These results suggest that a reduction in Aβ levels might be attainable via overexpression of any of several proteases, thus increasing the number of targets available for therapeutic evaluation. In contrast to the positive results obtained in mice, it is notable that upregulation of an IDE homologue had no effect in a *D. melanogaster* model of Aβ toxicity, whereas upregulation of a neprilysin homologue significantly decreased Aβ levels and inhibited Aβ42-induced neurodegeneration (Finelli et al., 2004). However, in this model, Aβ42, rather than APP, was overexpressed, in this case directly in the secretory pathway, by fusing the Aβ42 peptide directly to a signal peptide. Thus, the disparate results obtained with IDE in this fly model may indicate that IDE targets a different pool of Aβ than NEP (which is clearly present in the secretory pathway), or may reflect differences in the subcellular localization or tissue distribution of IDE homologues in invertebrate organisms.

Transcriptional upregulation of IDE remains a theoretical possibility, but the prospects for this approach may be more limited by comparison to the prospects for NEP. Treatment of primary hippocampal neurons with insulin resulted in a phophoinositol-3 kinase-dependent increase in IDE protein levels (Zhao et al. 2004). However, this increase was relatively modest ($\sim 25\%$) compared to the large effects known activators of NEP expression can exert (several fold). Several characteristics of IDE resemble those seen for known "housekeeping" genes, including (i) a relatively even tissue distribution, with no tissues known to lack IDE; and (ii) the presence of a prominent "CpG island" (i.e., a regions of chromosomal DNA containing relatively a low frequency of methylated cytosines) in IDE's promoter region (Farris et al. 2005). However, these observations do not exclude the possibility that activators of IDE expression may yet be found, perhaps through compound screening. It should be noted that overall IDE upregulation would result in increased degradation of cytosolic IDE substrates represented by AICD in addition to extracellular ones such as Aβ. Therefore, a more attractive approach may be to search for compounds that promote the translocation of IDE to extracytoplasmic compartments. In addition, while IDE's overall mRNA levels appear to be relatively stable, functionally distinct variants might nonetheless be differentially expressed. In this regard, recent work has identified a novel splice isoform of IDE with significantly reduced catalytic efficiency (Farris et al. 2005). Further research may uncover additional transcriptional variants with functional consequences.

Some work has examined the possibility of pharmacological augmentation of IDE activity. While the activation of enzymatic activity by a small molecule is not unprecedented (e.g., Efanov et al. 2005), screening of over 150,000 small molecules using a novel Aβ degradation assay (Leissring et al. 2003) failed to identify any direct activators of IDE-mediated Aβ degradation (M.L. and D. Selkoe, unpublished observations). A more realistic approach might be to identify an endogenous inhibitor of IDE and screen for compounds that disrupt the inhibitor-protease interaction. Several candidate endogenous inhibitors have been reported, including adenosine triphosphate (ATP; Camberos et al. 2001), long-chain fatty acids (Hamel et al. 2003) and ubiquitin (Saric et al. 2003). Interestingly, whereas ATP was initially reported to inhibit the degradation of insulin by IDE, it was later found to dramatically *activate* the hydrolysis of short fluorogenic peptide substrates, by as much as 75-fold, while having little effect on the hydrolysis of larger substrates such as Aβ or glucagon (Song et al. 2004; M.L. and D. Selkoe, unpublished observations). Other work by Hersh and colleagues (Song et al. 2003) found that the hydrolysis of Aβ and other substrates (but not insulin) could be activated by low levels of a second substrate; in light of evidence that IDE exists as a homodimer (Safavi et al. 1996; Song et al. 2003; Farris et al. 2005), this effect may represent an example of positive intersubunit cooperativity.

Knowledge about the physiological and pathophysiological roles NEP and ECEs, and their implication in the catabolism of Aβ has depended critically on the widespread availability of potent and selective inhibitors (Iwata et al. 2000; Eckman et al. 2001; Newell et al. 2003; Dolev et al. 2004). However, only very non-specific and low potency inhibitors (e.g., zinc chelators, thiol-alkylating agents) are currently available for IDE or indeed for any other members of its extended superfamily. Insulin is often used as a competitive substrate, but as mentioned above, its potent bioactivity and disulfide-bonded nature places limits on its utility, particularly *in vivo* paradigms. Bacitracin, a cyclic peptide used extensively for its antibiotic properties, is another inhibitor of IDE (Roth 1981), but its potency in *in vitro* experiments with recombinant

IDE is relatively low (M.L. personal observations) and little is known about its potential to inhibit proteases other than IDE. One of us (M.L.) has recently developed peptidic hydroxamate inhibitors of IDE that may fill the important need for potent and selective IDE inhibitors (unpublished observations). These reagents may also aid in the development of novel pharmacological tools, such as photoaffinity probes and improved activity assays, that could help to answer several outstanding questions about the biology and pharmacology of IDE, and facilitate in the search for small-molecule activators of IDE activity.

5.   Plasmin

Plasmin (EC 3.4.21.1, a.k.a. fibrinolysin) is a serine protease that is the ultimate effector in the fibrinolytic cascade. Plasmin is generated from an inactive zymogen (plasminogen) through proteolysis by urokinase- or tissue-type plasminogen activators (uPA and tPA, respectively). Interest in the potential involvement of plasmin in AD pathogenesis was prompted by a report by Tucker and colleagues (Tucker et al. 2000), showing that plasmin could degrade both monomeric and fibrillar Aβ, reducing the toxicity of both, and was fueled by the discovery of genetic linkage between late-onset AD and a region of Chr. 10 containing the gene for uPA (*PLAU;* Bertram et al. 2000; Ertekin-Taner et al. 2000; Myers et al. 2000; see Chapter __). Neither uPA nor tPA directly degrades Aβ; instead, both exert their effects via proteolytic conversion of plasminogen to plasmin (Van Nostrand and Porter 1999; Tucker et al. 2000; Tucker et al. 2002). Interestingly, tPA appears to be activated generally by ligands that contain cross-beta structure, which include assembled forms of Aβ (van Nostrand et al. 1999; Kranenburg et al. 2002; Kranenburg et al. 2005).

Despite a compelling array of *in vitro* and cell culture experiments supporting a potential role for plasmin in the catabolism of Aβ, animal modeling modeling studies have not yielded straightforward results. On the one hand, the clearance of Aβ injected into the hippocampus of mice lacking tPA or plasminogen was significantly attenuated relative to wild-type animals (Melchor et al. 2003). On the other hand, cerebral Aβ was not found to be elevated in mice lacking plasminogen (Tucker et al. 2004), and similarly, no elevation in brain Aβ was observed in uPA knockout mice, despite significant elevations of Aβ in plasma (Ertekin-Taner et al. 2005; see Table 1). This pattern of results suggests that the plasmin system may not be involved in the regulation of cerebral Aβ under normal conditions, but might instead be engaged only under pathological circumstances.

From a therapeutic perspective, plasmin compares favorably to IDE, NEP and the ECEs because its activity is tightly regulated by endogenous inhibitors, such as plasminogen-activator inhibitor 1 (PAI-1) and α-2 plasmin inhibitor.

Thus, compounds that disrupt these endogenous inhibitors, by activating plasmin, could tilt the balance of Aβ metabolism in favor of increased clearance. In this regard, administration of PAI-1 inhibitors to APP transgenic mice was recently reported to lower both plasma and brain Aβ levels (Pangalos et al. 2005). Plasmin is also an appealing therapeutic target because it degrades oligomeric forms of Aβ known to exert eletrophysiological and mnemonic disturbances at very low concentrations (Walsh et al. 2002; Cleary et al. 2005) more avidly than IDE or NEP (Leissring et al. 2003; D.M. Walsh and M.L., unpublished observations). Of course, the therapeutic effects of activating the fibrinolytic cascade will have to be balanced by the risk that such an approach might hold for inducing excessive bleeding or hemorrhagic stroke (van Nostrand et al. 1999).

6.   Other Aβ-degrading proteases

While the importance of *in vivo* studies to the field of Aβ degradation has been emphasized throughout this chapter, many seminal discoveries in this field were nonetheless initially made *in vitro* or in cell culture. An important goal for future research will be to confirm or confute the *in vivo* relevance of known Aβ-degrading proteases, and also to search for others that remain to be identified. This section briefly summarizes the evidence implicating a few candidate Aβ-degrading proteases.

Two matrix-metalloproteases (MMPs) have been shown to degrade Aβ *in vitro*, namely MMP-2 and MMP-9 (Roher et al. 1994; Backstrom et al. 1996). These candidates are interesting from a therapeutic perspective because, like plasmin, they normally exist as latent, inactive proenzymes. Notably, conversion of the MMP proenzymes to their active forms is mediated by a unique "cysteine switch" mechanism (Van Wart and Birkedal-Hansen 1990). Activation by this mechanism is normally mediated by proteolysis, but it is notable that it can also be triggered by small molecules (Van Wart et al. 1990). While compounds currently known to activate MMPs (e.g., mercurials) are quite toxic and therefore of little therapeutic value, it nonetheless seems plausible that small molecules might be identified with increased potency and specificity that could be used to selectively activate this pool of Aβ-degrading proteases.

Angiotensin-converting enzyme is yet another metalloprotease that has been shown to degrade Aβ *in vitro* (Hu et al. 2001). ACE is an interesting candidate in view of genetic evidence suggesting an association with risk for late-onset AD (e.g., Kehoe et al. 2004). In view of the widespread clinical use of ACE inhibitors, it is of considerable interest to establish whether ACE does in fact play an important role in AD pathogenesis. However, recent studies found no detectable increases in the brains of ACE knockout mice (Takaki and Saido, unpublished observations; Table 1).

The final protease to be considered is cathepsin D, a lysosomal aspartyl protease. Rather surprisingly, cathepsin D was actually shown to be by far the major degrader of Aβ in brain homogenates when degradation was assayed across a range of pH values (McDermott and Gibson, 1996). In addition to its well known localization within lysosomes, cathepsin D was recently demonstrated to exist in endosomes, where it may be involved in insulin catabolism (Authier et al. 2002). While significant alterations in cathepsin D expression levels have been observed in AD brain (Cataldo et al. 1996), the relevance of these observations to Aβ catabolism remains to be established.

7.   Future directions

As this chapter attests, enormous strides have been made in our understanding of the significance of Aβ catabolism for AD pathogenesis. Nonetheless, research in this area is in its infancy, and a number of important issues remain to be addressed. This section attempts to identify a few key areas that are especially deserving of future investigation.

An obvious and common question is: which Aβ-degrading proteases are most important to Aβ metabolism? At this stage we have only crude metrics with which to judge the relative contribution of different proteases. One widely accepted parameter is the relative increase in cerebral Aβ levels measured in mice with genetic deletion of different Aβ-degrading proteases, as summarized in Table 1. While there is some merit to this approach, several important caveats should be kept in mind. First, Aβ most likely exists in separate pools within the brain that are influenced differentially by different proteases. Thus, one may speculate that a given increase in overall cerebral Aβ levels might reflect a moderate increase in global Aβ levels or, alternatively, might reflect a comparatively large increase in a restricted pool of Aβ. Absent knowledge about which pools are targeted by different proteases, it makes little sense to regard total brain Aβ levels as a quantitatively meaningful measure by which to compare the relative importance of different Aβ-degrading proteases, particularly when different methods of extracting Aβ are frequently used (see Table 1). Accordingly, future experiments seeking to compare levels of brain Aβ in different protease knockout mice, should pay greater attention to biochemically distinguishable pools of Aβ, comparing different mouse brain regions (Iwata et al. 2000; Iwata et al. 2004) side by side on the same ELISA plate together with appropriate controls such as extracts from APP knockout mice. A second difficulty with gene-targeted animals is that life-long, pan-cellular deletion of a protease may have developmental or compensatory

Table 1. Comparison of increases in endogenous cerebral Aβ levels in gene-targeted mice lacking candidate Aβ-degrading proteases.

Gene	Study	Extraction method	n per genotype	% increase cerebral Aβ[a,b] (±SEM)[c]	
				Aβ40	Aβ42
NEP	Iwata et al. 2001	GuHCl	9	100.6 ± 12.0***	113.8 ± 24.9***
	Iwata et al. 2004	Tris-saline	6	100.3 ± 15.3***	82.3 ± 19.6***
		GuHCl		107.1 ± 18.2***	100.9 ± 17.3***
	Farris et al. 2003	Diethylamine	2–3	63.8 ± 26.2*	ND
	Ertekin-Taner et al. 2005	Formic acid	3	44.0 ± 8.4[d],*	32.8 ± 6.6[d] (n.s.)
IDE	Farris et al. 2003	Diethylamine	6–7	63.5 ± 4.7***	ND
			6–7[e]	13.1 ± 1.8[e],***	7.7 ± 2.1[e],*
	Miller et al. 2003	Diethylamine	3–4	54.7 ± 14.5***	35.7 ± 7.1***
ECE1	Eckman et al. 2003	Diethylamine	12–13	20.7 ± 2.7[f],***	19.1 ± 4.3[f],*
ECE2	Eckman et al. 2003	Diethylamine	9–10	30.5 ± 3.8***	18.9 ± 5.7**
PLG	Tucker et al. 2004	Formic acid	8	5.3[g,h] (n.s.)	−3.3[g,h] (n.s.)
PLAU	Ertekin-Taner et al. 2005	Formic acid	11–13	3.7 ± 2.7[d,i] (n.s.)	−0.4 ± 1.7[d,i] (n.s.)
ACE	Takaki et al. (unpublished data)	GuHCl	6	−14.4 ± 12.9 (n.s.)	−9.3 ± 6.9 (n.s.)

a Unless otherwise indicated, data reflect measurements from homozygous null mice relative to wild-type mice.

b All measurements performed using ELISA developed by N. Suzuki (BNT77/BA27/BC05; Scheuner et al. 1996).

c SEM values were calculated based on the assumption that data were normally distributed and that the numbers of data sets were identical between wild-type and KO groups.

d Quantitative data and *P* values estimated from graphical data in published study.

e Determined in a separate set of 6-month-old mice (Farris et al. 2003)

f Data from ECE1$^{+/-}$ mice relative to wild-type controls (homozygous deletion is lethal).

g Data from Plg$^{-/-}$ mice relative to Plg$^{+/-}$ mice in genetic background homozygous null for fibrin.

h Standard error not provided.

i 3- to 6-month-old and 11-month-old mice were used.

* = P value < 0.05; ** = P value <0.01; *** = P value < 0.001; n.s. = not significant; ND = not determined.

alterations that influence brain Aβ levels indirectly, rather than directly, as is often assumed. Finally, it should be kept in mind that most proteases (including β- and γ-secretases) have multiple substrates, and it is well-established that some peptides can exert powerful influences on brain Aβ levels (e.g., Carro et al. 2002). Future research with inducible and tissue-specific knockout animals and with judicious use of selective pharmacological agents, should help to clarify matters.

The question of which pools of Aβ are most relevant to disease is another topic of fundamental importance to AD research. It seems likely that much light could be shed on this issue by directly comparing the consequences of genetic or pharmacological manipulation of different Aβ-degrading proteases in appropriate AD mouse models, given the evidence that they target different pools of Aβ. In this regard, the use of proteases genetically engineered to be targeted to different subcellular compartments, as employed successfully in cultured cells (Hama et al. 2004), would seem to be an especially informative *in vivo* approach. The difficulty here is knowing which endpoints are most disease-relevant. For instance, it is conceivable that a given Aβ protease could have no effect on extracellular plaques, and yet nonetheless yield improvements in other endpoints, such as dendritic arborization, tau hyperphosphorylation, or behavioral or electrophysiological outcomes. It seems likely that future research in this field will inform our understanding of AD pathogenesis generally, and possibly lead to improved animal models that are more faithful to relevant features of the disease in man.

A final set of questions relates to the potential therapeutic use of Aβ-degrading proteases, either as targets for drug development, or potentially, as therapeutic agents. Here again, a number of important issues relate to the existence of multiple pools of Aβ, specifically what separates these pools and how Aβ traffics among them. Notably, peripheral administration of Aβ-binding agents that cannot cross the blood-brain barrier have been shown to lead to reductions in brain Aβ in APP transgenic mice (Deane et al. 2003; Matsuoka et al. 2003; see also Chapter 10). This has led to the hypothesis that pools of Aβ in the brain and in the periphery may exist in a state of dynamic equilibrium. While this concept has not been firmly established in humans, if it is confirmed, then peripheral administration of Aβ-degrading proteases would appear to offer certain advantages over the use of antibodies, because proteases would act catalytically rather than stochiometrically to eliminate Aβ, thus requiring significantly lower quantities to exert a therapeutic benefit. Intriguingly, it may not even be necessary to administer any therapeutic agent, if a way can be found to divert brain Aβ into compartments with excess clearance capacity, an effect that might be mediated, for example, by increasing the transport of Aβ across the blood-brain barrier (see Chapter 10). Consideration of these therapeutic possibilities serves as a reminder that different Aβ clearance mechanisms do not act in isolation, but are instead likely to be highly interdependent. Future research into the interplay between proteolytic degradation of Aβ, cell-mediated clearance, and the transport of Aβ into and out of the brain, may be the most fruitful road to the development of useful clearance-based therapies to combat AD.

Let us end this chapter by reflecting on the pragmatic issue of drug development aimed at proteolytic degradation of Aβ. Within contemporary pharmaceutical science, approximately 60% and 30% of successful medications targeted at proteins are receptor agonists/antagonists and enzyme inhibitors, respectively. Because of the very high incidence of AD in the aged populations, there is a strong socio-economical demand for cost-effective and preventive measures to combat AD. In this respect, treatment that requires surgical operations or recombinant proteins would probably be too costly as a medication for the treatment of the majority of the millions of AD patients worldwide. One possible solution is to develop low- molecular-weight receptor ligands that can activate Aβ-degrading activity specifically in the brain (Saito et al. 2003; Saito et al. 2005). The combination of such a medication with others targeting secretases or more down-stream events (i.e., inflammation, oxidative stress, etc.) would exert not only preventive but also therapeutic effects both in presymptomatic and symptomatic patients. If reduced activity of Aβ-degrading enzyme(s) in brain is indeed a cause for sporadic AD, detection of such a change in a region-specific manner by, for instance, non-invasive imaging (Sato et al. 2004; Higuchi et al. 2005) may provide a new opportunities to develop presymptomatic diagnoses and to quantify prognosis, thereby facilitating the timely treatment of this devastating neurodegenerative disease.

References

Authier F, Metioui M, Fabrega S, Kouach M, Briand G (2002) J Biol Chem 277(11):9437–9446.

Backstrom JR, Lim GP, Cullen MJ, Tokes ZA (1996) J Neurosci 16(24):7910–7919.

Bae SJ, Matsunaga Y, Takenaka M, Tanaka Y, Hamazaki Y, Shimizu K, Katayama I (2002) Int Arch Allergy Immunol 127(4):316–321.

Barnes K, Turner Kenny AJ (1992) J Neurochem 58(6):2088–2096.

Becker AB, Roth RA (1992) Proc Natl Acad Sci USA 89(9):3835–3839.

Bertram L, Blacker D, Mullin K, Keeney D, Jones J, Basu S, Yhu S, McInnis MG, Go RC, Vekrellis K, Selkoe DJ, Saunders AJ, Tanzi RE (2000) Science 290(5500):2302–2303.

Caccamo A, Oddo S, Sugarman MC, Akbari Y, LaFerla FM (2005) Neurobiol Aging 26(5):645–654.

Camberos MC, Perez AA, Udrisar DP, Wanderley MI, Cresto JC (2001) Exp Biol Med (Maywood) 226(4):334–341.

Carpentier M, Robitaille Y, DesGroseillers L, Boileau G, Marcinkiewicz M (2002) J Neuropathol Exp Neurol 61(10):849–856.

Carro E, Trejo JL, Gomez-Isla T, LeRoith D, Torres-Aleman I (2002) Nat Med 8(12):1390–1397.

Cataldo AM, Hamilton DJ, Barnett JL, Paskevich PA, Nixon RA (1996) J Neurosci 16(1):186–199.

Cleary JP, Walsh DM, Hofmeister JJ, Shankar GM, Kuskowski MA, Selkoe DJ, Ashe KH (2005) Nat Neurosci 8(1):79–84.

Cook DG, Leverenz JB, McMillan PJ, Kulstad JJ, Ericksen S, Roth RA, Schellenberg GD, Jin LW, Kovacina KS, Craft S (2003) Am J Pathol 162(1):313–319.

D'Adamio L, Shipp MA, Masteller EL, Reinherz EL (1989) Proc Natl Acad Sci USA 86(18):7103–7107.

Deane R, Du Yan S, Submamaryan RK, LaRue B, Jovanovic S, Hogg E, Welch D, Manness L, Lin C, Yu J, Zhu H, Ghiso J, Frangione B, Stern A, Schmidt AM, Armstrong DL, Arnold B, Liliensiek B, Nawroth P, Hofman F, Kindy M, Stern D, Zlokovic B (2003) Nat Med 9(7):907–913.

Dolev I, Michaelson DM (2004) Proc Natl Acad Sci USA 101(38):13909–13914.

Duckworth WC, Bennett RG, Hamel FG (1998) Endocr Rev 19(5):608–624.

Eckman EA, Reed DK, Eckman CB (2001) J Biol Chem 276(27):24540–24548.

Eckman EA, Watson M, Marlow L, Sambamurti K, Eckman CB (2003) J Biol Chem 278(4):2081–2084. Epub 2002 Dec 02.

Edbauer D, Willem M, Lammich S, Steiner H, Haass C (2002) J Biol Chem 277(16):13389–13393. Epub 2002 Jan 23.

Efanov AM, Barrett DG, Brenner MB, Briggs SL, Delaunois A, Durbin JD, Giese U, Guo H, Radloff M, Gil GS, Sewing S, Wang Y, Weichert A, Zaliani A, Gromada J (2005) Endocrinology 146(9):3696–3701.

Ertekin-Taner N, Graff-Radford N, Younkin LH, Eckman C, Baker M, Adamson J, Ronald J, Blangero J, Hutton M, Younkin SG (2000) Science 290(5500):2303–2304.

Ertekin-Taner N, Ronald J, Feuk L, Prince J, Tucker M, Younkin L, Hella M, Jain S, Hackett A, Scanlin L, Kelly J, Kihiko-Ehman M, Neltner M, Hersh L, Kindy M, Markesbery W, Hutton M, de Andrade M, Petersen RC, Graff-Radford N, Estus S, Brookes AJ, Younkin SG (2005) Hum Mol Genet 14(3):447–460.

Fakhrai-Rad H, Nikoshkov A, Kamel A, Fernstrom M, Zierath JR, Norgren S, Luthman H, Galli J (2000) Hum Mol Genet 9(14):2149–2158.

Farris W, Leissring MA, Hemming ML, Chang AY, Selkoe DJ (2005) Biochemistry 44(17):6513–6525.

Farris W, Mansourian S, Chang Y, Lindsley L, Eckman EA, Frosch MP, Eckman CB, Tanzi RE, Selkoe DJ.Guenette S (2003) Proc Natl Acad Sci USA 100(7):4162–4167.

Farris W, Mansourian S, Leissring MA, Eckman EA, Bertram L, Eckman CB, Tanzi RE, Selkoe DJ (2004) Am J Pathol 164(4):1425–1434.

Finelli A, Kelkar A, Song HJ, Yang H, Konsolaki M (2004) Mol Cell Neurosci 26(3):365–375.

Fukami S, Watanabe K, Iwata N, Haraoka J, Lu B, Gerard NP, Gerard C, St. George-Hyslop P, Saido TC (2002). Neurosci Res 43:39–56.

Funato H, Yoshimura M, Kusui K, Tamaoka A, Ishikawa K, Ohkoshi N, Namekata K, Okeda R, Ihara Y (1998) Am J Pathol 152(6):1633–1640.

Gasparini L, Gouras GK, Wang R, Gross RS, Beal MF, Greengard PXu, H (2001) J Neurosci 21(8):2561–2570.

Haass C, Schlossmacher MG, Hung AY, Vigo-Pelfrey C, Mellon A, Ostaszewski BL, Lieberburg I, Koo EH, Schenk D, Teplow DB, et al. (1992) Nature 359(6393):322–325.

Hama E, Saido TC (2005) Med Hypotheses 65(3):498–500.

Hama E, Shirotani K, Iwata N, Saido TC (2004) J Biol Chem 279(29):30259–30264.

Hamel FG, Upward JL, Bennett RG (2003) Endocrinology 144(6):2404–2408.

Higuchi, M.

Ho L, Qin W, Pompl PN, Xiang Z, Wang J, Zhao Z, Peng Y, Cambareri G, Rocher A, Mobbs CV, Hof PR, Pasinetti GM (2004) Faseb J 18(7):902–904.

Howell S, Nalbantoglu J, Crine P (1995) Peptides 16(4):647–652.

Hu J, Igarashi A, Kamata M, Nakagawa H (2001) J Biol Chem 276(51):47863–47868.

Huang J, Guan H, Booze RM, Eckman CB, Hersh LB (2004) Neurosci Lett 367(1):85–87.

Iwata N, Higuchi M, Saido TC (2005) Pharmacol Ther

Iwata N, Mizukami H, Shirotani K, Takaki Y, Muramatsu S, Lu B, Gerard NP, Gerard C, Ozawa K, Saido TC (2004) J Neurosci 24(4):991–998.

Iwata N, Takaki Y, Fukami S, Tsubuki S, Saido TC (2002) J Neurosci Res 70(3):493–500.

Iwata N, Tsubuki S, Takaki Y, Shirotani K, Lu B, Gerard NP, Gerard C, Hama E, Lee HJ, Saido TC (2001) Science 292(5521):1550–1552.

Iwata N, Tsubuki S, Takaki Y, Watanabe K, Sekiguchi M, Hosoki E, Kawashima-Morishima M, Lee HJ, Hama E, Sekine-Aizawa Y, Saido TC (2000) Nat Med 6(2):143–150.

Iwata S, Lee JW, Okada K, Lee JK, Iwata M, Rasmussen B, Link TA, Ramaswamy S, Jap BK (1998) Science 281(5373):64–71.

Kanemitsu H, Tomiyama T, Mori H (2003) Neurosci Lett 350(2):113–116.

Karamohamed S, Demissie S, Volcjak J, Liu C, Heard-Costa N, Liu J, Shoemaker CM, Panhuysen CI, Meigs JB, Wilson P, Atwood LD, Cupples LA, Herbert A (2003) Diabetes 52(6):1562–1567.

Kawarabayashi T, Younkin LH, Saido TC, Shoji MK, Hsiao A, Younkin SG (2001) J Neurosci 21, 372–381.

Kehoe PG, Katzov H, Andreasen N, Gatz M, Wilcock GK, Cairns NJ, Palmgren J, de Faire U, Brookes AJ, Pedersen NL, Blennow K, Prince JA (2004) Hum Genet 114(5):478–483.

Kranenburg O, Bouma B, Kroon-Batenburg LM, Reijerkerk A, Wu YP, Voest EE, Gebbink MF (2002) Curr Biol 12(21):1833–1839.

Kranenburg O, Gent YY, Romijn EP, Voest EE, Heck AJ, Gebbink MF (2005) Neuroscience 131(4):877–886.

Kurochkin IV (1998) FEBS Lett 427(2):153–156.

Kurochkin IV, Goto S (1994) FEBS Lett 345(1):33–37.

LaFerla FM, Tinkle BT, Bieberich CJ, Haudenschild CC, Jay G (1995) Nat Genet 9(1):21–30.

Lazarov O, Lee M, Peterson DA, Sisodia SS (2002) J Neurosci 22(22):9785–9793.

Lazarov O, Robinson J, Tang YP, Hairston IS, Korade-Mirnics Z, Lee VM, Hersh LB, Sapolsky RM, Mirnics K, Sisodia SS (2005) Cell 120(5):701–713.

Leissring MA, Farris W, Chang AY, Walsh DM, Wu X, Sun X, Frosch MP, Selkoe DJ (2003) Neuron 40(6):1087–1093.

Leissring MA, Farris W, Wu X, Christodoulou DC, Haigis MC, Guarente L, Selkoe DJ (2004) Biochem J 383(Pt. 3):439–446.

Leissring MA, Lu A, Condron MM, Teplow DB, Stein RL, Farris W, Selkoe DJ (2003) J Biol Chem 278(39):37314–37320.

Lemere CA, Lopera F, Kosik KS, Lendon CL, Ossa J, Saido TC, Yamaguchi H, Ruiz A, Martinez A, Madrigal L, Hincapie L, Arango JC, Anthony DC, Koo EH, Goate AM, Selkoe DJ (1996) Nat Med 2(10):1146–1150.

Lu T, Pan Y, Kao SY, Li C, Kohane I, Chan J, Yankner BA (2004) Nature 429(6994):883–891.

Makarova KS, Grishin NV (1999) Protein Sci 8(11):2537–2540.

Marr RA, Rockenstein E, Mukherjee A, Kindy MS, Hersh LB, Gage FH, Verma IM, Masliah E (2003) J Neurosci 23(6):1992–1996.

Maruyama M, Higuchi M, Takaki Y, Matsuba Y, Tanji H, Nemoto M, Tomita N, Matsui T, Iwata N, Mizukami H, Muramatsu S, Ozawa K, Saido TC, Arai H, Sasaki H (2005) Ann Neurol 57(6):832–842.

Matsuoka Y, Saito M, LaFrancois J, Saito M, Gaynor K, Olm V, Wang L, Casey E, Lu Y, Shiratori C, Lemere C, Duff K (2003) J Neurosci 23(1):29–33.

McDermott JR, Gibson AM (1996) Neuroreport 7(13):2163–2166.

Melchor JP, Pawlak R, Strickland S (2003) J Neurosci 23(26):8867–8871.

Melzig MF, Janka M (2003) Phytomedicine 10(6–7):494–498.

Miller BC, Eckman EA, Sambamurti K, Dobbs N, Chow KM, Eckman CB, Hersh LB, Thiele DL (2003) Proc Natl Acad Sci USA 100(10):6221–6226.

Mohajeri MH, Kuehnle K, Li H, Poirier R, Tracy J, Nitsch RM (2004) FEBS Lett 562(1–3):16–21.

Mohajeri MH, Wollmer MA, Nitsch RM (2002) J Biol Chem 277(38):35460–35465.

Mucke L, Masliah E, Yu GQ, Mallory M, Rockenstein EM, Tatsuno G, Hu K, Kholodenko D, Johnson-Wood, K.McConlogue, L. (2000) J Neurosci 20(11):4050–4058.

Murakami K, Irie K, Morimoto A, Ohigashi H, Shindo M, Nagao M, Shimizu T, Shirasawa T (2002) Biochem Biophys Res Commun 294(1):5–10.

Myers A, Holmans P, Marshall H, Kwon J, Meyer D, Ramic D, Shears S, Booth J, DeVrieze FW, Crook R, Hamshere M, Abraham R, Tunstall N, Rice F, Carty S, Lillystone S, Kehoe P, Rudrasingham V, Jones L, Lovestone S, Perez-Tur J, Williams J, Owen MJ, Hardy J, Goate AM (2000) Science 290(5500):2304–2305.

Newell AJ, Sue LI, Scott S, Rauschkolb PK, Walker DG, Potter PE, Beach TG (2003) Neurosci Lett 350(3):178–180.

Ott A, Stolk RP, van Harskamp F, Pols HA, Hofman A, Breteler MM (1999) Neurology 53(9):1937–1942.

Pangalos MN, Jacobsen SJ, Reinhart PH (2005) Biochem Soc Trans 33(Pt 4):553–558.

Pardossi-Piquard R, Petit A, Kawarai T, Sunyach C, Alves da Costa C, Vincent B, Ring S, D'Adamio L, Shen J, Muller U, St George Hyslop P, Checler F (2005) Neuron 46(4):541–554.

Qiu WQ, Walsh DM, Ye Z, Vekrellis K, Zhang J, Podlisny MB, Rosner MR, Safavi A, Hersh LB, Selkoe DJ (1998) J Biol Chem 273(49):32730–32708.

Qiu WQ, Ye Z, Kholodenko D, Seubert P, Selkoe DJ (1997) J Biol Chem 272(10):6641–6646.

Roher AE, Kasunic TC, Woods AS, Cotter RJ, Ball MJ, Fridman R (1994) Biochem Biophys Res Commun 205(3):1755–1761.

Roth RA (1981) Biochem Biophys Res Commun 98(2):431–438.

Safavi A, Miller BC, Cottam L, Hersh LB (1996) Biochemistry 35(45):14318–14325.

Saido TC (1998) Neurobiol Aging 19(1 Suppl):S69–75.

Saito T, Iwata N, Tsubuki S, Takaki Y, Takano J, Huang SM, Suemoto T, Higuchi M, Saido TC (2005) Nat Med 11(4):434–439.

Saito T, Takaki Y, Iwata N, Trojanowski J, Saido TC (2003) Sci Aging Knowl Environ 2003(3):PE1.

Saric T, Muller D, Seitz HJ, Pavelic K (2003) Mol Cell Endocrinol 204(1–2):11–20.

Sato K, Higuchi M, Iwata N, Saido TC, Sasamoto K (2004) Eur J Med Chem 39(7):573–578.

Scheuner D, Eckman C, Jensen M, Song X, Citron M, Suzuki N, Bird TD, Hardy J, Hutton M, Kukull W, Larson E, Levy-Lahad E, Viitanen M, Peskind E, Poorkaj P, Schellenberg G, Tanzi R, Wasco W, Lannfelt L, Selkoe D, Younkin S (1996) Nat Med 2(8):864–870.

Schmitz A, Schneider A, Kummer MP, Herzog V (2004) Traffic 5(2):89–101.

Seta KA, Roth RA (1997) Biochem Biophys Res Commun 231(1):167–171.

Shi J, Zhang S, Tang M, Ma C, Zhao J, Li T, Liu X, Sun Y, Guo Y, Han H, Ma Y, Zhao Z (2005) J Gerontol A Biol Sci Med Sci 60(3):301–306.

Shirotani K, Tsubuki S, Iwata N, Takaki Y, Harigaya W, Maruyama K, Kiryu-Seo S, Kiyama H, Iwata H, Tomita T, Iwatsubo T, Saido TC (2001) J Biol Chem 276(24):21895–21901.

Solano DC, Sironi M, Bonfini C, Solerte SB, Govoni S, Racchi M (2000) Faseb J 14(7):1015–1022.

Song ES, Juliano MA, Juliano L, Fried MG, Wagner SL, Hersh LB (2004) J Biol Chem 279(52):54216–54220.

Song ES, Juliano MA, Juliano L, Hersh LB (2003) J Biol Chem 278(50):49789–49794.

Takahashi RH, Nam EE, Edgar M, Gouras GK (2002) Histol Histopathol 17(1):239–246.

Takaki Y, Iwata N, Tsubuki S, Taniguchi S, Toyoshima S, Lu B, Gerard NP, Gerard C, Lee HJ, Shirotani K, Saido TC (2000) J Biochem (Tokyo) 128(6):897–902.

Taylor AB, Smith BS, Kitada S, Kojima K, Miyaura H, Otwinowski Z, Ito A, Deisenhofer J (2001) Structure (Camb) 9(7):615–625.

Tsubuki S, Takaki Y, Saido TC (2003) Lancet 361(9373):1957–1958.

Tucker HM, Kihiko-Ehmann M, Estus S (2002) J Neurosci Res 70(2):249–255.

Tucker HM, Kihiko-Ehmann M, Wright S, Rydel RE, Estus S (2000) J Neurochem 75(5):2172–2177.

Tucker HM, Kihiko M, Caldwell JN, Wright S, Kawarabayashi T, Price D, Walker D, Scheff S, McGillis JP, Rydel RE Estus S (2000) J Neurosci 20(11):3937–3946.

Tucker HM, Simpson J, Kihiko-Ehmann M, Younkin LH, McGillis JP, Younkin SG, Degen JL, Estus S (2004) Neurosci Lett 368(3):285–289.

Turner AJ, Isaac RE, Coates D (2001) Bioessays 23(3):261–269.

Turner AJ, Tanzawa K (1997) Faseb J 11(5):355–364.

Van Nostrand WE, Porter M (1999) Biochemistry 38(35):11570–11576.

Van Wart HE, Birkedal-Hansen H (1990) Proc Natl Acad Sci USA 87(14):5578–5582.

Vekrellis K, Ye Z, Qiu WQ, Walsh D, Hartley D, Chesneau V, Rosner MR, Selkoe DJ (2000) J Neurosci 20(5):1657–65.

von Rotz RC, Kohli BM, Bosset J, Meier M, Suzuki T, Nitsch RM, Konietzko U (2004) J Cell Sci 117(Pt 19):4435–4448.

Walsh DM, Klyubin I, Fadeeva JV, Cullen WK, Anwyl R, Wolfe MS, Rowan MJ, Selkoe DJ (2002) Nature 416(6880):535–539.

Wang DS, Iwata N, Hama E, Saido TC, Dickson DW (2003) Biochem Biophys Res Commun 310(1):236–241.

Wang DS, Lipton RB, Katz MJ, Davies P, Buschke H, Kuslansky G, Verghese J, Younkin SG, Eckman C, Dickson DW (2005) J Neuropathol Exp Neurol 64(5):378–385.

Watson GS, Peskind ER, Asthana S, Purganan K, Wait C, Chapman D, Schwartz MW, Plymate S, Craft S (2003) Neurology 60(12):1899–1903.

Yasojima K, Akiyama H, McGeer EG, McGeer PL (2001) Neurosci Lett 297(2):97–100.

Zhang Y, McLaughlin R, Goodyer C, LeBlanc A (2002) J Cell Biol 156(3):519–529.

Zhao L, Teter B, Morihara T, Lim GP, Ambegaokar SS, Ubeda OJ, Frautschy SA, Cole GM (2004) J Neurosci 24(49):11120–11126.

Chapter 11

Role of Aβ Transport and Clearance in the Pathogenesis and Treatment of Alzheimer's Disease

David M. Holtzman[1] and Berislav Zlokovic[2]

[1]*Departments of Neurology, Molecular Biology & Pharmacology*
Hope Center for Neurological Disorders
Washington University School of Medicine
660 S. Euclid Ave, Box 8111, St. Louis, MO 63110
Email: holtzman@neuro.wustl.edu

[2]*Department of Neurosurgery, University of Rochester*
601 Elmwood Avenue, Box 670 Rm G-9613, Rochester NY 14642
Email: berislav_zlokovic@urmc.rochester.edu

1. Introduction

Dementia is a major public health problem and Alzheimer's disease (AD) is the most common cause of dementia in the United States. Data from genetic, biochemical, and animal modeling studies strongly suggests that the aggregation and buildup of the 38-43 amino acid amyloid-β (Aβ) peptide is a central event in AD pathogenesis and also a target for treatment (Golde et al., 2000). There are relatively rare, early-onset autosomal dominant forms of AD that are caused by mutations in one of three genes: *APP*, *PS1*, and *PS2* (see Chapter 1). Mutations in these genes lead to one common phenotype, overproduction of all Aβ species or specifically of Aβ$_{42}$ (Selkoe, 2001b). There are also mutations in the Aβ sequence itself that usually lead to a disorder known as cerebral amyloid angiopathy (CAA) (Levy et al., 1990) and sometimes AD (Nilsberth et al., 2001). These mutations appear to affect the ability of Aβ to aggregate (Nilsberth et al., 2001) or its clearance (Deane et al., 2004b). In the more common form of AD, late-onset AD, that occurs after the age of 60 and accounts for greater than 99% of cases, there is not compelling evidence that Aβ or Aβ$_{42}$ is over-produced. Thus, understanding factors that control

Aβ metabolism after it is cleaved from the amyloid precursor protein (APP) may be critical in determining whether or not it will ultimately aggregate, build up in the brain and cerebral arterioles (CAA), and lead to neural and vascular toxicity. After Aβ is released by cells into the extracellular space, it can potentially be degraded by enzymes. This is discussed in Chapter 9 of this volume. When Aβ is in the extracellular space either inside or outside the brain, it can also be transported between different compartments (e.g. brain to blood, blood to brain) and can also be cleared locally by cells. The process of transport and clearance can be influenced by molecules (chaperones) that bind Aβ such as apolipoprotein E (apoE). Interestingly, *APOE* genotype is a strong genetic risk factor for AD (Strittmatter et al., 1993a) and remains the only proven genetic risk factor for AD as well as for CAA in hundreds of different studies (Strittmatter and Roses, 1996) (see Chapter 1). Given the likely importance of Aβ transport and clearance in regulating Aβ metabolism and whether it will ultimately aggregate and build up in the brain, this first section of this chapter will focus on data examining Aβ transport between brain and blood and *vice-versa*. The second section will discuss studies suggesting that endogenous Aβ chaperones such as apoE influence Aβ metabolism after it is released by cells and influence Aβ aggregation, clearance, and transport.

2.   Aβ transport between brain and blood

The blood-brain barrier (BBB) does not normally allow free exchange of polar solutes such as Aβ between brain and blood or blood and brain owing to the presence of tight junctions between brain endothelial cells that form a continuous cellular monolayer. On the other hand, a non-specific bulk flow of brain interstitial fluid (ISF) seems to be responsible for 10–15% of Aβ clearance from normal brain (Shibata et al., 2000). Because studies suggest that the rate of Aβ clearance from out of the brain is faster and larger than can be accounted for only by bulk ISF and CSF flow, a carrier-mediated or receptor-mediated transport system(s) for Aβ must exist at the BBB to remove Aβ from the CNS into circulation shortly after its physiological production or to shuttle circulating Aβ into the CNS, for review see (Tanzi et al., 2004; Zlokovic, 2005).

In 1993, Zlokovic and colleagues suggested that carrier and/or receptor-mediated transport across the BBB regulates brain Aβ (Zlokovic et al., 1993). A series of papers from different groups have verified the validity of the BBB transport hypothesis for Aβ (Maness et al., 1994; Zlokovic et al., 1994; Banks and Kastin, 1996; Ghersi-Egea et al., 1996; Ghilardi et al., 1996; Martel et al., 1996; Zlokovic et al., 1996; Martel et al., 1997; Poduslo et al., 1997; Shayo et al., 1997; Mackic et al., 1998; Shibata et al., 2000; Zlokovic et al., 2000;

DeMattos et al., 2001; Lam et al., 2001; Poduslo and Curran, 2001; Bading et al., 2002; DeMattos et al., 2002b; DeMattos et al., 2002c; Mackic et al., 2002; Monro et al., 2002; Deane et al., 2003; Deane et al., 2004b) after this initial report (Zlokovic et al., 1993). These studies have demonstrated that transport exchange across the BBB critically influences the concentration of soluble Aβ in the CNS which is central to formation of neurotoxic oligomeric Aβ species (Hardy and Selkoe, 2002) and vasculotoxic aggregated forms of Aβ (Paris et al., 2004). Under physiological conditions, the concentration of free Aβ in the brain ISF is estimated to be about 100-fold higher than in plasma (DeMattos et al., 2002a; Cirrito et al., 2003), but the total amount of free Aβ in the extracellular body fluid and plasma is 10-fold greater than in brain ISF and CSF (Zlokovic, 2004). Increased basal Aβ levels in plasma have been reported in APP transgenic mice (Kawarabayashi et al., 2001) or after active Aβ immunization in non-human primates (Lemere et al., 2004) and/or treatment of different Alzheimer's transgenic lines with Aβ-peripheral binding agents (DeMattos et al., 2002b; Deane et al., 2003; Matsuoka et al., 2003).

Transport of soluble Aβ across the BBB appears to be regulated by the receptor for advanced glycation end products (RAGE) (Deane et al., 2003) and the low-density lipoprotein receptor related protein-1 (LRP1) (Shibata et al., 2000; Deane et al., 2004b). In addition, other receptors such as megalin (LRP2) (Zlokovic et al., 1996) and P-glycoprotein (Cirrito et al., 2005) may contribute to transport of Aβ across the BBB. Aβ-binding proteins that including α2-macroglobulin, apoE and apoJ also appear to influence transport of Aβ at the BBB (Zlokovic, 2005). Besides the vascular clearance and BBB transport of Aβ, proteolytic degradation of the peptide (Iwata et al., 2001; Selkoe, 2001a), its oligomerization, aggregation, and of course the production (Hardy and Selkoe, 2002; Tanzi et al., 2004; Zlokovic, 2004) and clearance of different forms of Aβ (fibrillar vs. soluble) by other cells of the neurovascular capillary unit, including astrocytes (Wyss-Coray et al., 2003; Koistinaho et al., 2004; Zlokovic, 2005), may have a major role in determining the levels and the form of Aβ brain accumulation and associated neuronal and vascular toxicity (Fig. 1).

2.1 Plasma Aβ transport

The autosomal dominant mutations that cause early-onset AD all increase Aβ$_{42}$ in plasma and brain (Citron et al., 1992; Cai et al., 1993; Suzuki et al., 1994; Borchelt et al., 1996; Scheuner et al., 1996). A late onset AD locus on chromosome 10 acts to increase plasma Aβ (Ertekin-Taner et al., 2000). Only a few studies have analyzed the levels of plasma Aβ in AD patients vs. age-matched controls suggesting no changes (Mehta et al., 2000). However, two studies suggest an increased risk for AD in cognitively normal elderly

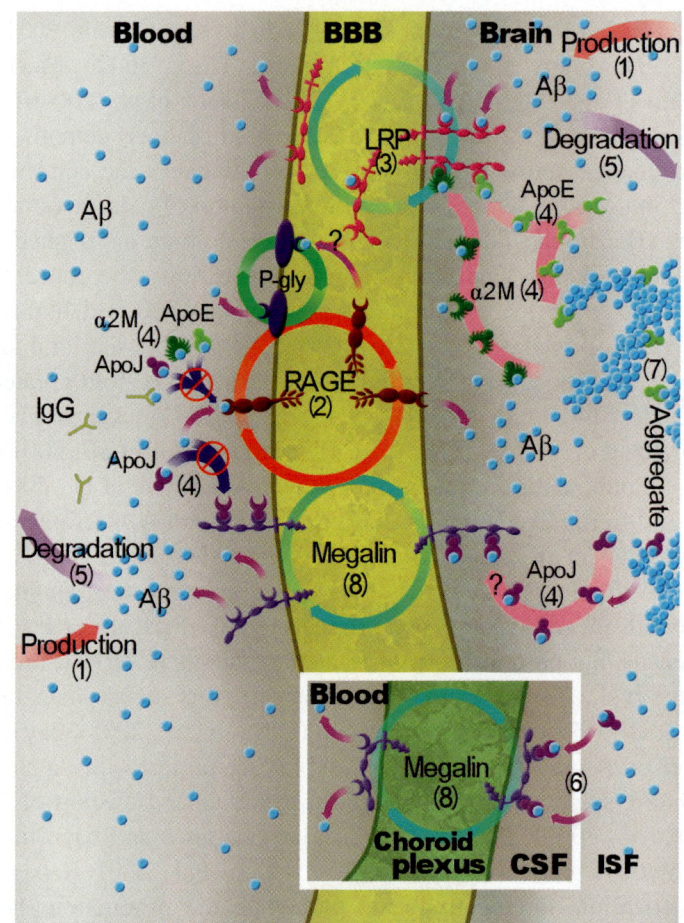

Figure 1. Brain Aβ homeostasis is controlled by numerous pathways: (**1**) peripheral and central production; (**2**) rapid receptor-mediated transport of soluble forms across the BBB from blood to brain via RAGE, and (**3**) from brain to blood via LRP; (**4**) binding to transport proteins, e.g., apoE, apoJ, α2-macroglobulin (α2M), which may influence Aβ sequestration in plasma, brain ISF and CSF, as well as the form of its accumulation in brain, i.e., soluble vs. fibrillar, and/or transport across the BBB and blood-CSF barrier; (**5**) proteolytic degradation by enkephalinase, insulinase, plasmin, tissue plasminogen activator, or matrix metalloproteinases, astrocyte-mediated degradation and microglia-mediated degradation; (**6**) slow removal via the ISF-CSF bulk flow; (**7**) oligomerization and aggregation; (**8**) transport in and out via megalin (LRP2) (Taken with permission from Zlokovic, 2005).

individuals with high levels of plasma Aβ (Mayeux et al., 1999; Mayeux et al., 2003). APP transgenic mice expressing the APPsw mutation leading to a form a familial AD (Tg-2576 mice) have high plasma levels of Aβ40 and 42, i.e., 4 nM and 0.4 nM, respectively, between 3 and 9 months of age (Kawarabayashi et al., 2001). In contrast, PDAPP mice and PD-hAPP mice have significantly lower baseline values of plasma Aβ (~200 pg/ml), but these levels can be significantly increased by 40-fold after a single i.v. injection of an anti-Aβ antibody (DeMattos et al., 2001) or soluble form of RAGE, sRAGE (Deane et al., 2003). Studies suggest that free Aβ in the plasma can be transported across the BBB into the CNS and that blocking this transport via a peripheral Aβ sequestering agent that does not cross the BBB such as an anti-Aβ antibody or sRAGE have an impact on the amount of soluble Aβ present in the CNS. Studies in these models suggest that that this peripheral trapping of Aβ shifts the plasma-CNS equilibrium of Aβ. According to the proposed transport model in Fig. 1, agents that do not penetrate the BBB but bind Aβ in plasma may promote efflux of rapidly mobilized pool of brain Aβ by creating a peripheral "sink" for Aβ.

2.2 Vascularly based Aβ-lowering strategies

Peripheral Aβ-binding agents such as anti-Aβ antibody (DeMattos et al., 2001; Sigurdsson et al., 2001; DeMattos et al., 2002b) and non-immune approaches with gelsolin and GM1 (Matsuoka et al., 2003), sRAGE (Deane et al., 2003) or soluble forms of LRP-1, sLRP-1 fragments (Deane et al., 2004b) may promote clearance of brain-derived Aβ thereby reducing Aβ levels and amyloid load in the CNS in different APP over-expressing mice. Serum amyloid P (SAP) component can be removed from human amyloid deposits in the tissues by drugs that are competitive inhibitors of SAP and may provide its rapid clearance (Pepys et al., 2002). It has been also suggested that insulin-like growth factor I may induce clearance of brain Aβ probably by enhancing transport of Aβ carrier proteins such as albumin and transthyretin into the CNS (Carro et al., 2002). Our ongoing studies indicate that soluble LRP-1 clusters II and IV, which do not penetrate from blood to brain but are high affinity Aβ binding agents (Deane et al., 2004b), may potently reduce the levels of endogenous Aβ in wild type mice (Zlokovic et al., unpublished observations).

2.3 RAGE, an influx transport receptor for Aβ

RAGE, a multiligand receptor in the immunoglobulin superfamily, binds a broad repertoire of ligands including products of non-enzymatic glycooxidation (AGE), Aβ, the S100/calgranulin family of pro-inflammatory cytokine-like mediators, and the high mobility group 1 DNA binding protein amphoterin (Hori et al., 1995; Yan et al., 1996; Hofmann et al., 1999; Kislinger et al., 1999; Stern et al., 2002). RAGE biology is largely dictated by the expression or accumulation of its ligands. Thus, in mature animals there is relatively little expression of RAGE in most tissues, whereas deposition of ligands triggers receptor expression (Stern et al., 2002). When pathogenic Aβ species accumulate in AD (Yan et al., 1996; Lue et al., 2001) or transgenic models of β-amyloidosis, i.e. Tg-2576 mice (Deane et al., 2003; LaRue et al., 2004) or Thy-1 APP triple mutant APP Swedish/ Dutch/Iowa mice (Davis et al., 2004), RAGE expression increases in affected cerebral vessels, neurons or microglia. In contrast to suppression of receptors observed with the LDL receptors in a lipoprotein-rich environment (Herz and Strickland, 2001) or LRP-1 in an $A\beta$ rich environment (Shibata et al., 2000; Deane et al., 2004b), RAGE is up-regulated by its ligands. This mechanism provides the potential for exacerbating cellular dysfunction due to RAGE-ligand interaction, as increasing expression of the receptor allows for more profound RAGE-mediated induction of cellular dysfunction.

RAGE binds soluble $A\beta$ in the nanomolar range, and mediates pathophysiologic cellular responses consequent to ligation by $A\beta$ (Mackic et al., 1998; Stern et al., 2002; Deane et al., 2004a; LaRue et al., 2004). In view of the upregulation of RAGE in AD brain vasculature (Yan et al., 1996) and RAGE-dependent binding and transport of $A\beta$ in a human model of the BBB (Mackic et al., 1998), Zlokovic and colleagues have recently demonstrated that RAGE mediates transport of physiologically relevant and pathophysiologically relevant concentrations of plasma $A\beta$ across the BBB, and that transport of pathophysiologically relevant concentrations of plasma $A\beta$ results in expression of proinflammatory cytokines in neurovascular cells and elaboration of endothelin-1, causing suppression of cerebral blood flow in response to whisker stimulation in mice (Deane et al., 2003). In addition, it has been also shown that treatment with sRAGE significantly reduces accumulation of $A\beta$ in brain of PD-hAPP mice (Deane et al., 2003). As an alternative approach, library screening for small molecules which prevent $A\beta$ binding to RAGE may offer potential new therapies to reduce $A\beta$ accumulation in the brain and associated cellular stress.

2.4   Brain to blood transport of $A\beta$

Studies in PDAPP mice have demonstrated that a single peripheral injection of the m266 monoclonal anti-$A\beta$ antibody increases plasma $A\beta$ from baseline levels of 200 pg/ml to 5–30 ng/ml within 24 hrs (DeMattos et al., 2001; DeMattos et al., 2002b). Given the similarity in plasma and CSF levels of $A\beta$ in humans (DeMattos et al., 2002a) and PDAPP mice (DeMattos et al., 2002c), it has been suggested that $A\beta$ efflux measurements from brain to plasma after challenge with an anti-$A\beta$ antibody may be useful for quantifying brain amyloid burden in patients at risk for or those who have been diagnosed with AD (DeMattos et al., 2002b). Development of plaques in primate models of β-amyloidosis and transgenic models shift a transport equilibrium for $A\beta$ between the CNS and plasma due to binding of soluble $A\beta$ from brain and plasma onto amyloid deposits in the CNS and around blood vessels (Bading et al., 2002; DeMattos et al., 2002b; Mackic et al., 2002). A study in squirrel monkeys, a primate model of CAA, has confirmed rapid elimination of Aβ from brain into plasma across the BBB (Bading et al., 2002). Studies in primates indicated an age-dependent decline in Aβ clearance via the BBB route that correlates with increases in Aβ 40/42 cerebrovascular immunoreactivity and amyloid deposition (Bading et al., 2002; Mackic et al., 2002) and down regulation of LRP-1, a major Aβ clearance receptor at the BBB.

2.5 LRP-1, a clearance receptor for Aβ

LRP-1, a multifunctional scavenger and signaling receptor, is a member of the LDL receptor family (Herz and Strickland, 2001; Green and Bu, 2002). LRP-1 was first recognized as a large endocytic receptor central to transport and metabolism of cholesterol and large apoE-containing lipoproteins. The heavy chain of LRP-1 (515 kDa) contains four ligand-binding domains (clusters I-IV) that bind a diverse array of structurally unrelated ligands, e.g., apoE, α2-macroglobulin, tissue plasminogen activator, plasminogen activator inhibitor-1, APP, factor VIII, lactoferrin (Herz and Strickland, 2001), and Aβ wild type and mutant peptides as recently shown by Zlokovic and colleagues (Deane et al., 2004b). The light chain of LRP-1 (85 kDa) contains a transmembrane domain and cytoplasmic tail that can be phosphorylated on serine and tyrosine residues (Li et al., 2001; van der Geer, 2002). Phosphorylation of LRP-1 in response to nerve growth factor accelerates endocytic activity of the receptor (Bu et al., 1998). The effects of platelet-derived growth factor-phosphorylation of the LRP-1 cytoplasmic tail (Loukinova et al., 2002) on endocytosis are less well understood (Green and Bu, 2002).

We have recently shown that LRP-1 functions as a clearance receptor for Aβ at the BBB (Shibata et al., 2000; Deane et al., 2004b). The LRP-1-mediated Aβ transcytosis is initiated at the ablumenal (brain) side of the endothelium and is therefore directly responsible for eliminating Aβ from brain ISF into blood (Fig. 1). We have reported that LRP-1-mediated transcytosis of Aβ across the BBB is substantially reduced by the high β-sheet content in Aβ and deletion of the receptor-associated protein gene. Despite low Aβ production in the brain, transgenic mice expressing low LRP-1-clearance mutant Aβ develop robust Aβ cerebral accumulation much earlier than Tg-2576 Aβ-overproducing mice (Davis et al., 2004; Deane et al., 2004b). While Aβ does not affect LRP-1 internalization and synthesis, it promotes proteasome-dependent LRP-1 degradation in brain endothelium at concentrations > 1 μM consistent with reduced brain capillary LRP levels in Aβ-accumulating transgenic mice, AD patients, and patients with cerebrovascular β-amyloidosis Dutch type. Thus, a low affinity LRP-1/Aβ interaction and/or Aβ-induced LRP-1 loss at the BBB appears to be important in brain accumulation of neurotoxic Aβ.

A recent study points to a link of the *MEOX2* (mesenchyme homebox 2) gene to neurovascular dysfunction in AD which may provide new mechanistic and therapeutic insights into this illness (Wu et al., 2005). It has been shown that expression of *MEOX2*, a regulator of vascular differentiation, is low in AD, and that restoring the expression of the protein it encodes, GAX, in brain endothelium (a site of the BBB) in individuals with AD stimulates angiogenesis, transcriptionally suppresses AFX1 forkhead transcription factor-mediated apoptosis, and increases the levels of LRP-1 at the BBB. In mice, deletion of *Meox2*, resulted in reductions in brain capillary density and resting cerebral blood flow, loss of the angiogenic response to hypoxia in the brain and an impaired Aβ efflux from brain caused by reduced LRP-1 levels (Wu et al., 2005). Thus, strategies to increase GAX levels with small molecules may result in improvement of neurovascular dysfunction in AD associated with enhanced LRP-1-mediated Aβ clearance from brain. Alternatively, we have also reported that statins, proteasome inhibitors and blockade of RAGE result in increased levels of LRP-1 at the BBB *in vitro* (Deane et al., 2004b).

3.   Regulation of Aβ transport, clearance, and aggregation by apoE

3.1   ApoE and Aβ

ApoE, a 299 amino acid lipid transport protein, has 3 common isoforms in humans, apoE2 (cys^{112}, cys^{158}), apoE3 (cys^{112}, arg^{158}), and apoE4 (arg^{112}, arg^{158}), that are products of alleles at a single gene locus on chromosome 19 (Mahley, 1988; Plump and Breslow, 1995). ApoE is expressed in several organs with highest levels in the liver and next highest expression in the brain where it is produced primarily by astrocytes (Boyles et al., 1985; Pitas et al., 1987b; Pitas et al., 1987a) and also by microglia (Stone et al., 1997). ApoE within the brain is derived exclusively from within the blood-brain-barrier (BBB) (Linton et al., 1991) where it is secreted in unique high density lipoprotein (HDL)-like lipoprotein particles and is present in the cerebrospinal fluid (CSF) at concentrations of $\sim$5 μg/ml (Pitas et al., 1987b; Pitas et al., 1987a; LaDu et al., 1998; Fagan et al., 1999). A potential connection between apoE and AD was first suggested when apoE-immunoreactivity (IR) was shown to be co-localized with extracellular amyloid deposits in subjects with AD and other types of amyloidoses (Namba et al., 1991; Wisniewski and Frangione, 1992). It was suggested that apoE may be a pathological chaperone for amyloidogenic proteins. It was of great interest when it was first reported by Strittmatter, Roses and colleagues that the ε4 allele of apoE was a genetic risk factor for AD (Strittmatter et al., 1993a). These investigators were searching for proteins in human CSF that bound to synthetic, immobilized Aβ peptide. One of the proteins that avidly bound to Aβ was apoE. Subsequently, they looked for whether there were any differences in the frequency of the different known isoforms of human apoE in AD vs. control subjects. They found that the allele frequency of $\varepsilon 4$ was ~0.15 in a control population but ~0.50 in a population of subjects with late-onset AD. Subsequently, numerous reports have confirmed that the $\varepsilon 4$ allele is an AD risk factor (particularly in the age range of 60–80) and that the $\varepsilon 2$ allele appears to be protective against AD (Corder et al., 1993; Mayeux et al., 1993; Corder et al., 1994; Poirer, 1994; Strittmatter and Roses, 1996). *APOE* isoforms also modify age of onset of AD in subjects with Down syndrome and in a large group of subjects with an FAD *PS-1* mutation (Hardy et al., 1994; Lai et al., 1999; Pastor et al., 2003). Pathological studies support the idea that apoE/Aβ interactions are somehow involved in AD in that at autopsy, individuals with AD have an *APOE4* dose-dependent increase in the density of Aβ deposits and CAA (Rebeck et al., 1993; Schmechel et al., 1993; Hyman et al., 1995). When investigators isolated fibrillar Aβ from AD brain, they found that it was stably associated with apoE (Naslund et al., 1995; Wisniewski et al., 1995). In addition to fibrillar Aβ, apoE was also found to be associated with soluble Aβ in CSF, plasma, and in both normal and AD brain homogenates *in vivo* (Koudinov et al., 1994; Koudinov et al., 1996; Permanne et al., 1997; Russo et al., 1998; Matsubara et al., 1999; Fagan et al., 2000). Further evidence that apoE/Aβ interactions are relevant to AD pathogenesis comes from studies of subjects with CAA. The $\varepsilon 4$ allele is a risk factor for the development of CAA (Schmechel et al., 1993; Greenberg et al., 1995; Greenberg et al., 1996; Premkuman et al., 1996). While there are several mechanisms as to how apoE could influence the pathogenesis of AD (e.g. effects on Aβ, neural plasticity, and/or behavior (Nathan et al., 1994; Holtzman et al., 1995; Arendt et al., 1997; Buttini et al., 1999; Buttini et al., 2000; Raber et al., 2000), the association of apoE with 2 diseases involving Aβ deposition (AD and CAA) strengthens the evidence that apoE/Aβ interactions are important in disease pathogenesis. Some, but not all, studies have also found that polymorphisms in the *APOE* promoter are associated with altered risk for AD (Bullido et al., 1998; Lambert et al., 1998; Rebeck et al., 1999). This suggests that apoE expression level may be an important AD risk modulator though further studies assessing apoE levels in the CNS as a function of promoter polymorphism are needed to determine if the promoter differences actually result in altered apoE levels in humans.

3.2 Potential role of apoE in Aβ clearance and metabolism

Several groups have performed *in vitro* experiments to investigate the ability of apoE to bind Aβ and influence its aggregation, conformation, and clearance. Studies have shown that lipid-free and lipid-associated apoE can increase (Ma et al., 1994; Sanan et al., 1994; Castano et al., 1995; Golabek et al., 1996;

Soto and Castano, 1996) or decrease (Evans et al., 1994; Wood et al., 1996) $A\beta$ fibrillogenesis. The reason for different results between studies is not clear but may be due to the different preparations of apoE and $A\beta$ utilized. It has been shown that both lipid-free and lipid-associated apoE can bind $A\beta$ when the individual proteins are mixed together (Strittmatter et al., 1993b; LaDu et al., 1994; LaDu et al., 1995; Aleshkov et al., 1997; Yang et al., 1997; Morikawa et al., 2005). Marked differences in isoform-binding are seen with lipid-containing preparations (Tokuda et al., 2000; Morikawa et al., 2005). *In vitro* studies examining apoE/$A\beta$ interactions have been informative; however, the *in vivo* environment in the brain interstitial fluid (ISF) and CSF where apoE and $A\beta$ interact is likely very different. ApoE is in HDL-like lipoproteins in these locales along with many other $A\beta$-binding molecules (e.g. $\alpha_2 M$, clusterin, HSPG). Further, the metabolism and removal of $A\beta$ and apoE/$A\beta$ complexes from the ISF and CSF is poorly understood and could occur via 1) $A\beta$ degrading enzymes (Iwata et al., 2001; Farris et al., 2003); 2) apoE receptors on glia/neurons (Hammad et al., 1997; Urmoneit et al., 1997; Beffert et al., 1999; Yang et al., 1999); 3) receptor-mediated transport across the blood-brain-barrier (Shibata et al., 2000); and 4) bulk flow along CSF and ISF drainage pathways (Weller et al., 1998). While some of these processes cannot be modeled *in vitro*, it will likely require *in vivo* studies to understand how apoE is influencing $A\beta$ metabolism. Figure 2 presents a model by which apoE can influence the clearance, metabolism, and aggregation of $A\beta$ via interactions with specific molecules and cells in the CNS.

If apoE-containing lipoproteins in the CNS "sequester" soluble $A\beta$, they could alter the metabolic pathway by which $A\beta$ is transported, cleared, and degraded. This could occur locally via apoE receptors present on cells within the brain. Several known receptors for apoE could participate in this process including the low density lipoprotein receptor (LDLR), the LDLR-related protein (LRP), the very low density lipoprotein receptor, LR8/apoER2, and LR11/SorLA-1, which are all expressed in the brain (Swanson et al., 1988; Rebeck et al., 1993; Christie et al., 1996; Jacobsen et al., 1996; Kim et al., 1996; Yamazaki et al., 1996). *In vitro* evidence suggests that LDLR and LRP may be involved (Beffert et al., 1999; Yang et al., 1999). There is also *in vivo* evidence that an LDLR family member(s) is somehow involved (Van Uden et al., 2002). Our recent data indicates that LDLR is probably the primary apoE-receptor in vivo that regulates apoE levels (Fryer et al., 2005b). For example, we found that astrocyte-secreted apoE is bound, taken up, and degraded by cells expressing LDLR but not by cells expressing LRP, apoER2, or VLDLR. We also found that $LDLR^{-/-}$ mice had a 50% elevation of endogenous murine apoE in the CNS. Interestingly, after crossing human apoE knock-in mice, expressing apoE2, apoE3, and apoE4, onto an $LDLR^{-/-}$ background, the absence of LDLR resulted in a 2–4 fold increase in CSF levels of apoE3 and apoE4. There was no

Possible mechanism(s) by which apoE influences Aβ metabolism

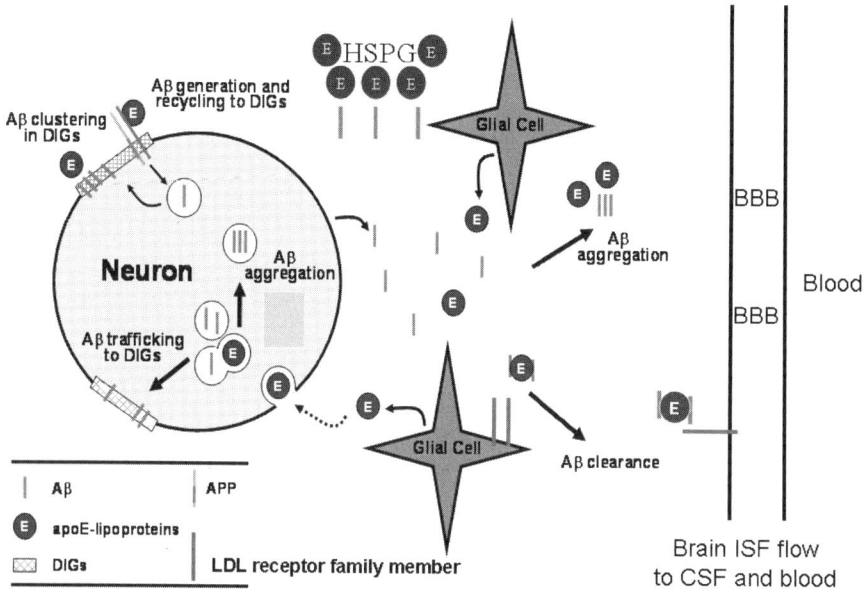

Figure 2. Model for the interactions between apoE and Aβ in the brain extracellular space. ApoE is secreted into the brain in interstitial fluid (ISF) from glial cell such as astrocytes and microglia. Aβ is secreted into the brain ISF mostly by neurons. There, apoE-containing high density lipoproteins (HDL) can interact with Aβ and can influence its clearance into cells via endocytic receptors such as LDL receptor family members such as LDLR. It can also facilitate retention of Aβ in the extracellular matrix of the brain and arterioles due to binding of apoE/Aβ complexes to heparin sulfate proteoglycans (HSPG). This may play a role in CAA formation. ApoE and Aβ have been shown to co-localize in the detergent-insoluble glycolipid rich membrane domains (DIGs) where their interaction may be promoted. In the ISF, apoE/Aβ interactions likely determine whether and when Aβ will aggregate. ApoE may also play a role in Aβ transport out of the brain via ISF/bulk flow as well as via the BBB through receptors such as LDLR or its family members. Taken with permission from (Fagan et al., 2002).

increase in apoE2 in the CSF of apoE2; LDLR$^{-/-}$ mice consistent with apoE2 being a poor ligand for LDLR. While these studies show that LDLR regulates the level of apoE in the CNS, the primary cells in the CNS responsible for this LDLR-mediated affect remain to be defined. LDLR has been shown to be expressed by astrocytes and neurons (Pitas et al., 1987b; Pitas et al., 1987a; Swanson et al., 1988; Rebeck et al., 1998).

In addition to apoE/Aβ interactions potentially influencing local Aβ clearance, it may be affecting transport of Aβ from the brain extracellular space back into the systemic circulation. As mentioned earlier in this chapter, studies have demonstrated that exogenously administered Aβ is rapidly transported from the CNS to plasma with an elimination half-life from brain of ≤ 30 minutes (Ghersi-Egea et al., 1996; Shibata et al., 2000; Ji et al., 2001). Further, evidence suggests that apoE may contribute to this process through a receptor-mediated transport mechanism, in part via LRP (Shibata et al., 2000). Whether LDLR plays a role in the efflux of apoE-Aβ transport out of the CNS remains to be defined.

3.3   Role of apoE/Aβ interactions in AD and CAA: Insights from *APP* transgenic mice

The most compelling data which demonstrate that apoE/Aβ interactions directly impact AD and CAA pathology come from studies utilizing APP transgenic mice. It was first shown that when PDAPP mice were bred onto an *apoE* knockout ($^{-/-}$) background, Aβ deposition still occurred, yet there were 2 major differences as compared to mice expressing murine apoE (Bales et al., 1997). First, there was not a delay in onset of Aβ deposition; however, there was a decrease in both Aβ deposition and Aβ levels once Aβ deposition began in PDAPP$^{+/+}$, *Apoe*$^{-/-}$ as compared to PDAPP$^{+/+}$, *Apoe*$^{+/+}$ mice. Second, PDAPP$^{+/+}$, *Apoe* $^{-/-}$ mice had diffuse but no fibrillar Aβ deposits (thioflavine-S positive-amyloid) until very old ages (Bales et al., 1999; Fagan et al., 2002). The only other Aβ binding molecule which when knocked out affects Aβ in a similar manner is clusterin (also called apolipoprotein J) (DeMattos et al., 2002d). Both of these strong effects of mouse apoE on Aβ fibril formation were gene dose-dependent. Follow-up studies in PDAPP, APPsw, and APP/PS1 mice have extended these findings and shown that the lack of murine apoE does not permanently block fibrillar Aβ formation but results in a delay and much less Aβ accumulation, neuritic plaque formation, and CAA (Bales et al., 1999; Holtzman et al., 1999; Holtzman et al., 2000b; Holtzman et al., 2000a; Fagan et al., 2002; Fryer et al., 2003; Costa et al., 2004). In fact, the absence of apoE has even more profound effects on CAA than on parenchymal Aβ. The absence of apoE in PDAPP and APPsw mice blocked CAA and associated brain hemorrhage through 2 years of age (Fryer et al., 2003). At time points prior to Aβ deposition (e.g. 3 months of age), the presence vs. the absence of murine apoE did not influence levels of Aβ as measured by ELISA in total brain tissue homogenates in initial reports. However, we have found that both murine and human apoE can modify extracellular levels of Aβ at times prior to Aβ deposition and that at least part of this affect is via Aβ clearance or transport (DeMattos et al., 2004; Fryer et al., 2005a). For example, we found that at 3 months of age, PDAPP, *Apoe*$^{-/-}$ mice have increased soluble brain and CSF Aβ levels with no evidence of increased Aβ production (DeMattos et al., 2004). Further, utilizing a newly developed technique to assess ISF Aβ by microdialysis (Cirrito et al., 2003), we found that the half-life of brain ISF Aβ

clearance was significantly decreased (by 50%) in the absence of apoE. These findings suggest that apoE binds to a pool of ISF Aβ *in vivo* that results in an alteration in its transport or clearance. In addition to these data, we have found that in 3 month old APPsw mice expressing apoE4 (prior to plaque deposition), there is an increase in the Aβ 40/42 ratio in soluble brain homogenates and a decrease in this ratio in the CSF relative to mice expressing murine apoE or apoE3 (Fryer et al., 2005a). This is further evidence that apoE is binding soluble Aβ and somehow influencing its transport within the CNS. In addition to effects on clearance of soluble Aβ, there is also recent evidence that apoE can influence the clearance of aggregated Aβ after its formation (Koistinaho et al., 2004). Work from the lab of Steven Paul showed that following *ex vivo* incubation of cultured astrocytes together with brain sections from APP transgenic mice that contained Aβ deposits, that the astroctyes could take up and degrade deposited Aβ. In contrast *Apoe*$^{-/-}$ astrocytes could not take up and degrade deposited Aβ. Experiments also showed that the ability of wild-type astrocytes to take and degrade Aβ was blocked by antibodies to apoE and also by the receptor-associated protein (RAP), a protein that blocks interactions with LDLR family members. Thus, under certain condition, astrocytes, in some way via apoE, may regulate the amount of deposited Aβ even after it converts from soluble to aggregated forms.

Overall, these and other results suggest that the elevated (but still relatively low and soluble) levels of human Aβ present from an early age in *APP* transgenic mice, over time, drives the conversion of soluble forms of Aβ to aggregated forms that have either a high β-sheet content (e.g. thioflavine-S positive amyloid plaques) or to forms that have a low β-sheet content (diffuse plaques). The presence of murine apoE somehow influences the probability that a % of the Aβ that converts to aggregated forms will convert into a β-sheet (i.e. it facilitates conformational change to β-sheet forms such as oligomers and fibrils with concomitant toxicity) through a variety of potential mechanisms. Mouse apoE is approximately 70% identical to human apoE at the amino acid level, and there are inherent functional differences between mouse and human apoE isoforms on plasma lipoprotein metabolism *in vivo* (Sullivan et al., 1997). Thus, it was possible that human apoE would have different effects on Aβ metabolism as compared to murine apoE. To explore the effect of human apoE isoforms on Aβ deposition *in vivo*, we produced and compared PDAPP, *Apoe*$^{+/+}$ and PDAPP, *Apoe*$^{-/-}$ mice to PDAPP mice expressing similar levels of human *APOE2*, *E3*, or *E4* under the control of an astrocyte-specific promoter (Sun et al., 1998). In addition, we have bred APPsw mice to both *Apoe*-/- mice and to apoE knock-in mice in which *APOE2*, *E3*, or *E4* expression is controlled by the endogenous murine apoE promoter (Fryer et al., 2005b; Fryer et al., 2005a). Overall, similar results have emerged from these experiments with these mice. First, PDAPP and APPsw mice in the presence or absence of mouse apoE all began to develop Aβ deposits by 9 months of age which increased dramatically by 15 months of age in both groups of animals ($apoE^{+/+}$>>apoE$^{-/-}$ in absolute level; $apoE^{-/-}$ with no fibrillar Aβ) (Holtzman et al., 1999; Holtzman et al., 2000b; Holtzman et al., 2000a). Second, and in contrast to the effects of mouse apoE or no apoE, human apoE2, apoE3 and apoE4, expressed at levels comparable to endogenous murine apoE, markedly delayed the onset of Aβ deposition until 15–24 months of age (Holtzman et al., 2000b; Fagan et al., 2002; Fryer et al., 2005a). Third, between 15–24 months of age, an isoform-specific effect of human apoE emerged such that earlier and greater Aβ deposition occurred with the order of effects being E4>E3>E2 (Holtzman et al., 2000b; Fagan et al., 2002; Fryer et al., 2005a). In addition, studies with APPsw mice expressing human apoE isoforms shows a strong effect of apoE4 favoring the formation of CAA over parenchymal plaques (Fryer et al., 2005a). In sum, in addition to the isoform-specific effects on Aβ deposition, the expression of human *APOE* isoforms in the brain of APP transgenic mice results in a delay in onset and less Aβ deposition than in APP transgenic mice expressing murine *Apoe* or no *Apoe*. Since evidence suggests that apoE does not influence Aβ production from APP (Biere et al., 1995; DeMattos et al., 2004) and both *in vitro* and *in vivo* evidence suggest that apoE can influence CNS Aβ transport and clearance, further work to sort out the detailed mechanisms underlying the ability of apoE to influence these processes will likely lead to important insights into AD pathogenesis.

4.   Summary

In addition to Aβ production, the factors that control Aβ degradation, clearance, transport, and fibrillogenesis play an important role in the process by which Aβ converts from soluble forms to both soluble and insoluble forms that aggregate in the brain and contribute to AD and CAA pathogenesis. Studies summarized herein suggest that transport of Aβ between brain and blood and blood and brain plays an important role in regulating Aβ metabolism and that specific molecules such as LRP, RAGE, and apoE regulate this process. In addition, apoE, acting as a chaperone for both soluble and fibrillar Aβ can modulate Aβ clearance, transport, and fibrillogenesis and in doing so, plays an important role in AD and CAA pathogenesis. Further studies to sort out the cell biology of these clearance mechanisms as well as which other molecules are involved in these processes will provide important additional insights into AD and CAA pathogenesis.

References

Aleshkov S, Abraham CR, Zannis VI (1997) Biochemistry 36:10571–10580.

Arendt T, Schindler C, Brückner MK, Eschrich K, Bigl V, Zedlick D, Marcova L (1997) J Neurosci 17:516–529.

Bading JR, Yamada S, Mackic JB, Kirkman L, Miller C, Calero M, Ghiso J, Frangione B, Zlokovic BV (2002) J Drug Target 10:359–368.

Bales KR, Verina T, Cummins DJ, Du Y, Dodel JC, Saura J, Fishman CE, DeLong CA, Piccardo P, Petegnief V, Ghetti B, Paul SM (1999) Proc Natl Acad Sci USA 96:15233–15238.

Bales KR, Verina T, Dodel RC, Du Y, Altstiel L, Bender M, Hyslop P, Johnstone EM, Little SP, Cummins DJ, Piccardo P, Ghetti B, Paul SM (1997) Nature Genet 17:263–264.

Banks WA, Kastin AJ (1996) Life Sci 59:1923–1943.

Beffert U, Aumont N, Dea D, Lussier-Cacan S, Davignon J, Poirier J (1999) Mol Brain Res 68:181–185.

Biere AL, Ostaszewski B, Zhao H, Gillespie S, Younkin SG, Selkoe DJ (1995) Neurobiol Dis 2:177–187.

Borchelt DR, Thinakaran G, Eckman CB, Lee MK, Davenport F, Ratovitsky T, Prada CM, Kim G, Seekins S, Yager D, Slunt HH, Wang R, Seeger M, Levey AI, Gandy SE, Copeland NG, Jenkins NA, Price DL, Younkin SG, Sisodia SS, Neuron. 17(5):1005–13 N (1996) Neuron 17:1005–1013.

Boyles JK, Pitas RE, Wilson E, Mahley RW, Taylor JM (1985) J Clin Invest 76:1501–1513.

Bu G, Sun Y, Schwartz AL, Holtzman DM (1998) J Biol Chem 273:13359–13365.

Bullido MJ, Artiga MJ, Recuero M, Sastre I, Garcia MA, Aldudo J, Lendon C, Han SW, Morris JC, Frank A, Vazquez J, Goate A, Valdivieso F (1998) Nature Genet 18:69–71.

Buttini M, Akeefe H, Lin C, Mahley RW, Pitas RE, Wyss-Coray T, Mucke L (2000) Neurosci 97:207–210.

Buttini M, Orth M, Bellosta S, Akeefe H, Pitas RE, Wyss-Coray T, Mucke L, Mahley RW (1999) J Neurosci 19:4867–4880.

Cai X-D, Golde TE, Younkin SG (1993) Science 259:514–516.

Carro E, Trejo JL, Gomez-Isla T, LeRoith D, Torres-Aleman I (2002) Nat Med 8:1390–1397.

Castano EM, Prelli F, Wisniewski T, Golabek A, Kumar RA, Soto C, Frangione B (1995) Biochem J 306:599–604.

Christie RH, Chung H, Rebeck GW, Strickland D, Hyman BT (1996) J Neuropath Exp Neurol 55:491–498.

Cirrito JR, May PC, O'Dell MA, Taylor JW, Parsadanian M, Cramer JW, Audia JE, Nissen JS, Bales KR, Paul SM, DeMattos RB, Holtzman DM (2003) J Neurosci 23:8844–8853.

Cirrito JR, Deane R, Fagan AM, Spinner ML, Parsadanian M, Finn MB, Jiang H, Prior JL, Sagare A, Bales KR, Paul SM, Zlokovic BV, Piwnica-Worms D, Holtzman DM (2005) J Clin Invest 115:3285–3290.

Citron M, Oltersdorf T, Haas C, McConlogue L, Hung AY, Seubert P, Vigo-Pelfrey C, Lieberburg I, Selkoe DJ (1992) Nature 360:672–674.

Corder EH, Saunders AM, Strittmatter WJ, Schmechel DE, Gaskell PC, Small GW, Roses AD, Haines JL, Pericak-Vance MA (1993) Science 261:921–923.

Corder EH, Saunders AM, Risch NJ, Strittmatter WJ, Schmechel DE, Gaskell PC, Rimmler JB, Locke PA, Conneally PM, Schmader KE, Small GW, Roses AD, Haines JL, Pericak-Vance MA (1994) Nature Genet 7:180–184.

Costa DA, Nilsson LN, Bales KR, Paul SM, Potter H (2004) J Alzheimers Dis 6:509–514.

Davis J, Xu F, Deane R, Romanov G, Previti ML, Zeigler K, Zlokovic BV, Van Nostrand WE (2004) J Biol Chem 279:20296–20306.

Deane R, Wu Z, Zlokovic BV (2004a) Stroke 35:2628–2631.

Deane R, Wu Z, Sagare A, Davis J, Du Yan S, Hamm K, Xu F, Parisi M, LaRue B, Hu HW, Spijkers P, Guo H, Song X, Lenting PJ, Van Nostrand WE, Zlokovic BV (2004b) Neuron 43:333–344.

Deane R, Du Yan S, Submamaryan RK, LaRue B, Jovanovic S, Hogg E, Welch D, Manness L, Lin C, Yu J, Zhu H, Ghiso J, Frangione B, Stern A, Schmidt AM, Armstrong DL, Arnold B, Liliensiek B, Nawroth P, Hofman F, Kindy M, Stern D, Zlokovic B (2003) Nat Med 9:907–913.

DeMattos RB, Bales KR, Paul SM, Holtzman DM (2002a) Potential role of endogenous and exogenous Aβ binding molecules in Aβ clearance and metabolism. In: Metabolic Regulation of Amyloid Beta and Alzheimer's Disease (Saido T, ed), pp 123–139. Georgetown, TX: Landes Bioscience.

DeMattos RB, Bales K, Cummins DJ, Paul SM, Holtzman DM (2002b) Science 295:2264–2267.

DeMattos RB, Bales KR, Cummins DJ, Dodart J-C, Paul SM, Holtzman DM (2001) Proc Natl Acad Sci USA 98:8850–8855:8810.1073/pnas.151261398.

DeMattos RB, Bales KR, Parsadanian M, Kierson ME, O'Dell MA, Foss EM, Paul SM, Holtzman DM (2002c) J Neurochem 81:229–236.

DeMattos RB, O'dell MA, Parsadanian M, Taylor JW, Harmony JAK, Bales KR, Paul SM, Aronow BJ, Holtzman DM (2002d) Proc Natl Acad Sci USA 10:10843–10848.

DeMattos RB, Cirrito JR, Parsadanian M, May PC, O'Dell MA, Taylor JM, Harmony JAK, Aronow BJ, Bales KR, Paul SM, Holtzman DM (2004) Neuron 41:193–202.

Ertekin-Taner N, Graff-Radford N, Younkin LH, Eckman C, Baker M, Adamson J, Ronald J, Blangero J, Hutton M, Younkin SG (2000) Science 290:2303–2304.

Evans KC, Berger EP, Cho C-G, Weisgraber KH, Lansbury PT (1994) Proc Natl Acad Sci USA 92:763–767.

Fagan AM, Younkin LH, Morris JC, Cole TG, Younkin SG, Holtzman DM (2000) Ann Neurol 48:201–210.

Fagan AM, Watson M, Parsadanian M, Bales KR, Paul SM, Holtzman DM (2002) Neurobiol Dis 9:305–318.

Fagan AM, Holtzman DM, Munson G, Mathur T, Schneider D, Chang LK, Getz GS, Reardon CA, Lukens J, Shah JA, LaDu MJ (1999) J Biol Chem 274:30001–30007.

Farris W, Mansourian S, Chang Y, Lindsley L, Elizabeth A. Eckman EA, Frosch MP, Eckman CB, Tanzi RE, J. SD, Guénette S (2003) Proc Natl Acad Sci USA 100:4162–4167.

Fryer JD, Taylor JW, DeMattos RB, Bales KR, Paul SM, Parsadanian M, Holtzman DM (2003) J Neurosci 23:7889–7896.

Fryer JD, Simmons K, Parsadanian M, Bales KR, Paul SM, Sullivan PM, Holtzman DM (2005a) J Neurosci 25:2803–2810.

Fryer JD, Demattos RB, McCormick LM, O'Dell MA, Spinner ML, Bales KR, Paul SM, Sullivan PM, Parsadanian M, Bu G, Holtzman DM (2005b) J Biol Chem 280:25754–25759.

Ghersi-Egea J-F, Gorevic PD, Ghiso J, Frangione B, Patlak CS, Fensternacher JD (1996) J Neurochem 67:880–883.

Ghilardi JR, Catton M, Stimson ER, Rogers S, Walker LC, Maggio JE, Mantyh PW (1996) Neuroreport 7:2607–2611.

Golabek AA, Soto C, Vogel T, Wisniewski T (1996) J Biol Chem 271:10602–10606.

Golde TE, Eckman CB, Younkin SG (2000) Biochim Biophys Acta 1502:172–187.

Green PS, Bu G (2002) Curr Opin Lipidol 13:569–572.

Greenberg SM, Rebeck GW, Vonsattel JPG, Gomez-Isla T, Hyman BT (1995) Ann Neurol 38:254–259.

Greenberg SM, Briggs ME, Hyman BT, Kokoris GJ, Takis C, Kanter DS, Kase CS, Pessin MS (1996) Stroke 27:1333–1337.

Hammad SM, Ranganathan S, Loukinova E, Twal WO, Argraves WS (1997) J Biol Chem 272:18644–18649.

Hardy J, Selkoe DJ (2002) Science 297:353–356.

Hardy J, Crook R, Perry R, Raghavan R, Roberts G (1994) Lancet 343:979–980.

Herz J, Strickland DK (2001) J Clin Invest 108:779–784.

Hofmann MA, Drury S, Fu C, Qu W, Taguchi A, Lu Y, Avila C, Kambham N, Bierhaus A, Nawroth P, Neurath MF, Slattery T, Beach D, McClary J, Nagashima M, Morser J, Stern D, Schmidt AM (1999) Cell 97:889–901.

Holtzman DM, Pitas RE, Kilbridge J, Nathan B, Mahley RW, Bu G, Schwartz AL (1995) Proc Natl Acad Sci USA 92:9480–9484.

Holtzman DM, Bales KR, Wu S, Bhat P, Parsadanian M, Fagan AM, Chang LK, Sun Y, Paul SM (1999) J Clin Invest 103:R15–R21.

Holtzman DM, Fagan AM, Mackey B, Tenkova T, Sartorius L, Paul SM, Bales K, Ashe KH, Irizzary MC, Hyman BT (2000a) Ann Neurol 47:739–747.

Holtzman DM, Bales KR, Tenkova T, Fagan AM, Parsadanian M, Sartorius LJ, Mackey B, Olney J, McKeel D, Wozniak D, Paul SM (2000b) Proc Natl Acad Sci USA 97:2892–2897.

Hori O, Brett J, Slattery T, Cao R, Zhang J, Chen JX, Nagashima M, Lundh ER, Vijay S, Nitecki D, et al. (1995) J Biol Chem 270:25752–25761.

Hyman BT, West ML, Rebeck GW, Buldyrev SV, Mantegna RN, Ukleja M, Havlin S, Stanley HE (1995) Proc Natl Acad Sci USA 92:3586–3590.

Iwata N, Tsubuki S, Takaki Y, Shirotani K, Bu G, Gerard NP, Gerard C, Hama E, Lee H-J, Saido TC (2001) Science 292:1550–1552.

Jacobsen L, Madsen P, Moestrup SK, Lund AH, Tommerup N, Nykjaer A, Sottrup-Jensen L, Gliemann J, Petersen CM (1996) J Biol Chem 271:31379–31383.

Ji Y, Permanne B, Sigurdsson EM, Holtzman DM, Wisniewski T (2001) J Alzheimer's Dis 3:23–30.

Kawarabayashi T, Younkin LH, Saido TC, Shoji M, Ashe KH, Younkin SG (2001) J Neurosci 21:372–381.

Kim DH, Iijima H, Goto K, Sakai J, Ishii H, Kim HJ, Suzuki H, Kondo H, Saeki S, Yamamoto T (1996) J Biol Chem 271:8373–8380.

Kislinger T, Fu C, Huber B, Qu W, Taguchi A, Du Yan S, Hofmann M, Yan SF, Pischetsrieder M, Stern D, Schmidt AM (1999) J Biol Chem 274:31740–31749.

Koistinaho M, Lin S, Wu X, Esterman M, Koger D, Hanson J, Higgs R, Liu F, Malkani S, Bales KR, Paul SM (2004) Nat Med 10:719–726.

Koudinov A, Matsubara E, Frangione B, Ghiso J (1994) Biochem Biophys Res Comm 205:1164–1171.

Koudinov AR, Koudinova NV, Kumar A, Beavis RC, Ghiso J (1996) Biochem Biophys Res Commun 223:592–597.

LaDu MJ, Falduto MT, Manelli AM, Reardon CA, Getz GS, Frail DE (1994) J Biol Chem 269:23404–23406.

LaDu MJ, Pederson TM, Frail DE, Reardon CA, Getz GS, Falduto MT (1995) J Biol Chem 270:9030–9042.

LaDu MJ, Gilligan SM, Lukens SR, Cabana VG, Reardon CA, Van Eldik LJ, Holtzman DM (1998) J Neurochem 70:2070–2081.

Lai F, Kammann E, Rebeck GW, Anderson A, Chen Y, Nixon RA (1999) Neurology 53:331–336.

Lam FC, Liu R, Lu P, Shapiro AB, Renoir JM, Sharom FJ, Reiner PB (2001) J Neurochem 76:1121–1128.

Lambert J-C, Pasquier F, Cottel D, Frigard B, Amouyel P, Chartier-Harlin M-C (1998) Hum Mol Genet 7:533–540.

LaRue B, Hogg E, Sagare A, Jovanovic S, Maness L, Maurer C, Deane R, Zlokovic BV (2004) J Neurosci Methods 138:233–242.

Lemere CA, Beierschmitt A, Iglesias M, Spooner ET, Bloom JK, Leverone JF, Zheng JB, Seabrook TJ, Louard D, Li D, Selkoe DJ, Palmour RM, Ervin FR (2004) Am J Pathol 165:283–297.

Levy E, Carman MD, Fernandez MIJ, Power MD, Lieberburg I, van Duinen SG, Bots GT, Luyendijk W, Frangione B (1990) Science 248:1124–1126.

Li Y, van Kerkhof P, Marzolo MP, Strous GJ, Bu G (2001) Mol Cell Biol 21:1185–1195.

Linton MF, Gish R, Hubl ST, Butler E, Esquivel C, Bry WI, Boyles JK, Wardell MR, Young SG (1991) J Clin Invest 88:270–281.

Loukinova E, Ranganathan S, Kuznetsov S, Gorlatova N, Migliorini MM, Loukinov D, Ulery PG, Mikhailenko I, Lawrence DA, Strickland DK (2002) J Biol Chem 277:15499–15506.

Lue LF, Walker DG, Brachova L, Beach TG, Rogers J, Schmidt AM, Stern DM, Yan SD (2001) Exp Neurol 171:29–45.

Ma J, Yee A, Brewer HB, Das S, Potter H (1994) Nature 372:92–94.

Mackic JB, Bading J, Ghiso J, Walker L, Wisniewski T, Frangione B, Zlokovic BV (2002) Vascul Pharmacol 38:303–313.

Mackic JB, Stins M, McComb JG, Calero M, Ghiso J, Kim KW, Yan SD, Stern D, Schmidt AM, Frangione B, Zlokovic BV (1998) J Clin Inverst 102:734–743.

Mahley RW (1988) Science 240:622–630.

Maness LM, Banks WA, Podlisny MB, Selkoe DJ, Kastin AJ (1994) Life Sci 55:1643–1650.

Martel CL, Mackic JB, McComb JG, Ghiso J, Zlokovic BV (1996) Neurosci Lett 206:157–160.

Martel CL, Mackic JB, Matsubara E, Governale S, Miguel C, Miao W, McComb JG, Frangione B, Ghiso J, Zlokovic BV (1997) J Neurochem 69:1995–2004.

Matsubara E, Ghiso J, Frangione B, Amari M, Tomidokoro Y, Ikeda Y, Harigaya Y, Okamoto K, Shoji M (1999) Ann Neurol 45:537–541.

Matsuoka Y, Saito M, LaFrancois J, Saito M, Gaynor K, Olm V, Wang L, Casey E, Lu Y, Shiratori C, Lemere C, Duff K (2003) J Neurosci 23:29–33.

Mayeux R, Honig LS, Tang MX, Manly J, Stern Y, Schupf N, Mehta PD (2003) Neurology 61:1185–1190.

Mayeux R, Stern Y, Ottman R, Tatemichi TK, Tang M-X, Maestre G, Ngai C, Tycko B, Ginsberg H (1993) Ann Neurol 34:752–754.

Mayeux R, Tang MX, Jacobs DM, Manly J, Bell K, Merchant C, Small SA, Stern Y, Wisniewski HM, Mehta PD (1999) Ann Neurol 46:412–416.

Mehta PD, Pirttila T, Mehta SP, Sersen EA, Aisen PS, Wisniewski HM (2000) Arch Neurol 57:100–105.

Monro OR, Mackic JB, Yamada S, Segal MB, Ghiso J, Maurer C, Calero M, Frangione B, Zlokovic BV (2002) Neurobiol Aging 23:405–412.

Morikawa M, Fryer JD, Sullivan PM, Christopher EA, Wahrle SE, DeMattos RB, O'Dell MA, Fagan AM, Lashuel HA, Walz T, Asai K, Holtzman DM (2005) Neurobiol Dis 19:66–76.

Namba Y, Tomonaga M, Kawasaki H, Otomo E, Ikeda K (1991) Brain Res 541:163–166.

Naslund J, Thyberg J, Tjernberg LO, Wernstedt C, Karlstrom AR, Bogdanovic N, Gandy SE, Lannfelt L, Terenius L, Nordstedt C (1995) Neuron 15:219–228.

Nathan BP, Bellosta S, Sanan DA, Weisgraber KH, Mahley RW, Pitas RE (1994) Science 264:850–852.

Nilsberth C, Westlind-Danielsson A, Eckman CB, Condron MM, Axelman K, Forsell C, Stenh C, Luthman J, Teplow DB, Younkin SG, Naslund J, Lannfelt L (2001) Nature Neurosci 4:887–893.

Paris D, Townsend K, Quadros A, Humphrey J, Sun J, Brem S, Wotoczek-Obadia M, DelleDonne A, Patel N, Obregon DF, Crescentini R, Abdullah L, Coppola D, Rojiani AM, Crawford F, Sebti SM, Mullan M (2004) Angiogenesis 7:75–85.

Pastor P, Roe CM, Villegas A, Bedoya G, Chakraverty S, García G, Tirado V, Norton J, Ríos S, Martínez M, Kosik KS, Lopera F, Goate AM (2003) Ann Neurol 54:63–69.

Pepys MB, Herbert J, Hutchinson WL, Tennent GA, Lachmann HJ, Gallimore JR, Lovat LB, Bartfai T, Alanine A, Hertel C, Hoffmann T, Jakob-Roetne R, Norcross RD, Kemp JA, Yamamura K, Suzuki M, Taylor GW, Murray S, Thompson D, Purvis A, Kolstoe S, Wood SP, Hawkins PN (2002) Nature 417:254–259.

Permanne B, Perez C, Soto C, Frangione B, Wisniewski T (1997) Biochem Biophys Res Commun 240:715–720.

Pitas RE, Boyles JK, Lee SH, Hui D, Weisgraber KH (1987a) J Biol Chem 262:14352–14360.

Pitas RE, Boyles JK, Lee SH, Foss D, Mahley RW (1987b) Biochim Biophys Acta 917:148–161.

Plump AS, Breslow JL (1995) Annu Rev Nutr 15:495–518.

Poduslo JF, Curran GL (2001) Neuroreport 12:3197–3200.

Poduslo JF, Curran GL, Haggard JJ, Biere AL, Selkoe DJ (1997) Neurobiol Dis 4:27–34.

Poirer J (1994) Trends Neurol Sci 17:525–530.

Premkuman DRD, Cohen DL, Hedera P, Friedland RP, Kalaria RN (1996) Am J Pathol 148:2083–2095.

Raber J, Wong D, Yu GQ, Buttini M, Mahley RW, Pitas RE, Mucke L (2000) Nature 404:352–354.

Rebeck GW, Reiter JS, Strickland DK, Hyman BT (1993) Neuron 11:575–580.

Rebeck GW, Alonzo NC, Berezovska O, Harr SD, Knowles RB, Growdon JH, Hyman BT, Mendez AJ (1998) Exp Neurol 149:175–182.

Rebeck GW, Cheung BS, Growdon WB, Deng A, Akuthota P, Locascio J, Greenberg SM, Hyman BT (1999) Neurosci Lett 272:155–158.

Russo C, Angelini G, Dapino D, Piccini A, Piombo G, Schettini G, Chen S, Teller JK, Zaccheo D, Gambetti P, Tabaton M (1998) Proc Natl Acad Sci USA 95:15598–15602.

Sanan DA, Weisgraber KH, Russel SJ, Mahley RW, Huang D, Saunders A, Schmechel D, Wisniewksi T, Frangione B, Roses B, Roses AD, Strittmatter WJ (1994) J Clin Invest 94:860–869.

Scheuner D, Eckman C, Jensen M, Song X, Citron M, Suzuki N, Bird TD, Hardy J, Hutton M, Kukull W (1996) Nature Med 2:864–852.

Schmechel DE, Saunders AM, Strittmattter WJ, Crain BJ, Hulette CM, Joo SH, Pericak-Vance MA, Goldgaber D, Roses AD (1993) Proc Natl Acad Sci USA 90:9649–9653.

Selkoe DJ (2001a) Neuron 32:177–180.

Selkoe DJ (2001b) Physiol Rev 81:741–766.

Shayo M, McLay RN, Kastin AJ, Banks WA (1997) Life Sci 60:PL115–118.

Shibata M, Yamada S, Kumar SR, Calero M, Bading J, Frangione B, Holtzman DM, Miller CA, Strickland DK, Ghiso J, Zlokovic BV (2000) J Clin Invest 106:1489–1499.

Sigurdsson EM, Scholtzova H, Mehta P, Frangione B, Wisniewski T (2001) Am J Pathol 159:439–447.

Soto C, Castano EM (1996) Biochem J 314:701–707.

Stern D, Yan SD, Yan SF, Schmidt AM (2002) Adv Drug Deliv Rev 54:1615–1625.

Stone DJ, Rozovsky I, Morgan TE, Anderson CP, Hajian H, Finch CE (1997) Exp Neurol 143:313–318.

Strittmatter WJ, Roses AD (1996) Annu Rev Neurosci 19:53–77.

Strittmatter WJ, Saunders AM, Schmechel D, Pericak-Vance M, Enghild J, Salvesen GS, Roses AD (1993a) Proc Natl Acad Sci USA 90:1977–1981.

Strittmatter WJ, Weisgraber KH, Huang DY, Dong L-Y, Salvesen GS, Pericak-Vance M, Schmechel D, Saunders AM, Goldgaber D, Roses AD (1993b) Proc Natl Acad Sci USA 90:8098–8102.

Sullivan PM, Mezdour H, Aratani Y, Knouff C, Najib J, Reddick RL, Quarfordt SH, N. M (1997) J Biol Chem 272:17972–17980.

Sun Y, Wu S, Bu G, Onifade MK, Patel SN, LaDu MJ, Fagan AM, Holtzman DM (1998) J Neurosci 18:3261–3272.

Suzuki N, Cheung TT, Cai XD, Odaka A, Otvos LJ, Eckman C, Golde TE, Younkin SG (1994) Science 264:1336–1340.

Swanson LW, Simmons DM, Hofmann SL, Goldstein JL, Brown MS (1988) Proc Natl Acad Sci USA 85:9821–9825.

Tanzi RE, Moir RD, Wagner SL (2004) Neuron 43:605–608.

Tokuda T, Calero M, Matsubara E, Vidal R, Kumar A, Permanne B, Zlokovic B, Smith JD, Ladu MJ, Rostagno A, B. F, Ghiso J (2000) Biochem J 348:359–365.

Urmoneit B, Prikulis I, Wihl G, D'Urso D, Frank R, Heeren J, Beisiegel U, Prior R (1997) Lab Invest 77:157–166.

van der Geer P (2002) Trends Cardiovasc Med 12:160–165.

Van Uden E, Mallory M, Ieinbergs I, Alford M, Rockenstein E, Masliah E (2002) J Neurosci 22:9298–9304.

Weller RO, Massey A, Newman TA, Hutchings M, Kuo YM, Roher AE (1998) Am J Pathol 153:7225–7733.

Wisniewski T, Frangione B (1992) Neurosci Lett 135:235–238.

Wisniewski T, Lalowski M, Golabek AA, Vogel T, Frangione B (1995) Lancet 345:956–958.

Wood SJ, Chan W, Wetzel R (1996) Biochem 35:12623–12628.

Wu Z, Guo H, Chow N, Sallstrom J, Bell RD, Deane R, Brooks AI, Kanagala S, Rubio A, Sagare A, Liu D, Li F, Armstrong D, Gasiewicz T, Zidovetzki R, Song X, Hofman F, Zlokovic BV (2005) Nat Med.

Wyss-Coray T, Loike JD, Brionne TC, Lu E, Anankov R, Yan F, Silverstein SC, Husemann J (2003) Nat Med 9:453–457.

Yamazaki H, Bujo H, Kusunoki J, Seimiya K, Kanaki T, Morisaki N, Schneider WJ, Saito Y (1996) J Biol Chem 271:24761–24768.

Yan SD, Chen X, Fu J, Chen M, Zhu H, Roher A, Slattery T, Zhao L, Nagashima M, Morser J, Migheli A, Nawroth P, Stern D, Schmidt AM (1996) Nature 382:685–691.

Yang D-S, Smith JD, Zhou Z, Gandy S, Martins RN (1997) J Neurochem 68:721–725.

Yang DS, Small DH, Seydel U, Smith JD, Hallmayer J, Gandy SE, Martins RN (1999) Neurosci 90:1217–1226.

Zlokovic BV (2004) J Neurochem 89:807–811.

Zlokovic BV (2005) Trends Neurosci 28:202–208.

Zlokovic BV, Yamada S, Holtzman D, Ghiso J, Frangione B (2000) Nat Med 6:718–719.

Zlokovic BV, Ghiso J, Mackic JB, McComb JG, Weiss MH, Frangione B (1993) Biochem Biophys Res Comm 197:1034–1040.

Zlokovic BV, Martel CL, Mackic JB, Matsubara E, Wisniewski T, McComb JG, Frangione B, Ghiso J (1994) Biochem Biophys Res Commun 205:1431–1437.

Zlokovic BV, Martel CL, Matsubara E, McComb JG, Zheng G, McCluskey RT, Frangione B, Ghiso J (1996) Proc Natl Acad Sci USA 93:4229–4234.

Chapter 12

Tau Gene Mutations in FTDP-17 Syndromes*

Matthew J. Winton, John Q. Trojanowski and Virginia M-Y. Lee

Center for Neurodegenerative Disease Research and Institute on Aging
Department of Pathology and Laboratory Medicine
University of Pennsylvania School of Medicine
Philadelphia, PA 19104, USA

1.   Introduction

Frontal temporal dementias (FTDs) are a heterogeneous group of sporadic and familial neurodegenerative disorders that affect the frontal and anterior temporal lobes of the cerebral cortex, often with additional subcortical changes (for review see (Lee et al., 2001). FTDs usually develop between the ages of 40 and 75 years and are characterized clinically by the gradual loss of executive function, pronounced changes in personal and social behavior and language dysfunction (Neary et al., 1998). In addition extra-pyramidal signs, including tremors, rigidity, and slowness of movement can also be observed. Pathologically, this assorted collection of neurological diseases are diagnosed by severe neuron loss resulting in atrophy of the frontal and temporal lobes, gliosis in both white and gray matter, spongiosis, and the presence of abnormal intracellular inclusions. Currently, it is estimated that approximately 50% of all FTDs contain prominent tau pathology in neurons, glial or both and together with other neurodegenerative disorders that also possess abnormal deposits of hyperphosphorylated tau, these diseases are collectively termed tauopathies (Lee et al., 2001). Other types of FTD patients that do not develop tau lesions may develop ubiquitin positive inclusions that are tau and α-synclein negative and their identity remain to be elucidated. Within FTDs that develop tau lesions, several clinical subtypes have been identified and they include corticobasal degeneration (CBD), progressive supranuclear palsy (PSP), Pick's

---

* **Correspondence to:** V. M-Y. Lee, Email: vmylee@mail.med.upenn.edu

disease (PiD), autosomal dominantly inherited frontal temporal dementia and parkinsonism linked to chromosome 17 (FTDP-17) and others (Lee et al., 2001). However, the presence of filamentous tau inclusions in such diseases may differ in morphology, isoform composition, and brain location. Tau pathology isolated from affected regions of CBD brains contain both straight and paired helical filaments (PHF) of hyperphosphorylated 4R-tau isoforms (Komori, 1999; Ksiezak-Reding et al., 1998), whereas, tau pathology in PiD is a mixture of wide, straight filaments and wide, long-period twisted filaments composed exclusively of 3R-tau isoforms (Dickson, 1998; Munoz-Garcia and Ludwin, 1984; Murayama et al., 1990).

2.   Tau biology and function

Tau is a low molecular weight microtubule associated protein (MAP) that is expressed across diverse species. In humans, tau is predominantly found in neurons of both the peripheral and central nervous system, where it is mainly expressed in axons, although very low levels of tau expression have also been reported in glial cells (Binder et al., 1985; Cleveland et al., 1977; Couchie et al., 1992; LoPresti et al., 1995; Shin et al., 1991). Since its initial discovery tau has been implicated in a wide variety of important biological functions, such as intracellular vesicular transport, organization of the actin cytoskeleton and anchoring of phosphatases and kinases (for review see (Buee et al., 2000; Lee et al., 2001). However, tau is best characterized for its ability to bind to, stabilize and promote the polymerization of microtubules (MTs) (Cleveland et al., 1977; Weingarten et al., 1975).

Human tau is encoded by a single gene located on chromosome 17q21-22 that consists of 16 exons with the central nervous system (CNS) isoforms generated by alternative mRNA splicing (Andreadis et al., 1992; Neve et al., 1986). In the adult human brain, alternative splicing of exons (E) 2 (E2), 3 (E3) and 10 (E10) result, in the production of 6 isoforms, ranging in length from 352 to 441 amino acids (Goedert et al., 1989a; Goedert et al., 1989b). These isoforms differ in the absence (0N), or presence of 1 (1N), or 2 (2N) N-terminal inserts of unknown function and either 3 (3R) or 4 (4R) imperfect 18 amino acid microtubule-binding repeats located at the C-terminal (Andreadis et al., 1992; Goedert et al., 1989a; Goedert et al., 1989b) (Figure 1). Several studies provide evidence that these 3R and 4R carboxy-terminal repeats, along with specifically identified adjacent sequences are responsible for the binding of tau to MT (Butner and Kirschner, 1991; Gustke et al., 1994; Lee et al., 1989). Interestingly, the inter-repeat region between the first and second MT binding domain has 2-fold the binding affinity of any of the individual MT binding repeats (Goode and Feinstein, 1994). This region, unique to 4R-tau, is believed to be responsible for the increased MT affinity of 4R-tau as compared to 3R-tau (Goedert and Jakes, 1990).

Furthermore, tau is a phosphoprotein with 79 potential serine or threonine (Ser/Thr) phosphorylation acceptor sites identified within the longest isoform of tau, almost all of which are located within or adjacent to the MT binding repeats. Tau phosphorylation is a normal physiological process which decreases tau's binding affinity for MTs, thereby acting as an important regulator of MT polymerization (Biernat et al., 1993; Bramblett et al., 1993; Drechsel et al., 1992; Yoshida and Ihara, 1993). These phosphorylation sites can be sub-divided into 2 groups: (i) residues that are phosphorylated by proline-directed kinases; and (ii) residues that are phosphorylated by non-proline-directed kinases (for review see (Billingsley and Kincaid, 1997; Buee et al., 2000). While initially tau was thought to be phosphorylated at abnormal sites in AD relative to the normal brain, this idea was later dispelled. To date, approximately 30 of these residues have been shown to be phosphorylated under physiological conditions, although little is currently known about the exact biological functions and specific mechanisms by which these different phosphorylation sites are regulated.

3.   Familial frontal temporal dementias

Although considerably less common then their sporadic counterparts, several familial forms of FTDs, including the pathogenic mutations responsible, have been identified. At a consensus conference in 1997, the term frontal temporal dementia and parkinsonism linked to chromosome 17 (FTDP-17) was first implemented to describe a group of then 13 families all of which presented with heterogeneous, autosomal-dominantly inherited forms of FTDs with parkinsonism genetically linked to chromosome 17 (i.e.17q21-22) (Foster et al., 1997). Despite the diverse and varied list of clinical symptoms associated with this group of diseases; on post-mortem examination, the brains of all FTDP-17 patients contain abundant hyperphosphorylated tau-positive inclusions located in neurons, glia and in both neurons and glia, (Crowther and Goedert, 2000; Spillantini et al., 1998a). Moreover, the absence of additional neuropathological lesions, such as amyloid plaques or Lewy bodies, in almost all patients with these diseases strongly suggests that abnormalities in tau homeostasis are sufficient to induce the onset and progression of neurodegenerative disease.

As the tau gene had been previously been localized to chromosome 17q21-22 (Goedert et al., 1988; Neve et al., 1986) it was an obvious candidate for the disease locus, however, unequivocal support of this theory was not provided until 1998, when several groups simultaneously identified a number

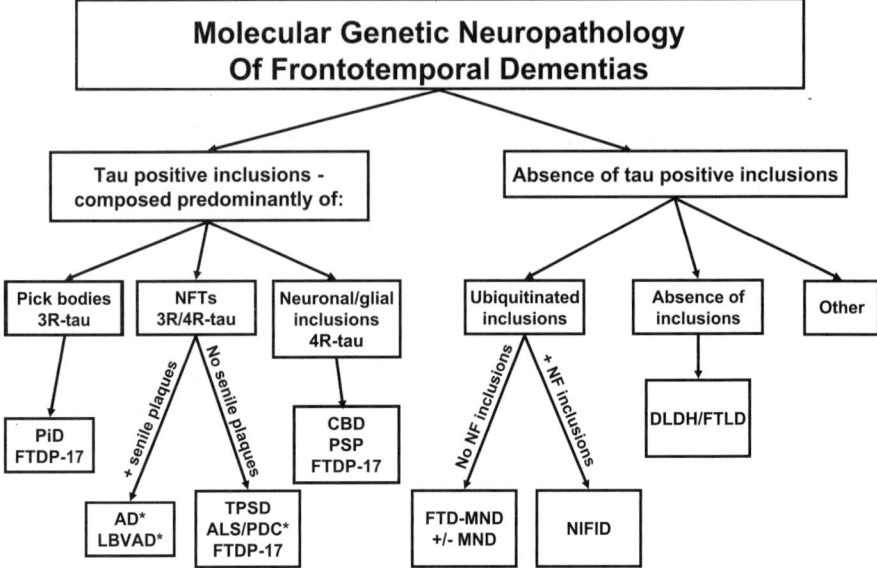

Figure 1. This alogorithm for the neuropathologic differential diagnosis of patients with different variants of FTD provides a nosology of these disorders that is grounded in their molecular genetics and neuropathology. While most FTDs are caused by tau pathologies,hence are classified as tauopathies, the neuropathology of other FTDs is very heterogeneous and the clinical manifestations of FTDs cannot predict the underlying neuropathology. While this algorithm represents an oversimplification of the definitions of these diverse FTDF variants it reflects an iterative effort to harmonize and rationalize current nomenclature for FTDs. Abbreviations: AD, Alzheimer's disease; ALS/PDC, Amyotrophic Lateral Sclerosis/Parkinson dementia complex of Guam; CBD, corticobasal degeneration; DLDH, dementia lacking distinctive histopathology; FTD-MND, FTD with motor neuron disease or motor neuron disease like pathology (i.e. ubiquitin positive but tau and α-synuclein negative inclusions); FTDP-17: frontotemporal dementia with parkinsonism linked to chromosome 17; FTLD, frontotemporal lobar degeneration, and alternative term for DLDH; LBVAD, LB variant of AD; NIFID, neuronal intermediate filament disease; PiD, Pick's disease; PSP, progressive supranuclear palsy; TPSD, tangle predominant dementia; 3R-tau, tau isoforms with 3 MT binding repeats; 4R-tau, tau isoforms with 4 MT binding repeats. * Identifies disorders that are double or triple brain amyloidoses because inclusions formed by multiple amyloidogenic proteins (e.g. SPs formed by fibrillar $A\beta$, NFTs formed by PHFtau, LBs formed by α-synuclein filaments) occur in varying degree in these diseases.

of FTDP-17 mutations which genetically mapped to the region of chromosome 17 containing the tau gene (Hutton et al., 1998; Poorkaj et al., 1998; Spillantini et al., 1998b; Clark et al., 1988). To date, more than 30 specific pathogenic mutations of tau have been identified in over 100 families (see Figure 2A and 2B).

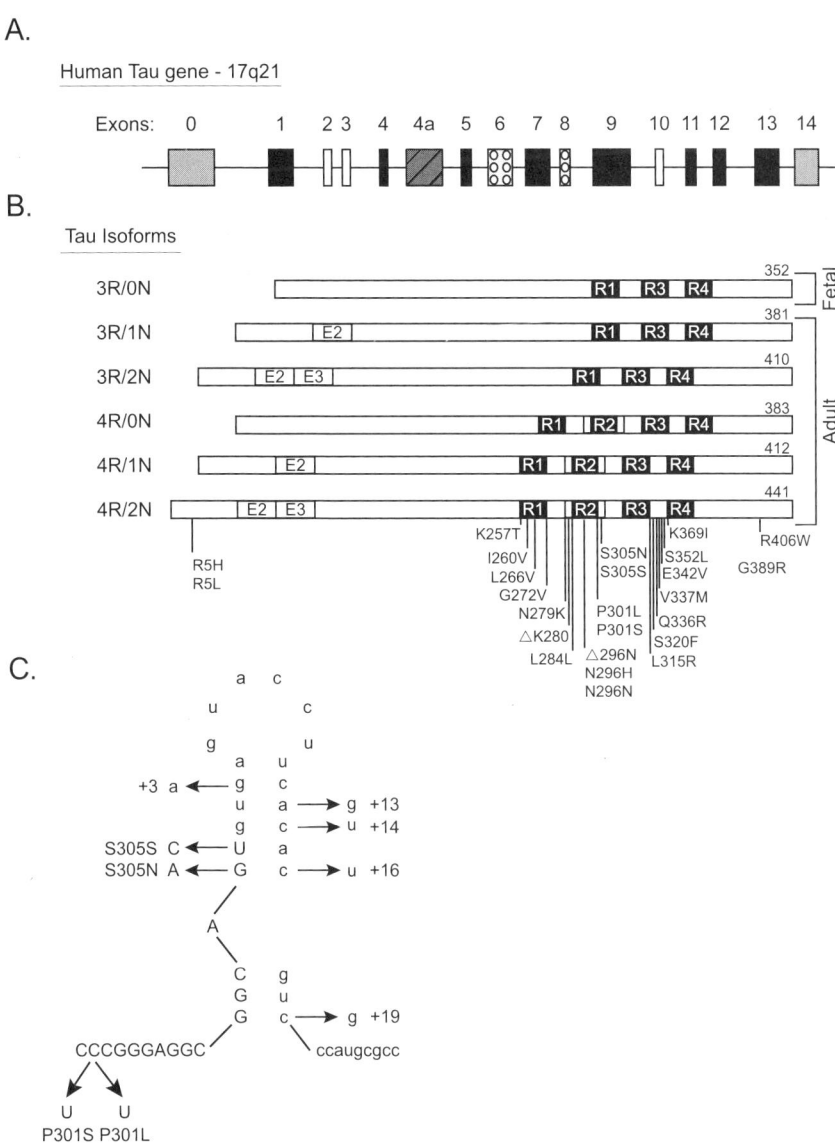

Figure 2. The genomic structure of the human tau gene and mutations in the tau gene associated with FTDP-17. (A) Human tau is encoded by a single gene located on chromosome 17q21-22 that consists of 16 exons. (B) Alternative splicing of exons 2, 3 and 10 (white boxes) results, in the production of 6 tau isoforms. Amino acid numbers are shown to the right. Twenty missense mutations, 3 silent mutation 2 deletion mutation in the coding region of tau are shown. Mutations are labeled according to nucleotide number and single letter amino acid code. (C) Predicted stem-loop structure located at the splice-donor site of the intron following the alternatively spliced exon 10. Nine mutations which increase the splicing of exon 10 are shown. Mutations are numbered with the first nucleotide of the splice-donor site designated +1. Exon sequences are shown in capital and intron sequences in lower-case letters.

These consist of 20 missense mutations [R5H (Hayashi et al., 2002), R5L (Poorkaj et al., 2002), K257T (Pickering-Brown et al., 2000; Rizzini et al., 2000), I260V (Grover et al., 2003), L266V (Hogg et al., 2003; Kobayashi et al., 2003b), G272V (Hutton et al., 1998; Rosso et al., 2003; Spillantini et al., 1998b), N279K (Delisle et al., 1999; Hasegawa et al., 1999; Wszolek et al., 2000; Yasuda et al., 2000), N296H (Iseki et al., 2001; Yoshida et al., 2002), P301L (Dumanchin et al., 1998; Heutink et al., 1997; Poorkaj et al., 2001; Rosso et al., 2003), P301S (Bugiani et al., 1999; Sperfeld et al., 1999), S305N (Hasegawa et al., 1999; Iijima et al., 1999), L315R (Rosso et al., 2003), S320F (Rosso et al., 2003), Q336R (Pickering-Brown et al., 2000), V337M (Hasegawa et al., 1998), E342V, S352L, K369I (Neumann et al., 2001), G389R (Murrell et al., 1999), and R406W (Hutton et al., 1998; Rosso et al., 2003)] 3 silent mutations [L284L (D'Souza et al., 1999), N296N (Spillantini et al., 2000a) and S305S (Stanford et al., 2000)] and 2 deletion mutations [ΔK280 and ΔN296 (D'Souza et al., 1999; Oliva and Pastor, 2004; Rizzu et al., 1999; Rosso et al., 2003; Yoshida et al., 2002)] (see Figure 2b). Mutations are numbered according to the longest CNS isoform of tau with 441 amino acids; 4R2N. The majority of these mutations are introduced in E9-13, except for 2 recently identified mutations which are present in E1 (R5H and R5L)(Hayashi et al., 2002; Poorkaj et al., 2002). In addition, 9 intronic mutations, clustered adjacent to the splice donor site of the intron directly following the alternatively spliced exon 10 have been identified and are located at positions +3 (Spillantini et al., 1998b), +11, +12, +13, +14 (Hutton et al., 1998), +16 (Goedert et al., 1999b), +19 +29, and +33 (Rizzu et al., 1999) (see Figure 2b). Mutations are numbered with the first nucleotide of the splice-donor site designated +1.

4.   The effects of mutations on tau function

Recent molecular and biochemical data suggests that FTDP-17 mutations in tau promote tau dysfunction and, in turn, lead to the development of intracellular aggregates via two main pathogenic mechanisms: (i) altering the mRNA splicing of exon 10; and (ii) decreasing tau-MT interactions. In addition, it has been suggested that a subset of these mutations may also increases tau filament formation.

4.1   Mutations affecting the splicing of exon 10

To date, the majority of known intronic mutations (Clark et al., 1998; Goedert et al., 1999b; Hulette et al., 1999; Hutton et al., 1998; Miyamoto et al., 2001; Morris et al., 1999; Spillantini et al., 1998b; Tolnay et al., 2000; Yoshida and Ihara, 1993), as well as, several coding region mutations located with in E10, such as N279K (Clark et al., 1998; Hasegawa et al., 1999), L284L

(D'Souza et al., 1999), N296N (Spillantini et al., 2000b), S305N (D'Souza et al., 1999; Hasegawa et al., 1999; Iijima et al., 1999), and S305S (Stanford et al., 2000), induce pathogenesis by altering the mRNA splicing of E10. As previously described, the alternative splicing of E10 produces 3 tau isoforms with 3 MT-binding repeats (3R-tau) and 3 isoforms with 4 MT-binding repeats (4R-tau). In the adult human brain the ratio of 3R tau to 4R tau is approximately 1:1, however, the biological significance of this ration is currently unknown. Tau mutations, which act at the RNA level, affect the alternative mRNA splicing of E10 and in doing so change the ratio of 4R- and 3R-tau isoforms. Therefore, with the exception of $\Delta 280K$ which has been shown to decrease E10 splicing (D'Souza et al., 1999), all mutations located within E10 itself or the intron following E10 lead to aggregation of predominantly 4R-tau. The regulation of E10 splicing in the tau gene appears to be exceedingly complex and is discussed in more detail in additional chapters. Briefly, it is believed to involve multiple cis-acting regulatory elements that either enhance or inhibit the utilization of the E10 5' splice site. For instance, the N279K mutation is suggested to magnify an exon splicing enhancer (ESE) (D'Souza and Schellenberg, 2000), whereas the silent mutation L284L may negate an exon splicing silencing element (ESS) believed to suppress exon 10 inclusion (D'Souza et al., 1999; Si et al., 1998). Sequences located at the boundary between exon 10 and the intron which follows, are suggested to inhibit the splicing due to the presence of a stem loops structure that limits access of the splicing machinery to the 5'-splice site (see Figure 2C) (Hutton et al., 1998; Spillantini et al., 1998b) . This potential pathogenic mechanism is supported by biochemical studies which demonstrate that sarkosyl-insoluble tau isolated from FTDP-17 brains with these mutations contain only 4R-tau isoforms (Clark et al., 1998; Goedert et al., 1999b; Hong et al., 1998; Hulette et al., 1999; Reed et al., 1998; Spillantini et al., 1997; Spillantini et al., 1998b; Yasuda et al., 2000) and further confirmed by exon-trapping studies which shown an increase in E10 splicing (Delisle et al., 1999; Goedert et al., 1999b; Hutton et al., 1998; Varani et al., 1999; Yasuda et al., 2000). The exact mechanisms by which changes in tau isoform expression leads to neuronal and glial cell dysfunction are currently unknown; however, it is believed that a specific ratio of 3R- to 4R-tau may be necessary for normal tau and MT function (Hong et al., 1998). In addition, the overproduction of 4R-tau results in an increase in free, cytoplasmic tau, which result in its hyperphosphorylation and filament formation.

The morphology of filaments produced from tau mutations that affect RNA splicing of E10 are described as wide, twisted ribbon-like filaments. These fibrillary lesions, which mirror those found in the brains of patients with sporadic 4R-tau disorders, such as CBD, are present in both neurons and glia and have been identified in individuals with +3, +11, +12, +13, and +16 intronic mutations (Goedert et al., 1999b; Miyamoto et al., 2001; Pickering-Brown et al., 2000; Spillantini et al., 1997; Yoshida and Ihara, 1993). Moreover, coding region mutations whose primary effects are at the RNA level (N279K, L284L, N296N, S305N, and S305S) also display abundant gray and white matter pathology, however, several studies report clinical variability amongst the different mutations, as well as, morphological differences in tau filaments. For instance, two neuropathological studies of the S305N mutation describe the presence of abundant NFT in one patient (Iijima et al., 1999) and PiD-like pathology in the second (Kobayashi et al., 2003a). In contrast, the N279K mutation, which has been identified 6 families, produces predominantly twisted tau filaments reminiscent of PSP (Delisle et al., 1999).

4.2   Mutations affecting tau-MT interactions

In contrast to mutations that affect RNA splicing, several tau mutations have been identified which disrupt tau-MT interactions. Specifically, K257T, G272V, Δ280, P301l, P301S, V337M, G389R and R406W all have been shown to reduce the ability of tau to directly bind MTs as determined by *in vitro* assays (Barghorn et al., 2000; Bugiani et al., 1999; D'Souza et al., 1999; Hasegawa et al., 1998; Hong et al., 1998; Murrell et al., 1999; Pickering-Brown et al., 2000; Rizzini et al., 2000; Rizzu et al., 1999). The results from these studies suggest that these mutations affect both the affinity of tau (K_d) for microtubules and its binding capacity (β_{max}). Furthermore, the overexpression of these mutations in several different cell types induces, to varying degrees, a reduction of MT binding, the disorganization of MT morphology and deficits in the over all assembly and stability of MTs (Arawaka et al., 1999; Dayanandan et al., 1999; Frappier et al., 1999; Matsumura et al., 1999; Sahara et al., 2000; Vogelsberg-Ragaglia et al., 2000). However, in two separate studies, the transfection of these coding region mutations were shown to induce only minor effects (DeTure et al., 2000; Sahara et al., 2000). The inconsistency between these and other studies may be due to differences in levels of tau expression and the methods by which tau levels and MT binding were quantified. It is believed that this deficit in tau-MT binding increases the levels of cytosolic tau, which may facilitate aggregation of tau into filamentous inclusions. That being said, even if these mutations induce only modest decreases in tau-MT binding, the potential cumulative effect of the lifespan of an individual may be sufficient to induce abnormal tau pathology. Interesting, besides having addition pathogenic effects, two missense mutations, S305N and Q336R have actually been shown to increase tau's ability to promote MT assembly, however the biological significance of this heightened affinity is unknown (Hasegawa et al., 1999; Pickering-Brown et al., 2000).

Two novel missense mutations located in exon 1, R5H (Hayashi et al., 2002) and R5L, (Hayashi et al., 2002; Poorkaj et al., 2002), which display neuronal and glial pathology, were recently identified. Although located at the N-terminus of tau, these mutations have also been reported to decrease MT assembly (Hayashi et al., 2002; Poorkaj et al., 2002). Recently, a conformation specific antibody, Alz-50, was shown to only recognizes tau when it adopts a specific conformation in which part of the extreme N-terminal comes into contact with the microtubule binding domain (Carmel et al., 1996). This change in conformation is believed to occur early in the NFT formation process and therefore it possible that the R5H and R5L mutations may enhance this interaction.

A subset of FTDP-17 mutations, including K257T, G272V, Δ280, P301L, P301S, V337M and R406W, have been documented to directly increase the tendency of tau to aggregate. This effect is particularly marked for the P301L and P301S mutations. Several studies have demonstrated that members of this subset of tau gene mutations directly increase heparin- or arachidonic-induced filament formation, as compared to wild-type tau (Arrasate et al., 1999; Barghorn et al., 2000; Gamblin et al., 2000; Goedert et al., 1999a; Rizzini et al., 2000). Moreover, the over expression of these mutants in intact cells produced insoluble amorphous and fibrillar tau aggregates (Vogelsberg-Ragaglia et al., 2000).

5. Conclusion

The identification of multiple autosomal-dominant tau gene mutations that give rise to a family of neurodegenerative disorders provides concrete evidence that abnormalities in tau are sufficient to induce the onset and progression of neurodegenerative disease. Although the majority of tauopathies are sporadic in nature, a greater understanding of the etiology and disease mechanisms involved in FTDP-17 syndromes will increase our current understanding of both the physiological and pathological functions of tau. Analogous to the identification of familial mutations involved in the pathogenesis of Alzheimer's disease, the discovery of such a large family of tau gene mutations which presents a wide range of phenotypes has lead to the development of more complex *in vitro* assays, better cell culture models and the development of multiple transgenic animal models. These biological models can be used to identify potential therapeutic targets and develop new mechanism based drugs aimed at stopping and/or prevented frontal temporal dementias and other tauopathies.

Acknowledgments

We thank our many colleagues at Penn and beyond for their contributions to the work summarized here which has been supported by grants from the NIH (AG10124, AG14382, AG17586), the Oxford Foundation, and the Marian S. Ware Alzheimer Program. VMYL is the John H. Ware 3rd Professor for Alzheimer's Disease Research and JQT is the William Maul Measy-Truman G. Schnabel Jr. M.D. Professor of Geriatric Medicine and Gerontology.

References

Andreadis A, Brown WM, Kosik KS (1992) Structure novel exons of the human tau gene. Biochemistry 31:10626–10633.

Arawaka S, Usami M, Sahara N, Schellenberg GD, Lee G, Mori H (1999) The tau mutation (val337met) disrupts cytoskeletal networks of microtubules. Neuroreport 10:993–997.

Arrasate M, Perez M, Armas-Portela R, Avila J (1999) Polymerization of tau peptides into fibrillar structures. The effect of FTDP-17 mutations. FEBS Lett 446:199–202.

Barghorn S, Zheng-Fischhofer Q, Ackmann M, Biernat J, von Bergen M, Mandelkow EM, Mandelkow E (2000) Structure, microtubule interactions,paired helical filament aggregation by tau mutants of frontotemporal dementias. Biochemistry 39:11714–11721.

Biernat J, Gustke N, Drewes G, Mandelkow E. M,Mandelkow E (1993) Phosphorylation of Ser262 strongly reduces binding of tau to microtubules: distinction between PHF-like immunoreactivity and microtubule binding. Neuron 11:153–163.

Billingsley ML, Kincaid RL (1997) Regulated phosphorylation and dephosphorylation of tau protein: effects on microtubule interaction, intracellular traffickingneurodegeneration. Biochem J 323 (Pt 3):577–591.

Binder, L. I, Frankfurter A, Rebhun LI (1985) The distribution of tau in the mammalian central nervous system. J Cell Biol 101:1371–1378.

Bramblett GT, Goedert M, Jakes R, Merrick SE, Trojanowski JQ, Lee VM-Y (1993) Abnormal tau phosphorylation at Ser396 in Alzheimer's disease recapitulates developmentcontributes to reduced microtubule binding. Neuron 10:1089–1099.

Buee L, Bussiere T, Buee-Scherrer V, Delacourte A, Hof PR (2000) Tau protein isoforms, phosphorylationrole in neurodegenerative disorders. Brain Res Brain Res Rev 33:95–130.

Bugiani O, Murrell JR, Giaccone G, Hasegawa M, Ghigo G, Tabaton M, Morbin M, Primavera A, Carella F, Solaro C, et al. (1999) Frontotemporal dementiacorticobasal degeneration in a family with a P301S mutation in tau. J Neuropathol Exp Neurol 58:667–677.

Butner KA, Kirschner MW (1991) Tau protein binds to microtubules through a flexible array of distributed weak sites. J Cell Biol 115:717–730.

Carmel G, Mager EM, Binder LI, Kuret J (1996) The structural basis of monoclonal antibody Alz50's selectivity for Alzheimer's disease pathology. J Biol Chem 271:32789–32795.

Clark L N, Poorkaj P, Wszolek Z, Geschwind DH, Nasreddine ZS, Miller B, Li D, Payami H, Awert F, Markopoulou K, et al. (1998) Pathogenic implications of mutations in the tau gene in pallido-ponto-nigral degenerationrelated neurodegenerative disorders linked to chromosome 17. Proc Natl Acad Sci USA 95:13103–13107.

Cleveland DW, Hwo SY, Kirschner MW (1977) Purification of tau, a microtubule-associated protein that induces assembly of microtubules from purified tubulin. J Mol Biol 116:207–225.

Couchie D, Mavilia C, Georgieff IS, Liem RK, Shelanski ML, Nunez J (1992) Primary structure of high molecular weight tau present in the peripheral nervous system. Proc Natl Acad Sci USA 89:4378–4381.

Crowther RA, Goedert M (2000) Abnormal tau-containing filaments in neurodegenerative diseases. J Struct Biol 130:271–279.

D'Souza I, Poorkaj P, Hong M, Nochlin D, Lee VM-Y, Bird TD, Schellenberg GD (1999) Missensesilent tau gene mutations cause frontotemporal dementia with parkinsonism-chromosome 17 type, by affecting multiple alternative RNA splicing regulatory elements. Proc Natl Acad Sci USA 96:5598–5603.

D'Souza I, Schellenberg GD (2000) Determinants of 4-repeat tau expression. Coordination between enhancinginhibitory splicing sequences for exon 10 inclusion. J Biol Chem 275:17700–17709.

Dayanandan R, Van Slegtenhorst M, Mack TG, Ko L, Yen SH, Leroy K, Brion JP, Anderton BH, Hutton M, Lovestone S (1999) Mutations in tau reduce its microtubule binding properties in intact cellsaffect its phosphorylation. FEBS Lett 446:228–232.

Delisle MB, Murrell JR, Richardson R, Trofatter JA, Rascol O, Soulages X, Mohr M, Calvas P, Ghetti B (1999) A mutation at codon 279 (N279K) in exon 10 of the Tau gene causes a tauopathy with dementiasupranuclear palsy. Acta Neuropathol (Berl) 98:62–77.

DeTure M, Ko LW, Yen S, Nacharaju P, Easson C, Lewis J, van Slegtenhorst M, Hutton M, Yen SH (2000) Missense tau mutations identified in FTDP-17 have a small effect on tau-microtubule interactions. Brain Res 853:5–14.

Dickson DW (1998) Pick's disease: a modern approach. Brain Pathol 8:339–354.

Drechsel DN, Hyman AA, Cobb MH, Kirschner MW (1992) Modulation of the dynamic instability of tubulin assembly by the microtubule-associated protein tau. Mol Biol Cell 3:1141–1154.

Dumanchin C, Camuzat A, Campion D, Verpillat P, Hannequin D, Dubois B, Saugier-Veber P, Martin C, Penet C, Charbonnier F, et al. (1998) Segregation of a missense mutation in the microtubule-associated protein tau gene with familial frontotemporal dementiaparkinsonism. Hum Mol Genet 7:1825–1829.

Foster NL, Wilhelmsen K, Sima AA, Jones MZ, D'Amato CJ, Gilman S (1997) Frontotemporal dementiaparkinsonism linked to chromosome 17: a consensus conference. Conference Participants. Ann Neurol 41:706–715.

Frappier T, Liang NS, Brown K, Leung CL, Lynch T, Liem RK, Shelanski ML (1999) Abnormal microtubule packing in processes of SF9 cells expressing the FTDP-17 V337M tau mutation. FEBS Lett 455:262–266.

Gamblin TC, King ME, Kuret J, Berry RW, Binder LI (2000) Oxidative regulation of fatty acid-induced tau polymerization. Biochemistry 39:14203–14210.

Goedert M, Jakes R (1990) Expression of separate isoforms of human tau protein: correlation with the tau pattern in braineffects on tubulin polymerization. Embo J 9:4225–4230.

Goedert M, Jakes R, Crowther RA (1999a) Effects of frontotemporal dementia FTDP-17 mutations on heparin-induced assembly of tau filaments. FEBS Lett 450:306–311.

Goedert M, Spillantini MG, Crowther RA, Chen SG, Parchi P, Tabaton M, Lanska DJ, Markesbery WR, Wilhelmsen KC, Dickson DW, et al. (1999b) Tau gene mutation in familial progressive subcortical gliosis. Nat Med 5:454–457.

Goedert M, Spillantini MG, Jakes R, Rutherford D, Crowther RA (1989a) Multiple isoforms of human microtubule-associated protein tau: sequenceslocalization in neurofibrillary tangles of Alzheimer's disease. Neuron 3:519–526.

Goedert M, Spillantini MG, Potier, MC, Ulrich J, Crowther RA (1989b) Cloningsequencing of the cDNA encoding an isoform of microtubule-associated protein tau containing four tandem repeats: differential expression of tau protein mRNAs in human brain. Embo J 8:393–399.

Goedert M, Wischik CM, Crowther RA, Walker JE, Klug A (1988) Cloningsequencing of the cDNA encoding a core protein of the paired helical filament of Alzheimer disease: identification as the microtubule-associated protein tau. Proc Natl Acad Sci USA 85, 4051–4055.

Goode BL, Feinstein SC (1994) Identification of a novel microtubule bindingassembly domain in the developmentally regulated inter-repeat region of tau. J Cell Biol 124:769–782.

Grover A, England E, Baker M, Sahara N, Adamson J, Granger B, Houlden H, Passant U, Yen SH, DeTure M, Hutton M (2003) A novel tau mutation in exon 9 (1260V) causes a four-repeat tauopathy. Exp Neurol 184:131–140.

Gustke N, Trinczek B, Biernat J, Mandelkow EM, Mandelkow E (1994) Domains of tau proteininteractions with microtubules. Biochemistry 33:9511–9522.

Hasegawa M, Smith MJ, Goedert M (1998) Tau proteins with FTDP-17 mutations have a reduced ability to promote microtubule assembly. FEBS Lett 437:207–210.

Hasegawa M, Smith MJ, Iijima M, Tabira T, Goedert M (1999) FTDP-17 mutations N279KS305N in tau produce increased splicing of exon 10. FEBS Lett 443:93–96.

Hayashi S, Toyoshima Y, Hasegawa M, Umeda Y, Wakabayashi K, Tokiguchi S, Iwatsubo T, Takahashi H (2002) Late-onset frontotemporal dementia with a novel exon 1 (Arg5His) tau gene mutation. Ann Neurol 51:525–530.

Heutink P, Stevens M, Rizzu P, Bakker E, Kros JM, Tibben A, Niermeijer MF, van Duijn CM, Oostra BA, van Swieten JC (1997) Hereditary frontotemporal dementia is linked to chromosome 17q21-q22: a geneticclinicopathological study of three Dutch families. Ann Neurol 41:150–159.

Hogg M, Grujic ZM, Baker M, Demirci S, Guillozet AL, Sweet AP, Herzog LL, Weintraub S, Mesulam MM, LaPointe NE, et al. (2003) The L266V tau mutation is associated with frontotemporal dementiaPick-like 3R4R tauopathy. Acta Neuropathol (Berl) 106:323–336.

Hong M, Zhukareva V, Vogelsberg-Ragaglia V, Wszolek Z, Reed L, Miller BI, Geschwind DH, Bird TD, McKeel D, Goate A, et al. (1998) Mutation-specific functional impairments in distinct tau isoforms of hereditary FTDP-17. Science 282:1914–1917.

Hulette CM, Pericak-Vance MA, Roses AD, Schmechel DE, Yamaoka LH, Gaskell PC, Welsh-Bohmer KA, Crowther RA, Spillantini MG (1999) Neuropathological features of frontotemporal dementiaparkinsonism linked to chromosome 17q21-22 (FTDP-17): Duke Family 1684. J Neuropathol Exp Neurol 58:859–866.

Hutton M, Lendon CL, Rizzu P, Baker M, Froelich S, Houlden H, Pickering-Brown S, Chakraverty S, Isaacs A, Grover A, et al. (1998) Association of missense5'-splice-site mutations in tau with the inherited dementia FTDP-17. Nature 393:702–705.

Iijima M, Tabira T, Poorkaj P, Schellenberg GD, Trojanowski JQ, Lee VM-Y, Schmidt ML, Takahashi K, Nabika T, Matsumoto T, et al. (1999) A distinct familial presenile dementia with a novel missense mutation in the tau gene. Neuroreport 10:497–501.

Iseki E, Matsumura T, Marui W, Hino H, Odawara T, Sugiyama N, Suzuki K, Sawada H, Arai T, Kosaka K (2001) Familial frontotemporal dementiaparkinsonism with a novel N296H mutation in exon 10 of the tau genea widespread tau accumulation in the glial cells. Acta Neuropathol (Berl) 102:285–292.

Kobayashi K, Kidani T, Ujike H, Hayashi M, Ishihara T, Miyazu K, Kuroda S, Koshino Y (2003a) Another phenotype of frontotemporal dementiaparkinsonism linked to chromosome-17 (FTDP-17) with a missense mutation of S305N closely resembling Pick's disease. J Neurol 250:990–992.

Kobayashi T, Ota S, Tanaka K, Ito Y, Hasegawa M, Umeda Y, Motoi Y, Takanashi M, Yasuhara M, Anno M, et al. (2003b) A novel L266V mutation of the tau gene causes frontotemporal dementia with a unique tau pathology. Ann Neurol 53:133–137.

Komori T (1999) Tau-positive glial inclusions in progressive supranuclear palsy, corticobasal degenerationPick's disease. Brain Pathol 9:663–679.

Ksiezak-Reding H, Yang G, Simon M, Wall JS (1998) Assembled tau filaments differ from native paired helical filaments as determined by scanning transmission electron microscopy (STEM). Brain Res 814:86–98.

Lee G, Neve RL, Kosik KS (1989) The microtubule binding domain of tau protein. Neuron 2:1615–1624.

Lee VM-Y, Goedert M, Trojanowski JQ (2001) Neurodegenerative tauopathies. Annu Rev Neurosci 24:1121–1159.

LoPresti P, Szuchet S, Papasozomenos SC, Zinkowski RP, Binder LI (1995) Functional implications for the microtubule-associated protein tau: localization in oligodendrocytes. Proc Natl Acad Sci USA 92:10369–10373.

Matsumura N, Yamazaki T, Ihara Y (1999) Stable expression in Chinese hamster ovary cells of mutated tau genes causing frontotemporal dementiaparkinsonism linked to chromosome 17 (FTDP-17). Am J Pathol 154:1649–1656.

Miyamoto K, Kowalska A, Hasegawa M, Tabira T, Takahashi K, Araki W, Akiguchi I, Ikemoto A (2001) Familial frontotemporal dementiaparkinsonism with a novel mutation at an intron 10+11-splice site in the tau gene. Ann Neurol 50:117–120.

Morris HR, Perez-Tur J, Janssen JC, Brown J, Lees AJ, Wood NW, Hardy J, Hutton M, Rossor MN (1999) Mutation in the tau exon 10 splice site region in familial frontotemporal dementia. Ann Neurol 45:270–271.

Munoz-Garcia D, Ludwin SK (1984) Classicgeneralized variants of Pick's disease: a clinicopathological, ultrastructural, and immunocytochemical comparative study. Ann Neurol 16:467–480.

Murayama S, Mori H, Ihara Y, Tomonaga M (1990) Immunocytochemicalultrastructural studies of Pick's disease. Ann Neurol 27:394–405.

Murrell JR, Spillantini MG, Zolo P, Guazzelli M, Smith MJ, Hasegawa M, Redi F, Crowther RA, Pietrini P, Ghetti B, Goedert M (1999) Tau gene mutation G389R causes a tauopathy with abundant pick body-like inclusionsaxonal deposits. J Neuropathol Exp Neurol 58:1207–1226.

Neary D, Snowden JS, Gustafson L, Passant U, Stuss D, Black S, Freedman M, Kertesz A, Robert PH, Albert M, et al. (1998) Frontotemporal lobar degeneration: a consensus on clinical diagnostic criteria. Neurology 51:1546–1554.

Neumann M, Schulz-Schaeffer W, Crowther RA, Smith MJ, Spillantini MG, Goedert M, Kretzschmar HA (2001) Pick's disease associated with the novel Tau gene mutation K369I. Ann Neurol 50:503–513.

Neve RL, Harris P, Kosik KS, Kurnit, DM, Donlon TA. (1986) Identification of cDNA clones for the human microtubule-associated protein tauchromosomal localization of the genes for taumicrotubule-associated protein 2. Brain Res 387:271–280.

Oliva R, Pastor P (2004) Tau gene delN296 mutation, Parkinson's disease,atypical supranuclear palsy. Ann Neurol 55:448–449.

Pickering-Brown S, Baker M, Yen SH, Liu WK, Hasegawa M, Cairns N, Lantos PL, Rossor M, Iwatsubo T, Davies Y, et al. (2000) Pick's disease is associated with mutations in the tau gene. Ann Neurol 48:859–867.

Poorkaj P, Bird TD, Wijsman E, Nemens E, Garruto RM, Anderson L, Andreadis A, Wiederholt WC, Raskind M, Schellenberg GD (1998) Tau is a candidate gene for chromosome 17 frontotemporal dementia. Ann Neurol 43:815–825.

Poorkaj P, Grossman M, Steinbart E, Payami H, Sadovnick A, Nochlin D, Tabira T, Trojanowski JQ, Borson S, Galasko D, et al. (2001) Frequency of tau gene mutations in familial and sporadic cases of non-Alzheimer dementia. Arch Neurol 58:383–387.

Poorkaj P, Muma NA, Zhukareva V, Cochran EJ, Shannon KM, Hurtig H, Koller WC, Bird TD, Trojanowski JQ, Lee VM-Y, Schellenberg G D (2002) An R5L tau mutation in a subject with a progressive supranuclear palsy phenotype. Ann Neurol 52:511–516.

Reed LA, Schmidt ML, Wszolek ZK, Balin BJ, Soontornniyomkij V, Lee VM-Y, Trojanowski JQ, Schelper RL (1998) The neuropathology of a chromosome 17-linked autosomal dominant parkinsonismdementia ("pallido-ponto-nigral degeneration"). J Neuropathol Exp Neurol 57:588–601.

Rizzini C, Goedert M, Hodges JR, Smith MJ, Jakes R, Hills R, Xuereb JH, Crowther RA, Spillantini MG (2000) Tau gene mutation K257T causes a tauopathy similar to Pick's disease. J Neuropathol Exp Neurol 59:990–1001.

Rizzu P, Van Swieten JC, Joosse M, Hasegawa M, Stevens M, Tibben A, Niermeijer MF, Hillebrand M, Ravid R, Oostra BA, et al. (1999) High prevalence of mutations in the microtubule-associated protein tau in a population study of frontotemporal dementia in the Netherlands. Am J Hum Genet 64:414–421.

Rosso SM, Donker Kaat L, Baks T, Joosse M, de Koning I, Pijnenburg Y, de Jong D, Dooijes D, Kamphorst W, Ravid R, et al. (2003) Frontotemporal dementia in The Netherlands: patient characteristicsprevalence estimates from a population-based study. Brain 126:2016–2022.

Sahara N, Tomiyama T, Mori H (2000) Missense point mutations of tau to segregate with FTDP-17 exhibit site-specific effects on microtubule structure in COS cells: a novel action of R406W mutation. J Neurosci Res 60:380–387.

Shin RW, Iwaki T, Kitamoto T, Tateishi J (1991) Hydrated autoclave pretreatment enhances tau immunoreactivity in formalin-fixed normalAlzheimer's disease brain tissues. Lab Invest 64:693–702.

Si ZH, Rauch D, Stoltzfus CM (1998) The exon splicing silencer in human immunodeficiency virus type 1 Tat exon 3 is bipartite and acts early in spliceosome assembly. Mol Cell Biol 18:5404–5413.

Sperfeld AD, Collatz MB, Baier H, Palmbach M, Storch A, Schwarz J, Tatsch K, Reske S, Joosse M, Heutink P, Ludolph AC (1999) FTDP-17: an early-onset phenotype with parkinsonismepileptic seizures caused by a novel mutation. Ann Neurol 46:708–715.

Spillantini MG, Bird TD, Ghetti B (1998a) Frontotemporal dementiaParkinsonism linked to chromosome 17: a new group of tauopathies. Brain Pathol 8:387–402.

Spillantini MG, Goedert M, Crowther RA, Murrell JR, Farlow MR, Ghetti B (1997) Familial multiple system tauopathy with presenile dementia: a disease with abundant neuronalglial tau filaments. Proc Natl Acad Sci USA 94:4113–4118.

Spillantini MG, Murrell JR, Goedert M, Farlow MR, Klug A, Ghetti B (1998b) Mutation in the tau gene in familial multiple system tauopathy with presenile dementia. Proc Natl Acad Sci USA 95:7737–7741.

Spillantini MG, Van Swieten JC, Goedert M (2000a) Tau gene mutations in frontotemporal dementiaparkinsonism linked to chromosome 17 (FTDP-17). Neurogenetics 2:193–205.

Spillantini MG, Yoshida H, Rizzini C, Lantos PL, Khan N, Rossor MN, Goedert M, Brown J (2000b) A novel tau mutation (N296N) in familial dementia with swollen achromatic neurons and corticobasal inclusion bodies. Ann Neurol 48:939–943.

Stanford PM, Halliday GM, Brooks WS, Kwok JB, Storey CE, Creasey H, Morris JG, Fulham MJ, Schofield PR (2000) Progressive supranuclear palsy pathology caused by a novel silent mutation in exon 10 of the tau gene: expansion of the disease phenotype caused by tau gene mutations. Brain 123 (Pt 5), 880–893.

Tolnay M, Grazia Spillantini M, Rizzini C, Eccles D, Lowe J, Ellison D (2000) A new case of frontotemporal dementia and parkinsonism resulting from an intron 10 +3-splice site mutation in the tau gene: clinicalpathological features. Neuropathol Appl Neurobiol 26:368–378.

Varani L, Hasegawa M, Spillantini MG, Smith MJ, Murrell JR, Ghetti B, Klug A, Goedert M, Varani G (1999) Structure of tau exon 10 splicing regulatory element RNAdestabilization by mutations of frontotemporal dementiaparkinsonism linked to chromosome 17. Proc Natl Acad Sci USA 96:8229–8234.

Vogelsberg-Ragaglia V, Bruce J, Richter-Landsberg C, Zhang B, Hong M, Trojanowski JQ, Lee VM-Y (2000) Distinct FTDP-17 missense mutations in tau produce tau aggregatesother pathological phenotypes in transfected CHO cells. Mol Biol Cell 11:4093–4104.

Weingarten MD, Lockwood AH, Hwo SY, Kirschner MW (1975) A protein factor essential for microtubule assembly. Proc Natl Acad Sci USA 72:1858–1862.

Wszolek ZK, Tsuboi Y, Uitti RJ, Reed L (2000) Two brothers with frontotemporal dementia-parkinsonism with an N279K mutation of the tau gene. Neurology 55:1939.

Yasuda M, Takamatsu J, D'Souza I, Crowther RA, Kawamata T, Hasegawa M, Hasegawa H, Spillantini MG, Tanimukai S, Poorkaj P, et al. (2000) A novel mutation at position +12 in the intron following exon 10 of the tau gene in familial frontotemporal dementia (FTD-Kumamoto). Ann Neurol 47:422–429.

Yoshida H, Crowther RA, Goedert M (2002) Functional effects of tau gene mutations deltaN296N296H. J Neurochem 80:548–551.

Yoshida H, Ihara Y (1993) Tau in paired helical filaments is functionally distinct from fetal tau: assembly incompetence of paired helical filament-tau. J Neurochem 61:1183–1186.

Chapter 13

Animal Models of Tauopathy

Karen Duff[1], Pavan Krishnamurthy[1], Emmanuel Planel[1]
and Michael Hutton[2]

[1] *Department of Psychiatry*
Department of Physiology & Neuroscience
New York University School of Medicine
Email: duff@nki.rfmh.org

[2]*Mayo Clinic College of Medicine*
4500 San Pablo, Jacksonville FL 32224
Email: hutton.michael@mayo.edu

1.   Introduction

The tauopathies are a group of neurodegenerative diseases that are characterized by the accumulation of intraneuronal filamentous aggregates of the microtubule associated protein, tau. The tauopathies includes over twenty different conditions including Alzheimer's disease (AD), Progressive Supranuclear Palsy (PSP) and the familial tauopathy, Frontotemporal dementia and Parkinsonism linked to chromosome 17 (FTDP). Tau is a member of the family of Microtubule Associated Proteins (MAPs) and along with other MAPs, tau is normally involved in modulating microtubule assembly and maintaining microtubule stability within neurons (Matus, 1988). However, during the pathogenic cascade that leads to tauopathy, tau detaches from the microtubules, becomes hyperphosphorylated, and aggregates in filaments that eventually accumulate into the large somatodendritic neurofibrillary lesions that represent the histological hallmarks of this group of diseases. Despite much progress over the past ten years, the mechanism(s) by which this accumulation of abnormal tau species leads to neurodegeneration, and even the precise sequence of events that eventually results in the formation these neurofibrillary lesions is unclear. What we do know from human genetic studies is that tau dysfunction can be sufficient to cause neurodegeneration since mutations in the tau (*MAPT*) gene cause FTDP-17 and a common haplotype in *MAPT* is a robust genetic risk factor for the "sporadic" tauopathies, PSP and Cortical basal degeneration. As a result, considerable interest is currently focused on dissecting the pathogenic mechanism that leads to tau dysfunction, neurodegeneration and the clinical progression in the human tauopathies. To this end, the recent development of robust transgenic animal models that reproduce many of the neuropathological, biochemical and even clinical features of different human tauopathies represent an important advance that has already yielded sometimes surprising insights into this major group of human neurodegenerative diseases. Several recent reviews (Brandt et al., 2005; Lee et al., 2005b; Gotz et al., 2004) have described the salient features of the models, and the main purpose of this review is to identify how animal models have provided insight into the pathogenesis of tauopathy.

Transgenic models of tauopathy have been created in C. elegans (Kraemer et al., 2003) and Drosophila (Jackson et al., 2002b; Wittmann et al., 2001) as well as mice and rats. In addition to the tau transgenic models, models with varying degrees of tauopathy have been generated through overexpression of the cdk5 activator p25 (Cruz et al., 2003) or a cleaved form of ApoE (Brecht et al., 2004). Non transgenic models include animals subjected to hypothermia, either experimentally (Planel et al., 2004), or due to hibernation (Arendt et al., 2003).

2. Wild-type mice with neuronal pathology

The first attempt to generate a transgenic mouse model with tauopathy used a wild type tau transgene under the control of the human thy-1 promoter (Gotz et al., 1995). Expression of the transgene was low, and pathology was limited to somatodendritic localization and enhanced levels of phospho-tau in some neurons, without insoluble tau or NFT formation. A more pronounced phenotype was achieved using stronger promoters to drive wild type tau transgene expression (Ishihara et al., 1999; Probst et al., 2000; Spittaels et al., 1999). In these mice, motor neurons developed large numbers of enlarged axons with neurofilament- and tau-immunoreactive spheroids. This pathology is more reminiscent of amyotrophic lateral sclerosis (ALS) than tauopathy, and it likely results from high levels of tau expression in the motor neurons. Tau protein extracted from the brain and spinal cord of these transgenic mice becomes increasingly insoluble as the mice age, but NFTs were not observed except in mice at a very old age (Ishihara et al., 2001).

The generation of a transgenic mouse model overexpressing a cDNA of tau is complicated by the fact that there are six isoforms of human tau, and the investigator must choose only one to be overexpressed in the mouse. This limitation was overcome through the use of a genomic construct that contains all of the regulatory sequence necessary to allow alternative splicing, and the tau promoter allowing the correct temporal and spatial expression of all six isoforms of wild-type tau (Duff et al., 2000). The resulting mouse line had exceptionally high levels of human tau, but when the endogenous mouse tau was present, tau was distributed within the axons of cells and no abnormal tauopathy was observed. In contrast, when mouse tau was removed through cross breeding to a mouse with a dysfunctional tau gene, the resulting animal (known as the hTau line) developed progressive tauopathy that started with the human tau showing a normal (axonal) distribution. By three months of age, tau had been redistributed to the cell bodies in hippocampus and cortex, and neurons bearing pretangles with early phospho-epitopes were identified. Late phospho-epitopes and sarcosyl insoluble tau developed by 9 months. Tau filaments in the insoluble tau preparations that were visualized by electron microscopy, had a periodicity of 44.7 nm that is very similar to that of human AD paired helical filaments. By 15 months of age, thioflavin-S positive NFTs were observed and extensive cell loss was apparent in the cortex (Andorfer et al., 2005; Andorfer et al., 2003). Although there are differences between the hTau mouse and humans in terms of the ratio of 4R:3R tau isoforms, the tauopathy that develops is remarkably similar to that seen in human AD.

3. Mutant mice with neuronal pathology

Following the identification of mutations in the *MAPT* (tau) gene that cause FTDP-17 in 1998, the next generation of mice used mutant tau transgenes under a variety of promoters to achieve high level expression in neurons (Allen et al., 2002; Gotz et al., 2001a; Lewis et al., 2000; Tanemura et al., 2001; Tatebayashi et al., 2002) and glial cells (Gotz et al., 2001b; Gotz et al., 2001b; Higuchi et al., 2002). The first to be published (line JNPL3) expressed the shortest 4R tau isoform with the FTD-causing mutation P301L (Lewis et al., 2000). The mice developed hyperphosphorylated, sarcosyl-insoluble tau that formed twisted ribbons with a periodicity of 15–22 nm. Filamentous tau accumulated in cell bodies as mature NFTs that were positive for Gallyas silver stain and thioflavin S-fluorescent, mainly in spinal cord and brainstem. The mice underwent progressive degeneration of motor neurons leading to ataxia, dystonia and finally death by about 12 months of age. Recently, a mouse has been created that expresses mutant tau under the control of an inducible promoter (Santacruz et al., 2005). By switching the transgene off at certain points during pathology progression, the authors have been able to separate histochemically defined pathology development from neurodegeneration and cognitive decline.

4.   Mutant mice with glial pathology

Tauopathy in glial cells was achieved by expressing a tau construct with the G272V mutation under the control of a PrP-driven expression system that resulted in high expression in a subset of neurons and oligodendrocytes. Tau was hyperphosphorylated, and present as filaments. Thioflavin S-positive fibrillar inclusions were identified in oligodendrocytes and motor neurons in spinal cord (Gotz et al., 2001b). A second mouse over-expressed wild-type human tau in neurons and glial cells using the mouse Ta1 α-tubulin promoter. This led to glial pathology resembling the astrocytic plaques in CBD and the coiled bodies in CBD and PSP (Higuchi et al., 2002),

5.   Mechanisms of tauopathy: Insight from animal models

Several competing mechanisms have been proposed to attempt to explain how in AD and other tauopathies, insoluble tau and tangles form from normal, soluble tau. Most center on dysregulation of tau phosphorylation, either through increased activity of a protein kinase, or decreased activity of protein phosphatases, although other mechanisms have been implicated including glycosylation (Wang et al., 1996) and transglutamination (Norlund et al., 1999). Post-mortem studies of humans have shown that accumulation in cell bodies of tau phosphorylated at specific sites is an early event in neuronal degeneration as is conformational change recognized by the Alz50 or MC1 monoclonal antibodies (Trojanowski and Lee, 1994; Mandelkow et al., 1995; Iqbal and Grundke-Iqbal, 1991). It has been proposed that conformational change may allow tau to become a better substrate for kinases and/or worse substrate for phosphatases, but the relationship and timing of events is still not understood. (Hyman et al., 1988; Jicha et al., 1999a; Weaver et al., 2000c; Wolozin and Davies, 1987; Wolozin et al., 1986). Both events appear to occur before the formation of filamentous tau inclusions (Jicha et al., 1997; Vincent et al., 1998; Weaver et al., 2000b). Studies in brain tissue from early AD cases suggest that two phosphorylation events occur very early in the process of tangle formation: at serine 202, recognized by the AT8 (S202/T205), (Goedert et al., 1995) or CP13 monoclonal antibodies, and at threonine 231 recognized by the AT180 (Goedert et al., 1994) or TG3 monoclonal antibodies. Antibody PHF-1 recognizes tau phosphorylated at serine 396 and 404 (Otvos, Jr. et al., 1994), and it labels more mature, hyperphosphorylated forms of tau as seen in late stage tangles (Uboga and Price, 2000; Weaver et al., 2000a; Goedert et al., 1995; Goedert, 1993). Once filamentous tau aggregates have formed in neurons, as many as 21 different sites appear to be phosphorylated before the death of the neuron (Morishima and Ihara, 1994; Hasegawa et al., 1992). Many of these sites are phosphorylated to some extent in the normal brain (e.g. serines 202, 235 and 396/404, and threonine 231) (Matsuo et al., 1994a; Garver et al., 1994), while some are not (e.g. serines 214, 409 and perhaps 422) suggesting a stronger role in pathogenesis (Jicha et al., 1999b). In most tau transgenic mice, the principle determinant of the level of tau phosphorylation appears to be the amount of tau expressed as the amount of phosphorylation rises in parallel with the level of tau, and tau is not considered to be 'hyperphosphorylated'. As in humans, neurons with accumulated tau phosphorylated at epitopes such as serine 202 appear at an early stage, while late appearing epitopes such as S396/404, and conformation specific epitopes such as that recognized by MC1 only become abundant in cell bodies at later stages. Several phospho-epitopes are considered to be "abnormal" or "pathological", in the sense that they are usually disease-specific. Serine 422, and AT100 (T212 and S214), (Zheng-Fischhofer et al., 1998) are considered such pathological epitopes (Bussiere et al., 1999; Hasegawa et al., 1996; Matsuo et al., 1994b) and it is of note, that serine 422 has been suggested to be enhanced in tau transgenic mice exposed to high levels of Aβ (Gotz et al., 2001b).

Tau is the substrate for several kinases *in vivo* and *in vitro* and a large number of kinases have been reported to be associated with neurofibrillary tangles in the AD brain (Buee and Delacourte, 2001). In cultured neurons or in normal animal brain, the two major kinases involved in abnormal tau phosphorylation appear to be glycogen synthase kinase-3 (GSK3) and cyclin dependent kinase-5 (cdk5) (see Planel et al., 2002 and Maccioni et al., 2001 for comprehensive reviews). To examine the effect of elevated GSK-3β activity in the brain, two groups have created transgenic mice (Spittaels et al., 2000; Lucas et al., 2001). The latter study generated an inducible transgenic in which pretangles formed, but significant tauopathy did not develop. An interesting observation was made when a GSK-3β mouse was crossed to a mouse overexpressing human tau (Spittaels et al., 2000). Expression of the human tau transgene produced axonal abnormalities, including accumulation of tau and neurofilaments in the axon hillock region. However, these abnormalities were not found in the double transgenic mice. The authors suggested that accumulation of excess tau in the cell body of the single transgenic tau mice was due to inhibition of axonal transport, but that excess tau in the double transgenics was prevented from interfering with axonal transport following phosphorylation by GSK3. It has also been suggested that phosphorylation of "excess" tau protects against disruption of microtubule dynamics (Schneider et al., 1999; Stamer et al., 2002a) and the idea that abnormal phosphorylation may be a protective mechanism in general has also been proposed by Lee et al (Lee et al., 2005a). However, other studies have suggested the opposite. The double transgenic fly created by Jackson and colleagues (Jackson et al., 2002a) expressed human tau and *shaggy*, the fly homolog of GSK3. The double transgenic showed enhanced degeneration and NFT formation compared to the single transgenic tau fly. A very recent study has shown essentially the same results in a doubly transgenic mouse that co-expresses GSK-3β and a triple mutant tau transgene (line VLW) (Engel et al., 2005). Furthermore, pharmacological inhibition of GSK3 with two different drugs resulted in reduced insoluble tau, and axonal degeneration in spinal cord of JNPL3 mice (Noble et al., 2005). This suggested that hyperphosphorylated, insoluble tau is associated with neurodegeneration in the mouse, and that the formation, or stability of insoluble tau is affected by its phosphorylation state. Interestingly, in this model, NFT density did not appear to be decreased by GSK inhibitor treatment. These data support evidence from the invertebrate models (Jackson et al., 2002a; Wittmann et al., 2001), and the inducible mouse model (Santacruz et al., 2005) that there is a disconnect between the formation of mature NFTs and neurodegeneration.

The impact of GSK on tauopathy has also been implicated following the observation that presenilin-1 (PS1) can activate PI3K/Akt signaling leading to inactivation of GSK3, and that pathogenic mutations in PS1 inhibit the PS1-dependent PI3K/Akt activation. This then promotes GSK-3 activity and tau overphosphorylation at AD-related epitopes (Baki et al., 2004). Unfortunately, the implications of this for tauopathy development were not borne out in a double-transgenic mouse expressing mutant PS1 and human wild-type tau as tangle pathology was not seen even in very old mice (Boutajangout et al., 2002).

Overall, data from different model systems suggest that phosphorylation at some epitopes would prevent tau aggregation, while other epitopes would promote it. Recent data from tau pseudophosphorylation supports this concept (Necula and Kuret, 2004).

Neurodegeneration has been strongly associated with cdk5 activity through elevation of its activator, p25 (Patrick et al., 1999; Patrick et al., 2001). Several systems have been observed to be abnormal in AD such as calpains which cleave p35 to p25, that could promote increased activity of kinases such as cdk5 and GSK3 (Grynspan et al., 1997b; Grynspan et al., 1997a). There is one report of cdk5 transgenic mice, in which a mouse carrying three transgenes, human 4R tau, cdk5 and the p35 regulator of cdk5, was described (Van den Haute et al., 2001). This mouse had elevated cdk5 activity, some abnormalities in the distribution of neurofilaments, but little abnormal tau phosphorylation or pathology. There are two reports of mice transgenic for the p25 activator of cdk5. In both reports, there was some evidence for increased tau phosphorylation, but other than spheroids and axopathy, no additional tau pathology was seen although behavioral abnormalities were reported (Ahlijanian et al., 2000; Bian et al., 2002). A recent paper (Cruz et al., 2003) used an inducible p25 transgene to obtain high levels of expression postnatally. Interestingly, tangle pathology and neurodegeneration resulted, suggesting a significant role for p25/cdk5 in the induction of tau pathogenesis. Furthermore, co-expression of p25 in mice overexpressing mutant tau (JNPL3 line) showed enhanced insoluble tau levels and accelerated NFT formation (Noble et al., 2003) suggesting that enhanced cdk5 activity can also enhance ongoing pathology progression.

Overexpression of tau in transgenic animals leads to accumulation in the cell soma. Redistribution of tau from axonal to somatodendritic compartments is observed early in the pathogenic progression in human brains. In general, mice with higher levels of tau have a more severe phenotype, and animals with pathological mutations have accelerated pathology development compared to wild-type animals with equivalent levels of tau. In most transgenic mice, overexpression is a result of heterologous promoter activity. Unfortunately, targeted mouse models in which the mouse tau gene is mutated but expressed at normal endogenous levels have yet to be created so it is unknown whether the mutations will cause tauopathy in mice when present at physiological levels. It is likely that in the diseased brain, the normal mechanisms responsible for the clearance of tau break down, but it is not clear whether this is a cause, or effect of tauopathy. Little is known about the degradation of tau *in vivo*, but it has been shown that inhibition of proteasome activity by injection of the drug epoxomicin negates the clearance of tau, supporting the idea that tau is, in part, cleared by the proteasome (Oddo et al., 2004). In addition, new data implicates the co-chaperone heat-shock cognate (Hsc)70-interacting protein (CHIP), a ubiquitin protein ligase that can collaborate with molecular chaperones to selectively ubiquitinate denatured proteins thus facilitating protein folding and preventing protein aggregation. CHIP has been shown to regulate tau ubiquitination (Petrucelli et al., 2004) and it is upregulated by early NFT formation in humans and JNPL3 mice (Sahara et al., 2005). It remains to be determined, however, whether manipulation of CHIP or other chaperones has any effect on the levels of tau aggregation *in vivo* as data has been contradictory (Petrucelli et al., 2004; Sahara et al., 2005).

Some of the mutant tau mice have undergone behavioral assessment. The data suggest that tau aggregate formation, or the associated neuronal dysfunction and neurodegeneration, can significantly impair performance even in the absence of classic NFT formation (Arendash et al., 2004; Pennanen et al., 2004; Tanemura et al., 2002; Tatebayashi et al., 2002). A recent study has carefully examined when degeneration and cognitive decline occur relative to the appearance of abnormal tau forms including classically defined NFTs (Santacruz et al., 2005). Inducible mice were created in which a mutant (P301L) tau transgene could be switched off through the administration of doxycycline. In one line, the expression level was extremely high leading to robust NFT formation by approximately 6 months of age with massive cortico-limbic neuronal loss. Memory loss was present in all mice by 4 months of age as assessed by the Morris Water Maze test. By assessing neuropathological changes and cognitive performance before and after the transgene was switched off, the authors were able to investigate the reversibility of tau pathogenesis and the relationship between NFT formation, neuronal loss and memory function. Two things were apparent from this study: first, once tau has taken on a particular conformation and/or phosphorylation state, its aggregation is self-perpetuating, drawing in soluble tau species within the cell; and second, some form of tau (termed tau*) that precedes NFT formation correlates with cognitive decline and neurodegeneration, not the formation of mature NFT. This fits well with observations from the *Drosophila* models in which neurodegeneration occurred without NFT formation (Jackson et al., 2002b; Wittmann et al., 2001) and the hTau mice (Andorfer et al., 2005) in which neurons with severe morphological abnormalities do not always have the highest density of tau filaments. Indeed, several individual neurons displaying a dying nuclear morphology did not contain filaments, while many of those that had accumulated tau aggregates and filaments appeared to be intact in terms of nuclear morphology. Further studies are needed to assess the disconnect between filament formation, neurodegeneration and cognitive decline and whether neurotoxic tau intermediates such as tau fragments or oligomeric forms are involved, or whether degeneration is due to loss of normal tau function. Neurotoxic tau intermediates may be analogous to soluble oligomeric $A\beta$ that has been proposed to be a potent neurotoxic intermediate in the Alzheimer brain (Glabe, 2005). In general, more and more evidence suggests that NFTs and amyloid plaques are not the major damaging agents in the AD brain, although it is likely that mature lesions eventually contribute to cellular dysfunction (Tsai et al., 2005).

What exactly causes degeneration and cell death in the neurons of mice with tauopathy is not yet clear. Most of the data from tau transgenic lines suggests that apoptosis is not a primary mechanism for neuronal cell death, and this is in line with what is known from human tauopathies (Migheli et al., 1994; Atzori et al., 2001; Ferrer et al., 2001). Two mutant tau models with pathology primarily in the spinal cord failed to show apoptosis in tangle-bearing neurons using standard methods such as DNA fragmentation or immunostaining for cleaved α-fodrin or activated caspase-3 (Allen et al., 2002; Zehr et al., 2004). In hTau mice, features of both apoptosis and necrosis have been seen in cortical neurons where cell loss is quite significant (Andorfer et al., 2005). However, cell death in the hTau mouse model is not associated with induction of any caspases and incomplete, abnormal re-entry into the cell cycle has been proposed as a cell death mechanism (Andorfer et al., 2005). Whether the apparent discrepancies between the models indicate regional or severity differences remains to be seen.

Tau can be cleaved by proteases such as the executioner caspase, caspase 3, as well as other caspases, most likely at position 421 (Fasulo et al., 2000;

Gamblin et al., 2003). Studies from Carl Cotman's and Lester Binder's groups indicate that stimuli such as Aβ or oxidative stress can activate caspase 3 which can readily cleave tau at Asp 421, causing a conformation change detectable by the MC1 antibody. The resulting tau fragments (known as Δtau) can then act as an aggregation seed (Cotman et al., 2005). Recent studies have shown that a transgenic rat that overexpresses a Δtau construct has neurofibrillary degeneration that correlates with behavioral and cognitive deficits suggesting that these fragments can have a significant, detrimental effect *in vivo* (Hrnkova et al., 2005).

5.1   Interaction of Aβ and tau in mouse models

Alzheimer's disease is characterized not only by the development of tauopathy, but by the accumulation of amyloid into plaques. The main component of plaque amyloid is a heterogenous collection of peptides known as Aβ, which are derived from the proteolytic cleavage of the amyloid precursor protein, APP. AD causing mutations in APP and other genes including the presenilins increase the levels of Aβ42, and as a result, elevated Aβ42 peptide has been implicated as a major neurotoxin involved in the etiology of AD. The relevance of elevated Aβ in one of its forms (soluble monomers, oligomers or fibrils) and pathogenic tau to neurodegeneration has been debated extensively. A conclusive link between the two has only recently been shown *in vivo,* using mouse models of tauopathy exposed to elevated Aβ (Lewis et al., 2001; Gotz et al., 2001a; Oddo et al., 2004).

One of the possible links between Aβ elevation and tau pathogenesis may be through activation of neuronal signal transduction pathways, and much data has been generated to demonstrate this. Cell based studies have suggested a link between Aβ, cdk5 and tau (Town et al., 2002; Rank et al., 2002; Alvarez et al., 2001 and Otth et al., 2002) and cdk5-induced tau hyperphosphorylation has been shown in APP transgenic mouse models. Increased tau phosphorylation by cdk5 is sufficient to directly destabilize MTs and contribute to Aβ toxicity (Evans et al., 2000). In support of this, Faibushevich and colleagues (Li et al., 2003) have shown that taxol, a microtubule-stablizing agent, is able to prevent Aβ induced tau hyperphosphorylation in cortical neurons, and that administration of a taxol analog to mice inhibits Aβ induced cdk5 activation. Taxol did not inhibit cdk5 directly, but blocked Aβ-induced calpain activation, decreasing the cleavage of p35 to the cdk5 activator p25 and thus decreasing tau phosphorylation.

The accumulation of active GSK3β (phosphorylated at tyr 216) has been observed in dystrophic neurites in the brains of APP transgenic mice (Tomidokoro et al., 2001) and it colocalizes with neurons containing hyperphosphorylated tau in the brains of patients with AD (Pei et al., 1999) and in JNPL3 mice

(Ishizawa et al., 2003). In rabbits treated intracisternally with aggregated $A\beta$1-42, activation of caspase 3, tau phosphorylation and nuclear translocation of NFKβ and GSK-3β were observed (Ghribi et al., 2003). Lithium, a GSK3 inhibitor, reduced the levels of $A\beta$ in an APP mouse line (Phiel et al., 2003) and reduced insoluble tau levels (Perez et al., 2003) and axonal degeneration (Noble et al., 2005) in transgenic mice with tauopathy.

Both focal adhesion kinase (FAK) and fyn (a Src-family kinase) are possible kinase activators, the latter through tyrosine phosphorylation of the MAP kinases, GSK3 and cdk5. Both are associated with degenerating neurons (Shirazi and Wood, 1993) and dystrophic neurites around plaques (Grace and Busciglio, 2003) in AD brain, and APP transgenic mice (Tomidokoro et al., 2001). A number of groups suggest that elevation of $A\beta$ leads to an increase in tyrosine phosphorylation (Williamson et al., 2002a) via fyn and FAK (Tomidokoro et al., 2001) and upregulation of tau kinases such as GSK3 (Tomidokoro et al., 2001; Takashima et al., 1996). Willimason and colleagues (Williamson et al., 2002a) have blocked $A\beta$-induced tyrosine phosphorylation of tau by the addition of a Src family kinase inhibitor (PP2). These results also suggest that early signaling events involving fyn may be involved in neurodegenerative changes in response to $A\beta$ (Williamson et al., 2002b), an idea supported by Lambert et al. (Lambert et al., 1998) who have shown that brain slices from fyn-knockout mice are immune to oligomeric $A\beta$ toxicity.

In AD brain, activation of the three MAP kinases (JNK, p38 and ERK) has been demonstrated in neurons and dystrophic neurites (Hensley et al., 1999; Shoji et al., 2000; Perry et al., 1999; Zhu et al., 2001). Various studies have shown associations between APP, tau and the MAP kinase pathway: JNK binding-inhibitory protein binds to APP (Matsuda et al., 2001), JNK can phosphorylate APP (Standen et al., 2001) and presenilin activity has been shown to inhibit JNK (Kim et al., 2001). Savage et al. (Savage et al., 2002) have shown an upregulation of JNK and p38 activity in PS/APP transgenic mouse models of AD. In contrast to the *in vitro* models, this study detected a down-regulation of the ERK pathway in response to high levels of amyloid in AD animal models, results supported by the findings of Dineley et al. (Dineley et al., 2001). This is probably age/amyloid load related as young mice had increased ERK, and in AD brain, activated ERK is present during the initial stages of neurofibrillary degeneration in neurons in regions that are devoid of amyloid deposition (Pei et al., 2002).

5.2   Effect of elevated $A\beta$ on tauopathy

One of the primary questions in understanding Alzheimer's disease has been the relative contribution of amyloid plaques versus neurofibrillary tangles to degeneration and cognitive decline. The first transgenic models of AD used mutant APP transgenes to elevate the levels of APP, and thus Aβ peptides in the brain (Games et al., 1995; Hsiao et al., 1996). These mice make large amounts of Aβ1-40 and 42, and they form amyloid plaques at less than 12 months of age. Tauopathy was limited to enhanced staining with tau phosphorylation specific antibodies in the dystrophic neurites that cluster around plaques and neuropil threads or tangles have not been observed in any of the amyloid forming mouse models. Significantly, not only are neuropil threads and tangles absent in APP and APP/PS1 mice but there is also no evidence of overt neurodegeneration associated with amyloid plaque formation alone. Recently, a triple transgenic mouse overexpressing APP, PS1 and wild-type tau has been described (Boutajangout et al., 2004). Even with elevated levels of human tau, the mouse failed to develop more severe pathology in an environment of elevated Aβ. Thus it would appear that in mice, Aβ elevation and/or plaque formation is not sufficient to initiate tauopathy, or it does so too slowly to be seen in mice with wild-type tau.

Several published reports have shown that elevated Aβ does however have a pronounced effect on tauopathy progression in transgenic tau mice that have pre-existing tauopathy. Gotz and colleagues stereotaxically injected synthetic preparations of fibrillar Aβ42 into the somatosensory cortex and the CA1 region of P301L and wild-type human tau transgenic mice. Eighteen days following the injections, they observed a five-fold increase of NFT in the amygdala of P301L transgenic, but not wild-type tau transgenic or control mice (Gotz and Nitsch, 2001). To discount the non-specific effects of injection trauma on NFT formation, the investigators also injected the reverse peptide; NFT load was not affected suggesting that Aβ itself is a neurotoxic agent. NFT formation in the Aβ42-injected P301L mice was tightly correlated with phosphorylation of tau at epitope AT100 (T212/S214), and S422 but not AT8 (S202/T205) suggesting that these epitopes are particularly important in Aβ enhanced tangle formation. Progeny from a cross between the mutant APP transgenic line Tg2576 (Hsiao et al., 1996) and the FTDP-mutant tau line JNPL3 (Lewis et al., 2000) developed more pronounced tauopathy compared to singly transgenic JNPL3 littermates, and the pathology was more prominent in areas such as the olfactory bulb, the entorhinal cortex and the amygdala which lacked extensive pathology in the JNPL3 littermates (Lewis et al., 2001). In a similar experiment, Tg2576 was crossed with a different tau transgenic (line VLW) that expresses three FTDP causing mutations. At 9 months of age, the doubly transgenic mice, but not the tau-only VLW line, developed intracytoplasmic straight tau filaments composed of abnormally hyperphosphorylated tau in cortex and hippocampus, which became consistently present and numerous in older animals (Lim et al., 2001; Perez et al., 2005); In agreement with data from Gotz et al. (2001a), phosphorylation at two epitopes, serines 422 and 262 site was significantly enhanced in the double transgenic mice. At an age before NFTs had formed, sarkosyl insoluble tau aggregates from double transgenic mice contained tau filaments that were noticeably wider (about 10 nm in diameter) than in the singly transgenic VLW brains (about 2 nm in diameter), (Perez et al., 2005) suggesting an effect of Aβ on the process of tau aggregation. This supports data from tissue culture studies showing the importance of the serine 422 site in tau fibrillogenesis in response to Aβ (Ferrari et al., 2003).

The triple transgenic model of Oddo and colleagues overexpresses mutant human APP, PS1 and tau at comparable levels, and shows an early deficit in long-term synaptic plasticity that correlates with pathology progression. In these animals, intraneuronal Aβ accumulates, then forms amyloid plaques and NFTs in an age-dependent fashion (Oddo et al., 2003). Extracellular Aβ deposition precedes tau pathology by several months suggesting that elevated Aβ can enhance tauopathy development, but not vice-versa. Mice with just the mutant PS1 and tau transgenes generated in the same study do not develop NFT in the absence of APP overexpression and Aβ accumulation. Interestingly, injection of anti-Aβ antibodies into the brains of triple transgenic mice led to the rapid clearance of accumulated Aβ, and with it, the cell body accumulated tau. However, tau clearance only occurs in mice in which the tau is at an early stage in pathology progression (pretangles), before it becomes hyperphosphorylated, suggesting that only early tauopathy will be reversible. This is again consistent with the observations of Santacruz et al. (2005) that in the inducible tau mouse, NFTs and hyperphosphorylated insoluble tau were more stable than less pathogenic forms, and the observations of Noble and colleagues, that early stage, but not late stage tauopathy could be modulated by kinase inhibitor therapy (Noble et al., 2005).

Elevated Aβ only seems to enhance tauopathy in the mutant tau lines that have pre-existing tauopathy, not in lines with wild-type human tau, or endogenous tau. Thus Aβ seems to accelerate the development of tauopathy rather than cause it, at least in mice. For AD, this is an important point which differs from the traditional view that Aβ accumulation is the primary cause of neurodegeneration, with tauopathy being viewed largely as an epiphenomenon (Hardy and Selkoe, 2002). Indeed, the concept that Aβ would need some degree of tau malfunction to have its full toxic effects is supported by neuronal culture experiments which demonstrate that without tau, Aβ toxicity is greatly reduced (Rapoport et al., 2002). Clinical data also support this hypothesis because stage 1-2 NFTs usually begin to develop years before the first senile plaques (Braak and Braak, 1997). A recent study demonstrating that H1c tau haplotype carriers are more prone to AD also adds credibility to this hypothesis (Myers et al., 2005).

5.3 Effect of tauopathy on amyloidosis

Given the central role of tau in cellular transport and the maintenance of cellular integrity, it is feasible that tauopathy may initiate or enhance the amyloidogenic pathway. In support of this idea, Vitali (Vitali et al., 2004) have recently reported significant increases in the amount of soluble Aβ in the brains from patients with FTDP-17 in comparison to FTD lacking distinctive histopathology or normal controls. In terms of the mechanism involved, elevated levels of tau has been shown to interfere with the transport of several molecules including APP, out of the cell body (Stamer et al., 2002b) which could lead to aberrant trafficking of APP and increased processing into Aβ. Furthermore, mutant tau overexpression in the VLW line resulted in lysosomal abnormalities that included increased numbers of lysosomes with aberrant morphology (Lim et al., 2001) that mimic those found in human AD brains (Cataldo et al., 1996; Cataldo et al., 1994) and it is possible that a failure in the endosome-lysosomal machinery could enhance the imbalance between Aβ production and Aβ clearance. This suggestion is supported by data from doubly transgenic APP(Tg2576)-VLW mice. Neuropathological assessment of the double transgenic showed a significant increase in the amount of amyloid deposition in cortical and limbic areas compared to single transgenic APP mice, and in addition, significant neuronal loss in selective vulnerable areas like the entorhinal cortex and CA1 subfield of the hippocampus, increasing in an age-dependent fashion up to 36% at 16 months (Ribe et al., 2005). However, no enhancement of amyloid pathology was reported in a cross between APP(Tg2576) and a different tau line (JNPL3) (Lewis et al., 2001). The enhanced tauopathy in the VLW model compared to JNPL3 may account for the apparent discrepancy between the two sets of results.

6. Conclusion

Overall, animal models exist that replicate most, if not all of the pathological features of tauopathies. New models such as the inducible tau mouse will be especially useful to order events in pathogenesis, and mice that develop both amyloid plaques and tangles will be valuable tools for testing drugs.

References

Ahlijanian MK, Barrezueta NX, Williams RD, Jakowski A, Kowsz KP, McCarthy S, Coskran T, Carlo A, Seymour PA, Burkhardt JE, Nelson RB, McNeish JD (2000) Hyperphosphorylated tau and neurofilament and cytoskeletal disruptions in mice overexpressing human p25, an activator of cdk5. Proc Natl Acad Sci USA 97:2910–2915.

Allen B, Ingram E, Takao M, Smith MJ, Jakes R, Virdee K, Yoshida H, Holzer M, Craxton M, Emson PC, Atzori C, Migheli A, Crowther RA, Ghetti B, Spillantini MG, Goedert M (2002) Abundant tau filaments and nonapoptotic neurodegeneration in transgenic mice expressing human P301S tau protein. J Neurosci 22:9340–9351.

Alvarez A, Munoz JP, Maccioni RB (2001) A Cdk5-p35 stable complex is involved in the beta-amyloid-induced deregulation of Cdk5 activity in hippocampal neurons. Exp Cell Res 264:266–274.

Andorfer C, Acker CM, Kress Y, Hof PR, Duff K, Davies P (2005) Cell-cycle reentry and cell death in transgenic mice expressing nonmutant human tau isoforms. J Neurosci 25:5446–5454.

Andorfer C, Kress Y, Espinoza M, de Silva R, Tucker KL, Barde YA, Duff K, Davies P (2003) Hyperphosphorylation and aggregation of tau in mice expressing normal human tau isoforms. J Neurochem 86:582–590.

Arendash GW, Lewis J, Leighty RE, McGowan E, Cracchiolo JR, Hutton M, Garcia MF (2004) Multi-metric behavioral comparison of APPsw and P301L models for Alzheimer's disease: linkage of poorer cognitive performance to tau pathology in forebrain. Brain Res 1012:29–41.

Arendt T, Stieler J, Strijkstra AM, Hut RA, Rudiger J, Van der Zee EA, Harkany T, Holzer M, Hartig W (2003) Reversible paired helical filament-like phosphorylation of tau is an adaptive process associated with neuronal plasticity in hibernating animals. J Neurosci 23:6972–6981.

Atzori C, Ghetti B, Piva R, Srinivasan AN, Zolo P, Delisle MB, Mirra SS, Migheli A (2001) Activation of the JNK/p38 pathway occurs in diseases characterized by tau protein pathology and is related to tau phosphorylation but not to apoptosis. J Neuropathol Exp Neurol 60:1190–1197.

Baki L, Shioi J, Wen P, Shao Z, Schwarzman A, Gama-Sosa M, Neve R, Robakis NK (2004) PS1 activates PI3K thus inhibiting GSK-3 activity and tau overphosphorylation: effects of FAD mutations. EMBO J 23:2586–2596.

Bian F, Nath R, Sobocinski G, Booher RN, Lipinski WJ, Callahan MJ, Pack A, Wang KK, Walker LC (2002) Axonopathy, tau abnormalities, and dyskinesia, but no neurofibrillary tangles in p25-transgenic mice. J Comp Neurol 446:257–266.

Boutajangout A, Authelet M, Blanchard V, Touchet N, Tremp G, Pradier L, Brion JP (2004) Characterisation of cytoskeletal abnormalities in mice transgenic for wild-type human tau and familial Alzheimer's disease mutants of APP and presenilin-1. Neurobiol Dis 15:47–60.

Boutajangout A, Leroy K, Touchet N, Authelet M, Blanchard V, Tremp G, Pradier L, Brion JP (2002) Increased tau phosphorylation but absence of formation of neurofibrillary tangles in mice double transgenic for human tau and Alzheimer mutant (M146L) presenilin-1. Neurosci Lett 318:29–33.

Braak E, Braak H (1997) Alzheimer's disease: transiently developing dendritic changes in pyramidal cells of sector CA1 of the Ammon's horn. Acta Neuropathol (Berl) 93:323–325.

Brandt R, Hundelt M, Shahani N (2005) Tau alteration and neuronal degeneration in tauopathies: mechanisms and models. Biochim Biophys Acta 1739:331–354.

Brecht WJ, Harris FM, Chang S, Tesseur I, Yu GQ, Xu Q, Dee FJ, Wyss-Coray T, Buttini M, Mucke L, Mahley RW, Huang Y (2004) Neuron-specific apolipoprotein e4 proteolysis is associated with increased tau phosphorylation in brains of transgenic mice. J Neurosci 24:2527–2534.

Buee L, Delacourte A (2001) Tau phosphorylation. In: Functional Neurobiology of Aging (Hof PR, Mobbs CV, eds), pp 315–332. Academic Press.

Bussiere T, Hof PR, Mailliot C, Brown CD, Caillet-Boudin ML, Perl DP, Buee L, Delacourte A (1999) Phosphorylated serine422 on tau proteins is a pathological epitope found in several diseases with neurofibrillary degeneration. Acta Neuropathol (Berl) 97:221–230.

Cataldo AM, Hamilton DJ, Barnett JL, Paskevich PA, Nixon RA (1996) Properties of the endosomal-lysosomal system in the human central nervous system: disturbances mark most neurons in populations at risk to degenerate in Alzheimer's disease. J Neurosci 16:186–199.

Cataldo AM, Hamilton DJ, Nixon RA (1994) Lysosomal abnormalities in degenerating neurons link neuronal compromise to senile plaque development in Alzheimer disease. Brain Res 640:68–80.

Cotman CW, Poon WW, Rissman RA, Blurton-Jones M (2005) The role of caspase cleavage of tau in Alzheimer disease neuropathology. J Neuropathol Exp Neurol 64:104–112.

Cruz JC, Tseng HC, Goldman JA, Shih H, Tsai LH (2003) Aberrant Cdk5 activation by p25 triggers pathological events leading to neurodegeneration and neurofibrillary tangles. Neuron 40:471–483.

Dineley KT, Westerman M, Bui D, Bell K, Ashe KH, Sweatt JD (2001) Beta-amyloid activates the mitogen-activated protein kinase cascade via hippocampal alpha7 nicotinic acetylcholine receptors: In vitro and in vivo mechanisms related to Alzheimer's disease. J Neurosci 21:4125–4133.

Duff K, Knight H, Refolo LM, Sanders S, Yu X, Picciano M, Malester B, Hutton M, Adamson J, Goedert M, Burki K, Davies P (2000) Characterization of pathology in transgenic mice over-expressing human genomic and cDNA tau transgenes. Neurobiol Dis 7:87–98.

Engel T, Lucas JJ, Gomez-Ramos P, Moran MA, Avila J, Hernandez F (2005) Cooexpression of FTDP-17 tau and GSK-3beta in transgenic mice induce tau polymerization and neurodegeneration. Neurobiol Aging.

Evans DB, Rank KB, Bhattacharya K, Thomsen DR, Gurney ME, Sharma SK (2000) Tau Phosphorylation at Serine 396 and Serine 404 by Human Recombinant Tau Protein Kinase II Inhibits Tau's Ability to Promote Microtubule Assembly. J Biol Chem 275:24977–24983.

Fasulo L, Ugolini G, Visintin M, Bradbury A, Brancolini C, Verzillo V, Novak M, Cattaneo A (2000) The neuronal microtubule-associated protein tau is a substrate for caspase-3 and an effector of apoptosis. J Neurochem 75:624–633.

Ferrari A, Hoerndli F, Baechi T, Nitsch RM, Gotz J (2003) beta-Amyloid induces paired helical filament-like tau filaments in tissue culture. J Biol Chem 278:40162–40168.

Ferrer I, Blanco R, Carmona M, Ribera R, Goutan E, Puig B, Rey MJ, Cardozo A, Vinals F, Ribalta T (2001) Phosphorylated map kinase (ERK1, ERK2) expression is associated with early tau deposition in neurones and glial cells, but not with increased nuclear DNA vulnerability and cell death, in Alzheimer disease, Pick's disease, progressive supranuclear palsy and corticobasal degeneration. Brain Pathol 11:144–158.

Gamblin TC, Chen F, Zambrano A, Abraha A, Lagalwar S, Guillozet AL, Lu M, Fu Y, Garcia-Sierra F, LaPointe N, Miller R, Berry RW, Binder LI, Cryns VL (2003) Caspase cleavage of tau: linking amyloid and neurofibrillary tangles in Alzheimer's disease. Proc Natl Acad Sci USA 100:10032–10037.

Games D, Adams D, Alessandrini R, Barbour R, Berthelette P, Blackwell C, Carr T, Clemens J, Donaldson T, Gillespie F (1995) Alzheimer-type neuropathology in transgenic mice overexpressing V717F beta-amyloid precursor protein. Nature 373:523–527.

Garver TD, Harris KA, Lehman RA, Lee VM, Trojanowski JQ, Billingsley ML (1994) Tau phosphorylation in human, primate, and rat brain: evidence that a pool of tau is highly phosphorylated in vivo and is rapidly dephosphorylated in vitro. J Neurochem 63:2279–2287.

Ghribi O, Herman MM, Savory J (2003) Lithium inhibits Abeta-induced stress in endoplasmic reticulum of rabbit hippocampus but does not prevent oxidative damage and tau phosphorylation. J Neurosci Res 71:853–862.

Glabe CC (2005) Amyloid accumulation and pathogensis of Alzheimer's disease: significance of monomeric, oligomeric and fibrillar Abeta. Subcell Biochem 38:167–177.

Goedert M (1993) Tau protein and the neurofibrillary pathology of Alzheimer's disease. Trends Neurosci 16:460–465.

Goedert M, Jakes R, Crowther RA, Cohen P, Vanmechelen E, Vandermeeren M, Cras P (1994) Epitope mapping of monoclonal antibodies to the paired helical filaments of Alzheimer's disease: identification of phosphorylation sites in tau protein. Biochem J 301 (Pt 3):871–877.

Goedert M, Jakes R, Vanmechelen E (1995) Monoclonal antibody AT8 recognises tau protein phosphorylated at both serine 202 and threonine 205. Neurosci Lett 189:167–169.

Gotz J, Chen F, Barmettler R, Nitsch RM (2001a) Tau filament formation in transgenic mice expressing P301L tau. J Biol Chem 276:529–534.

Gotz J, Chen F, Van Dorpe J, Nitsch RM (2001b) Formation of neurofibrillary tangles in P301l tau transgenic mice induced by Abeta 42 fibrils. Science 293:1491–1495.

Gotz J, Nitsch RM (2001) Compartmentalized tau hyperphosphorylation and increased levels of kinases in transgenic mice. Neuroreport 12:2007–2016.

Gotz J, Probst A, Spillantini MG, Schafer T, Jakes R, Burki K, Goedert M (1995) Somatodendritic localization and hyperphosphorylation of tau protein in transgenic mice expressing the longest human brain tau isoform. EMBO J 14:1304–1313.

Gotz J, Streffer JR, David D, Schild A, Hoerndli F, Pennanen L, Kurosinski P, Chen F (2004) Transgenic animal models of Alzheimer's disease and related disorders: histopathology, behavior and therapy. Mol Psychiatry 9:664–683.

Grace EA, Busciglio J (2003) Aberrant activation of focal adhesion proteins mediates fibrillar amyloid beta-induced neuronal dystrophy. J Neurosci 23:493–502.

Grynspan F, Griffin WB, Mohan PS, Shea TB, Nixon RA (1997a) Calpains and calpastatin in SH-SY5Y neuroblastoma cells during retinoic acid-induced differentiation and neurite outgrowth: comparison with the human brain calpain system. J Neurosci Res 48:181–191.

Grynspan F, Griffin WR, Cataldo A, Katayama S, Nixon RA (1997b) Active site-directed antibodies identify calpain II as an early-appearing and pervasive component of neurofibrillary pathology in Alzheimer's disease. Brain Res 763:145–158.

Hardy J, Selkoe DJ (2002) The amyloid hypothesis of Alzheimer's disease: progress and problems on the road to therapeutics. Science 297:353–356.

Hasegawa M, Jakes R, Crowther RA, Lee VM, Ihara Y, Goedert M (1996) Characterization of mAb AP422, a novel phosphorylation-dependent monoclonal antibody against tau protein. FEBS Lett 384:25–30.

Hasegawa M, Morishima-Kawashima M, Takio K, Suzuki M, Titani K, Ihara Y (1992) Protein sequence and mass spectrometric analyses of tau in the Alzheimer's disease brain. J Biol Chem 267:17047–17054.

Hensley K, Floyd RA, Zheng NY, Nael R, Robinson KA, Nguyen X, Pye QN, Stewart CA, Geddes J, Markesbery WR, Patel E, Johnson GV, Bing G (1999) p38 kinase is activated in the Alzheimer's disease brain. J Neurochem 72:2053–2058.

Higuchi M, Ishihara T, Zhang B, Hong M, Andreadis A, Trojanowski J, Lee V (2002) Transgenic mouse model of tauopathies with glial pathology and nervous system degeneration. Neuron 35:433.

Hrnkova M, Zilka N, Filipcik P, Novak M (2005) Neurobiology of Aging; Vol. 25, Suppl. 2, pS233.

Hsiao K, Chapman P, Nilsen S, Eckman C, Harigaya Y, Younkin S, Yang F, Cole G (1996) Correlative memory deficits, Abeta elevation, and amyloid plaques in transgenic mice. Science 274:99–102.

Hyman BT, Van Hoesen GW, Wolozin BL, Davies P, Kromer LJ, Damasio AR (1988) Alz-50 antibody recognizes Alzheimer-related neuronal changes. Ann Neurol 23:371–379.

Iqbal K, Grundke-Iqbal I (1991) Ubiquitination and abnormal phosphorylation of paired helical filaments in Alzheimer's disease. Mol Neurobiol 5:399–410.

Ishihara T, Hong M, Zhang B, Nakagawa Y, Lee MK, Trojanowski JQ, Lee VM (1999) Age-dependent emergence and progression of a tauopathy in transgenic mice overexpressing the shortest human tau isoform. Neuron 24:751–762.

Ishihara T, Zhang B, Higuchi M, Yoshiyama Y, Trojanowski JQ, Lee VM (2001) Age-dependent induction of congophilic neurofibrillary tau inclusions in tau transgenic mice. Am J Pathol 158:555–562.

Ishizawa T, Sahara N, Ishiguro K, Kersh J, McGowan E, Lewis J, Hutton M, Dickson DW, Yen SH (2003) Co-localization of glycogen synthase kinase-3 with neurofibrillary tangles and granulovacuolar degeneration in transgenic mice. Am J Pathol 163:1057–1067.

Jackson GR, Wiedau-Pazos M, Sang TK, Wagle N, Brown CA, Massachi S, Geschwind DH (2002a) Human wild-type tau interacts with wingless pathway components and produces neurofibrillary pathology in Drosophila. Neuron 34:509–519.

Jackson RG, Wiedau-Pazos M, Sang TK, Wagle N, Brown CA, Massachi S, Geschwind D (2002b) Human wild-type tau interacts with wingless pathway components and produces neurofibrillary pathology in Drosophila. Neuron 34:509–519.

Jicha GA, Berenfeld B, Davies P (1999a) Sequence requirements for formation of conformational variants of tau similar to those found in Alzheimer's disease. J Neurosci Res 55:713–723.

Jicha GA, Lane E, Vincent I, Otvos L, Jr., Hoffmann R, Davies P (1997) A conformation- and phosphorylation-dependent antibody recognizing the paired helical filaments of Alzheimer's disease. J Neurochem 69:2087–2095.

Jicha GA, Weaver C, Lane E, Vianna C, Kress Y, Rockwood J, Davies P (1999b) cAMP-dependent protein kinase phosphorylations on tau in Alzheimer's disease. J Neurosci 19:7486–7494.

Kim JW, Chang TS, Lee JE, Huh SH, Yeon SW, Yang WS, Joe CO, Mook-Jung I, Tanzi RE, Kim TW, Choi EJ (2001) Negative regulation of the SAPK/JNK signaling pathway by presenilin 1. J Cell Biol 153:457–463.

Kraemer BC, Zhang B, Leverenz JB, Thomas JH, Trojanowski JQ, Schellenberg GD (2003) Neurodegeneration and defective neurotransmission in a Caenorhabditis elegans model of tauopathy. Proc Natl Acad Sci USA 100:9980–9985.

Lambert MP, Barlow AK, Chromy BA, Edwards C, Freed R, Liosatos M, Morgan TE, Rozovsky I, Trommer B, Viola KL, Wals P, Zhang C, Finch CE, Krafft GA, Klein WL (1998) Diffusible, nonfibrillar ligands derived from Abeta1-42 are potent central nervous system neurotoxins. Proc Natl Acad Sci USA 95:6448–6453.

Lee HG, Perry G, Moreira PI, Garrett MR, Liu Q, Zhu X, Takeda A, Nunomura A, Smith MA (2005a) Tau phosphorylation in Alzheimer's disease: pathogen or protector? Trends Mol Med 11:164–169.

Lee VM, Kenyon TK, Trojanowski JQ (2005b) Transgenic animal models of tauopathies. Biochim Biophys Acta 1739:251–259.

Lewis J, Dickson DW, Lin WL, Chisholm L, Corral A, Jones G, Yen SH, Sahara N, Skipper L, Yager D, Eckman C, Hardy J, Hutton M, McGowan E (2001) Enhanced neurofibrillary degeneration in transgenic mice expressing mutant tau and APP. Science 293:1487–1491.

Lewis J, McGowan E, Rockwood J, Melrose H, Nacharaju P, Van Slegtenhorst M, Gwinn-Hardy K, Paul MM, Baker M, Yu X, Duff K, Hardy J, Corral A, Lin WL, Yen SH, Dickson DW, Davies P, Hutton M (2000) Neurofibrillary tangles, amyotrophy and progressive motor disturbance in mice expressing mutant (P301L) tau protein. Nat Genet 25:402–405.

Li G, Faibushevich A, Turunen BJ, Yoon SO, Georg G, Michaelis ML, Dobrowsky RT (2003) Stabilization of the cyclin-dependent kinase 5 activator, p35, by paclitaxel decreases beta-amyloid toxicity in cortical neurons. J Neurochem 84:347–362.

Lim F, Hernandez F, Lucas JJ, Gomez-Ramos P, Moran MA, Avila J (2001) FTDP-17 mutations in tau transgenic mice provoke lysosomal abnormalities and Tau filaments in forebrain. Mol Cell Neurosci 18:702–714.

Lucas JJ, Hernandez F, Gomez-Ramon P, Moran MA, Hen R, Avila J (2001) Decreased nuclear beta-catenin, tau hyperphosphorylation and neurodegeneration in GSK-3beta conditional transgenic mice. EMBO J 20:27–39.

Maccioni RB, Otth C, Concha II, Munoz JP (2001) The protein kinase Cdk5. Structural aspects, roles in neurogenesis and involvement in Alzheimer's pathology. Eur J Biochem 268:1518–1527.

Mandelkow EM, Biernat J, Drewes G, Gustke N, Trinczek B, Mandelkow E (1995) Tau domains, phosphorylation, and interactions with microtubules. Neurobiol Aging 16:355–362.

Matsuda S, Yasukawa T, Homma Y, Ito Y, Niikura T, Hiraki T, Hirai S, Ohno S, Kita Y, Kawasumi M, Kouyama K, Yamamoto T, Kyriakis JM, Nishimoto I (2001) c-Jun N-terminal kinase (JNK)-interacting protein-1b/islet-brain-1 scaffolds Alzheimer's amyloid precursor protein with JNK. J Neurosci 21:6597–6607.

Matsuo ES, Shin RW, Billingsley ML, Van deVoorde A, O'Connor M, Trojanowski JQ, Lee VM (1994b) Biopsy-derived adult human brain tau is phosphorylated at many of the same sites as Alzheimer's disease paired helical filament tau. Neuron 13:989–1002.

Matsuo ES, Shin RW, Billingsley ML, Van deVoorde A, O'Connor M, Trojanowski JQ, Lee VM (1994a) Biopsy-derived adult human brain tau is phosphorylated at many of the same sites as Alzheimer's disease paired helical filament tau. Neuron 13:989–1002.

Matus A (1988) Microtubule-associated proteins: their potential role in determining neuronal morphology. Annu Rev Neurosci 11:29–44.

Migheli A, Cavalla P, Marino S, Schiffer D (1994) A study of apoptosis in normal and pathologic nervous tissue after in situ end-labeling of DNA strand breaks. J Neuropathol Exp Neurol 53:606–616.

Morishima M, Ihara Y (1994) Posttranslational modifications of tau in paired helical filaments. Dementia 5:282–288.

Myers AJ, Kaleem M, Marlowe L, Pittman AM, Lees A, Fung HC, Duckworth J, Leung D, Gibson A, Morris CM, de Silva R, Hardy J (2005) The H1c Haplotype at the MAPT Locus is associated with Alzheimer's Disease. Hum Mol Genet.

Necula M, Kuret J (2004) Pseudophosphorylation and glycation of tau protein enhance but do not trigger fibrillization in vitro. J Biol Chem 279:49694–49703.

Noble W, Olm V, Takata K, Casey E, Mary O, Meyerson J, Gaynor K, LaFrancois J, Wang L, Kondo T, Davies P, Burns M, Veeranna, Nixon R, Dickson D, Matsuoka Y, Ahlijanian M, Lau LF, Duff K (2003) Cdk5 is a key factor in tau aggregation and tangle formation in vivo. Neuron 38:555–565.

Noble W, Planel E, Zehr C, Olm V, Meyerson J, Suleman F, Gaynor K, Wang L, LaFrancois J, Feinstein B, Burns M, Krishnamurthy P, Wen Y, Bhat R, Lewis J, Dickson D, Duff K (2005) Inhibition of glycogen synthase kinase-3 by lithium correlates with reduced tauopathy and degeneration in vivo. Proc Natl Acad Sci USA 102:6990–6995.

Norlund MA, Lee JM, Zainelli GM, Muma NA (1999) Elevated transglutaminase-induced bonds in PHF tau in Alzheimer's disease. Brain Res 851:154–163.

Oddo S, Billings L, Kesslak JP, Cribbs DH, LaFerla FM (2004) Abeta immunotherapy leads to clearance of early, but not late, hyperphosphorylated tau aggregates via the proteasome. Neuron 43:321–332.

Oddo S, Caccamo A, Shepherd JD, Murphy MP, Golde TE, Kayed R, Metherate R, Mattson MP, Akbari Y, LaFerla FM (2003) Triple-transgenic model of Alzheimer's disease with plaques and tangles: intracellular Abeta and synaptic dysfunction. Neuron 39:409–421.

Otth C, Concha II, Arendt T, Stieler J, Schliebs R, Gonzalez-Billault C, Maccioni RB (2002) AbetaPP induces cdk5-dependent tau hyperphosphorylation in transgenic mice Tg2576. J Alzheimers Dis 4:417–430.

Otvos L, Jr., Feiner L, Lang E, Szendrei GI, Goedert M, Lee VM (1994) Monoclonal antibody PHF-1 recognizes tau protein phosphorylated at serine residues 396 and 404. J Neurosci Res 39:669–673.

Patrick GN, Zukerberg L, Nikolic M, de la MS, Dikkes P, Tsai LH (2001) reply: Neurobiolo-gyp25 protein in neurodegeneration. Nature 411:764–765.

Patrick GN, Zukerberg L, Nikolic M, de la MS, Dikkes P, Tsai LH (1999) Conversion of p35 to p25 deregulates Cdk5 activity and promotes neurodegeneration [see comments]. Nature 402:615–622.

Pei JJ, Braak E, Braak H, Grundke-Iqbal I, Iqbal K, Winblad B, Cowburn RF (1999) Distribution of active glycogen synthase kinase 3beta (GSK-3beta) in brains staged for Alzheimer disease neurofibrillary changes. J Neuropathol Exp Neurol 58:1010–1019.

Pei JJ, Braak H, An WL, Winblad B, Cowburn RF, Iqbal K, Grundke-Iqbal I (2002) Up-regulation of mitogen-activated protein kinases ERK1/2 and MEK1/2 is associated with the progression of neurofibrillary degeneration in Alzheimer's disease. Brain Res Mol Brain Res 109:45–55.

Pennanen L, Welzl H, D'Adamo P, Nitsch RM, Gotz J (2004) Accelerated extinction of conditioned taste aversion in P301L tau transgenic mice. Neurobiol Dis 15:500–509.

Perez M, Hernandez F, Lim F, Diaz-Nido J, Avila J (2003) Chronic lithium treatment decreases mutant tau protein aggregation in a transgenic mouse model. J Alzheimers Dis 5:301–308.

Perez M, Ribe E, Rubio A, Lim F, Moran MA, Ramos PG, Ferrer I, Isla MT, Avila J (2005) Characterization of a double (amyloid precursor protein-tau) transgenic: tau phosphorylation and aggregation. Neuroscience 130:339–347.

Perry G, Roder H, Nunomura A, Takeda A, Friedlich AL, Zhu X, Raina AK, Holbrook N, Siedlak SL, Harris PL, Smith MA (1999) Activation of neuronal extracellular receptor kinase (ERK) in Alzheimer disease links oxidative stress to abnormal phosphorylation. Neuroreport 10:2411–2415.

Petrucelli L, Dickson D, Kehoe K, Taylor J, Snyder H, Grover A, De LM, McGowan E, Lewis J, Prihar G, Kim J, Dillmann WH, Browne SE, Hall A, Voellmy R, Tsuboi Y, Dawson TM, Wolozin B, Hardy J, Hutton M (2004) CHIP and Hsp70 regulate tau ubiquitination, degradation and aggregation. Hum Mol Genet 13:703–714.

Phiel CJ, Wilson CA, Lee VM, Klein PS (2003) GSK-3alpha regulates production of Alzheimer's disease amyloid-beta peptides. Nature 423:435–439.

Planel E, Miyasaka T, Launey T, Chui DH, Tanemura K, Sato S, Murayama O, Ishiguro K, Tatebayashi Y, Takashima A (2004) Alterations in glucose metabolism induce hypothermia leading to tau hyperphosphorylation through differential inhibition of kinase and phosphatase activities: implications for Alzheimer's disease. J Neurosci 24:2401–2411.

Planel E, Sun X, Takashima A (2002) Role of GSK-3 beta in Alzheimer's disease pathology. Drug Development Research 56:491–510.

Probst A, Gotz J, Wiederhold KH, Tolnay M, Mistl C, Jaton AL, Hong M, Ishihara T, Lee VM, Trojanowski JQ, Jakes R, Crowther RA, Spillantini MG, Burki K, Goedert M (2000) Axonopathy and amyotrophy in mice transgenic for human four-repeat tau protein. Acta Neuropathol (Berl) 99:469–481.

Rank KB, Pauley AM, Bhattacharya K, Wang Z, Evans DB, Fleck TJ, Johnston JA, Sharma SK (2002) Direct interaction of soluble human recombinant tau protein with Abeta 1-42 results in tau aggregation and hyperphosphorylation by tau protein kinase II. FEBS Lett 514:263–268.

Rapoport M, Dawson HN, Binder LI, Vitek MP, Ferreira A (2002) Tau is essential to beta - amyloid-induced neurotoxicity. Proc Natl Acad Sci USA.

Ribe EM, Perez M, Puig B, Gich I, Lim F, Cuadrado M, Sesma T, Catena S, Sanchez B, Nieto M, Gomez-Ramos P, Moran MA, Cabodevilla F, Samaranch L, Ortiz L, Perez A, Ferrer I, Avila J, Gomez-Isla T (2005) Accelerated amyloid deposition, neurofibrillary degeneration and neuronal loss in double mutant APP/tau transgenic mice. Neurobiol Dis.

Sahara N, Murayama M, Mizoroki T, Urushitani M, Imai Y, Takahashi R, Murata S, Tanaka K, Takashima A (2005) In vivo evidence of CHIP up-regulation attenuating tau aggregation. J Neurochem 94:1254–1263.

Santacruz K, Lewis J, Spires T, Paulson J, Kotilinek L, Ingelsson M, Guimaraes A, Deture M, Ramsden M, McGowan E, Forster C, Yue M, Orne J, Janus C, Mariash A, Kuskowski M, Hyman B, Hutton M, Ashe KH (2005) Tau suppression in a neurodegenerative mouse model improves memory function. Science 309:476–481.

Savage MJ, Lin YG, Ciallella JR, Flood DG, Scott RW (2002) Activation of c-Jun N-Terminal Kinase and p38 in an Alzheimer's Disease Model Is Associated with Amyloid Deposition. J Neurosci 22:3376–3385.

Schneider A, Biernat J, von Bergen M, Mandelkow E, Mandelkow EM (1999) Phosphorylation that detaches tau protein from microtubules (Ser262, Ser214) also protects it against aggregation into Alzheimer paired helical filaments. Biochemistry 38:3549–3558.

Shirazi SK, Wood JG (1993) The protein tyrosine kinase, fyn, in Alzheimer's disease pathology. Neuroreport 4:435–437.

Shoji M, Iwakami N, Takeuchi S, Waragai M, Suzuki M, Kanazawa I, Lippa CF, Ono S, Okazawa H (2000) JNK activation is associated with intracellular beta-amyloid accumulation. Brain Res Mol Brain Res 85:221–233.

Spittaels K, Van den Haute C, Van Dorpe J, Geerts H, Mercken M, Bruynseels K, Lasrado R, Vandezande K, Laenen I, Boon T, van Lint J, Vandenheede J, Moechars D, Loos R, Van Leuven F (2000) Glycogen synthase kinase-3beta phosphorylates protein tau and rescues the axonopathy in the central nervous system of human four-repeat tau transgenic mice. J Biol Chem 275:41340–41349.

Spittaels K, Van den HC, Van Dorpe J, Bruynseels K, Vandezande K, Laenen I, Geerts H, Mercken M, Sciot R, Van Lommel A, Loos R, Van Leuven F (1999) Prominent axonopathy in the brain and spinal cord of transgenic mice overexpressing four-repeat human tau protein. Am J Pathol 155:2153–2165.

Stamer K, Vogel R, Thies E, Mandelkow E, Mandelkow EM (2002) Tau blocks traffic of organelles, neurofilaments, and APP vesicles in neurons and enhances oxidative stress. J Cell Biol 156:1051–1063.

Standen CL, Brownlees J, Grierson AJ, Kesavapany S, Lau KF, McLoughlin DM, Miller CC (2001) Phosphorylation of thr(668) in the cytoplasmic domain of the Alzheimer's disease amyloid precursor protein by stress-activated protein kinase 1b (Jun N-terminal kinase-3). J Neurochem 76:316–320.

Takashima A, Noguchi K, Michel G, Mercken M, Hoshi M, Ishiguro K, Imahori K (1996) Exposure of rat hippocampal neurons to amyloid beta peptide (25-35) induces the inactivation of phosphatidyl inositol-3 kinase and the activation of tau protein kinase I/glycogen synthase kinase-3 beta. Neurosci Lett 203:33–36.

Tanemura K, Akagi T, Murayama M, Kikuchi N, Murayama O, Hashikawa T, Yoshiike Y, Park JM, Matsuda K, Nakao S, Sun X, Sato S, Yamaguchi H, Takashima A (2001) Formation of filamentous tau aggregations in transgenic mice expressing V337M human tau. Neurobiol Dis 8:1036–1045.

Tanemura K, Murayama M, Akagi T, Hashikawa T, Tominaga T, Ichikawa M, Yamaguchi H, Takashima A (2002) Neurodegeneration with tau accumulation in a transgenic mouse expressing V337M human tau. J Neurosci 22:133–141.

Tatebayashi Y, Miyasaka T, Chui DH, Akagi T, Mishima K, Iwasaki K, Fujiwara M, Tanemura K, Murayama M, Ishiguro K, Planel E, Sato S, Hashikawa T, Takashima A (2002) Tau filament formation and associative memory deficit in aged mice expressing mutant (R406W) human tau. Proc Natl Acad Sci USA 99:13896–13901.

Tomidokoro Y, Ishiguro K, Harigaya Y, Matsubara E, Ikeda M, Park J, Yasutake K, Kawarabayashi T, Okamoto K, Shoji M (2001) Abeta amyloidosis induces the initial stage of tau accumulation in APP(Sw) mice. Neurosci Lett 299:169–172.

Town T, Zolton J, Shaffner R, Schnell B, Crescentini R, Wu Y, Zeng J, DelleDonne A, Obregon D, Tan J, Mullan M (2002) p35/Cdk5 pathway mediates soluble amyloid-beta peptide-induced tau phosphorylation in vitro. J Neurosci Res 69:362–372.

Trojanowski JQ, Lee VM (1994) Paired helical filament tau in Alzheimer's disease. The kinase connection. Am J Pathol 144:449–453.

Tsai JW, Chen Y, Kriegstein AR, Vallee RB (2005) LIS1 RNA interference blocks neural stem cell division, morphogenesis, and motility at multiple stages. J Cell Biol 170:935–945.

Uboga NV, Price JL (2000) Formation of diffuse and fibrillar tangles in aging and early Alzheimer's disease. Neurobiol Aging 21:1–10.

Van den Haute C, Spittaels K, Van Dorpe J, Lasrado R, Vandezande K, Laenen II, Geerts H, Van Leuven F (2001) Coexpression of Human cdk5 and Its Activator p35 with Human Protein Tau in Neurons in Brain of Triple Transgenic Mice. Neurobiol Dis 8:32–44.

Vincent I, Zheng JH, Dickson DW, Kress Y, Davies P (1998) Mitotic phosphoepitopes precede paired helical filaments in Alzheimer's disease. Neurobiol Aging 19:287–296.

Vitali A, Piccini A, Borghi R, Fornaro P, Siedlak SL, Smith MA, Gambetti P, Ghetti B, Tabaton M (2004) Soluble amyloid beta-protein is increased in frontotemporal dementia with tau gene mutations. J Alzheimers Dis 6:45–51.

Wang JZ, Grundke-Iqbal I, Iqbal K (1996) Glycosylation of microtubule-associated protein tau: an abnormal posttranslational modification in Alzheimer's disease. Nat Med 2:871–875.

Weaver CL, Espinoza M, Kress Y, Davies P (2000) Conformational change as one of the earliest alterations of tau in Alzheimer's disease. Neurobiol Aging 21:719–727.

Williamson R, Scales T, Clark BR, Gibb G, Reynolds CH, Kellie S, Bird IN, Varndell IM, Sheppard PW, Everall I, Anderton BH (2002a) Rapid tyrosine phosphorylation of neuronal proteins including tau and focal adhesion kinase in response to amyloid-beta peptide exposure: involvement of Src family protein kinases. J Neurosci 22:10–20.

Wittmann CW, Wszolek MF, Shulman JM, Salvaterra PM, Lewis J, Hutton M, Feany MB (2001) Tauopathy in Drosophila: neurodegeneration without neurofibrillary tangles. Science 293:711–714.

Wolozin B, Davies P (1987) Alzheimer-related neuronal protein A68: specificity and distribution. Ann Neurol 22:521–526.

Wolozin BL, Pruchnicki A, Dickson DW, Davies P (1986) A neuronal antigen in the brains of Alzheimer patients. Science 232:648–650.

Zehr C, Lewis J, McGowan E, Crook J, Lin WL, Godwin K, Knight J, Dickson DW, Hutton M (2004) Apoptosis in oligodendrocytes is associated with axonal degeneration in P301L tau mice. Neurobiol Dis 15:553–562.

Zheng-Fischhofer Q, Biernat J, Mandelkow EM, Illenberger S, Godemann R, Mandelkow E (1998) Sequential phosphorylation of Tau by glycogen synthase kinase-3beta and protein kinase A at Thr212 and Ser214 generates the Alzheimer-specific epitope of antibody AT100 and requires a paired-helical-filament-like conformation. Eur J Biochem 252:542–552.

Zhu X, Castellani RJ, Takeda A, Nunomura A, Atwood CS, Perry G, Smith MA (2001) Differential activation of neuronal ERK, JNK/SAPK and p38 in Alzheimer disease: the 'two hit' hypothesis. Mech Ageing Dev 123:39–46.

Chapter 14

Tau and Axonal Transport[*]

E.-M. Mandelkow, E. Thies and E. Mandelkow

Max-Planck-Unit for Structural Molecular Biology
22607 Hamburg, Germany

1. Properties and functions of tau protein

One of the characteristic features of AD is the loss of synapses at an early stage in the entorhinal region and hippocampus which corresponds to a loss of memory (Flood & Coleman, 1990; Terry et al., 1991). The triggers for these events are not understood, but factors such as inflammatory cytokines, oxidative stress, loss of growth factors, or the toxic effects of the Aβ peptide may be involved (Raff et al., 2002; Selkoe, 2002). The loss of synapses precedes the more conspicuous abnormal changes in the form of protein aggregates which are seen as senile plaques, neurofibrillary tangles, and others.

1.1 Traffic system of the cell

One clue for the early vulnerability of synapses comes from the highly asymmetric shape of neurons. Synapses and nerve terminals are located far from the cell body where most of the necessary components are synthesized, and therefore neurons require an efficient transport system. The distribution of material could conceivably take place by means of diffusion which might suffice for small local volumes and small molecules (e.g. neurotransmitters in a synaptic cleft), but this would lack efficiency and directionality over extended distances. To meet these criteria, cells have devised traffic systems in the form of cytoskeletal fibers which guide the transport of motor proteins and their cargoes (Hollenbeck & Saxton, 2005). Two distinct fiber systems for transport have evolved, the actin microfilaments and the microtubules. They interact with

---
[*] **Correspondence to:** Dr. Eva-Maria Mandelkow, Max-Planck-Unit for Structural Molecular Biology, Notkestrasse 85, 22607 Hamburg, Germany, Email: mandelkow@mpasmb.desy.de

numerous types of motor proteins which fall into three classes, the myosins (for the microfilament tracks) and the kinesins and dyneins (for microtubule tracks) (Hirokawa & Takemura, 2005). The two traffic systems often coexist in the same cell compartments. In mammalian cells, microtubules are mostly responsible for the long-haul traffic over large distances while microfilaments distribute material more locally. In both cases, the directionality of movement is determined by the intrinsic polarity of the tracks, combined with an in-built directionality of the motors (forward or reverse). Thus the "plus" end of microtubules usually points towards the periphery, and therefore "plus-end directed motors" such as most kinesins can support anterograde transport towards the nerve terminals, whereas minus-end directed motors such as dynein achieve retrograde movements back to the cell body. In both cases the fuel for the motors is ATP which is hydrolyzed to ADP and inorganic phosphate during the power strokes. Just as in a real railroad system, the traffic system of neurons requires not only tracks and motors, but also ties that keep the tracks stable. This role is fulfilled by ancillary proteins which in the case of microtubules are termed MAPs (for microtubule-associated proteins; review Cassimeris & Spittle, 2001). MAPs vary between cell types and cell compartments; in the case of neurons the most important ones are MAP2 (mostly dendritic), tau and MAP1b (mostly axonal). The association of MAPs with the microtubule tracks is regulated by phosphorylation which involves a variety of protein kinases and phosphatases (Stoothoff & Johnson, 2005). Furthermore, and unlike a railroad system, the microtubule cytoskeleton must be able to selfassemble and reorganize itself, and therefore the regulated disassembly is as important as assembly. The ability of microtubules to grow and shrink is in part an in-built property of the tubulin molecule and regulated by GTP turnover. However, in cells it is additionally controlled by microtubule stabilizers such as tau whose dissociation can induce microtubule breakdown, and by destabilizers such as katanin or kinesin-13 (MCAK) to ensure microtubule dynamics (Biernat et al., 2002; Baas & Qiang, 2005). This interplay is particularly important for neurite outgrowth and growth cone advance.

1.2   Tau in Alzheimer's disease

In the context of Alzheimer's disease, the cytoskeletal protein that has received most attention is the microtubule-associated protein tau. This is due to its conspicuous anomalous aggregation in the form of Alzheimer neurofibrillary tangles and neuropil threads which are made up of filaments of polymerized tau protein (paired helical filaments, PHFs, and a minority of straight filaments, Crowther & Goedert, 2000). In addition, this anomalous tau is highly phosphorylated, it is redistributed into the "wrong" compartment (from axonal to somatodendritic), and it is in part truncated by caspases and other proteases. This tau is detached from microtubules and has lost its ability to stabilize microtubules or promote microtubule growth. It is assumed that this change in properties leads to two major consequences, the destabilization of transport tracks and the pathological aggregation of tau in the cytosol. It is intuitively clear that both events have the ability to disrupt intracellular traffic. Another important property of tau in AD is that the aggregation has a well-defined pattern of spreading in the brain, starting from neurons in the transentorhinal region, to the hippocampus, and finally throughout the cortex. Thus the brain regions showing tau aggregation reflect the progression of clinical symptoms, from mild cognitive impairment to severe dementia (Braak stages, Braak & Braak, 1991).

1.3 Properties of tau

Tau is encoded by a gene on chromosome 17 which can be spliced into six main isoforms in the CNS with 352-441 aminoacid residues (Fig. 1) (Andreadis 2005). The repeat domain with 3 or 4 pseudo-repeats (~31 residues in length) and the flanking domains are important for the binding to microtubules. At the same time this domain is at the core of Alzheimer PHFs (Wille et al., 1992; Novak et al., 1993). This illustrates that the physiological and pathological functions of tau are closely related in the structure of the protein. Overall, tau

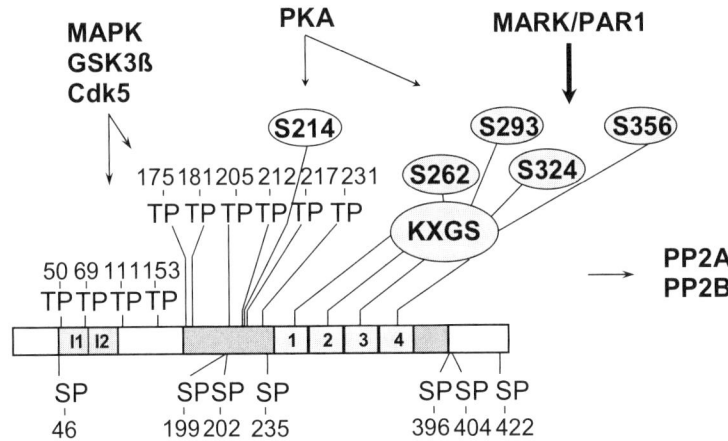

Figure 1. Diagram of tau protein: Domains, phosphorylation sites, and kinases. In CNS neurons, human tau occurs as 6 main isoforms derived from a single gene by alternative splicing (352-441 aminoacid residues). The 3 or 4 repeats in the C-terminal half (~31 residues each) constitute the center of the microtubule-binding domain, as well as the core of Alzheimer paired helical filaments. Tau contains a number of Ser and Thr residues, many of which show abnormally high phosphorylation in Alzheimer's disease and are diagnostic of Alzheimer tau. Phosphorylation sites within the repeats (at KXGS motifs) efficiently detach tau from microtubules.

has a basic and hydrophilic character due to the numerous lysine or arginine and polar residues. This makes tau unusually well soluble, with the consequence that the protein can be treated with heat or acids without loosing its biological function (Lee et al., 1988). A second consequence is that tau lacks the compact folding that is typical of most proteins; on the contrary it is a prototypic "natively unfolded" protein (Schweers et al., 1994). A number of mutations are known in the tau gene, many of which cause neurodegeneration (FTDP-17, Lee et al., 2001). The underlying cause may be either a change in tau protein function (e.g. lower microtubule binding or faster PHF aggregation) or an altered pattern of splice isoforms (D'Souza & Schellenberg, 2005).

1.4   Tau phosphorylation

In AD, tau can be phosphorylated to a high degree. The average content of phosphates rises $\sim$4-fold, from about 2 to 8 Pi per molecule (Kopke et al., 1993). These phosphates can be distributed over $\sim$30 different sites (Morishima-Kawashima et al., 1995). In the sequence, this corresponds to the numerous serines and threonines (45 and 35, 18% of the entire sequence) which are accessible to Ser/Thr directed kinases because of the open structure of tau. Thus, tau can be regarded as a sensitive reporter for the cytosolic balance between kinases and phosphatases, and indeed early stages of degeneration can be detected by means of phosphorylation-sensitive antibodies (Mandelkow & Mandelkow, 1998). Many of these sites occur in SP or TP motifs (7 and 10, resp.) which are preferred targets of proline-directed kinases (examples: MAP kinase, GSK-3β, Fig. 1). In addition tau contains 5 tyrosines (no. 18, 29, 197, 310, 394), some of which can be phosphorylated by Tyr-directed kinases (e.g. Y18 by the kinase fyn, Bhaskar et al., 2005). The biological consequences of tau phosphorylation are under debate but are heterogeneous. Phosphorylation can affect microtubule binding and/or PHF aggregation, others are functionally neutral. Of particular interest for tau interactions is the phosphorylation at the KXGS motifs in the repeat domain by the kinase MARK because this interrupts tau-microtubule binding and leads to dynamic microtubule and eventually breakdown (Biernat et al., 1993). This type of phosphorylation occurs early in the Alzheimer disease process (Augustinack et al., 2002) as well as in tau-inducible transgenic mice (unpublished data).

The interplay between tau and MARK becomes particularly noticeable on the level of neurite outgrowth. For example, N2a cells can be differentiated to develop neurites, but this requires inducers such as NGF or retinoic acid that allow microtubules to become dynamic (Brown et al., 1999). Transfection with tau greatly enhances the formation of neurites (since stable microtubules must support the neurite cytoskeleton), but outgrowth still requires differentiation inducers. On the other hand, transfection with MARK makes the cells independent of that requirement, so that the cells grow neurites spontaneously. Thus, formally the signalling of NGF and MARK have a similar effect on neurites to stimulate microtubule dynamics. The likely explanation is again the contradictory requirements on tau: Stable microtubules must be in place once the neurite is formed, but during the act of formation they must be dynamic. The specificity of the MARK-tau interaction is emphasized by the fact that kinase-dead MARK inhibits neurite outgrowth, even in the presence of differentiation inducers, and also does not phosphorylate tau at the KXGS motifs, or conversely KXGA mutants of tau (which cannot be phosphorylated by MARK) do not promote neurite outgrowth (Biernat et al., 2002; Timm et al., 2003).

1.5 Tau aggregation

Another property of tau important for its abnormal behavior in AD is the aggregation into fibers. This is a puzzling feature because of tau' excellent solubility which counteracts aggregation in physiological buffers. One explanation may be the interaction with polyanionic cellular structures (e.g. acidic proteins, RNA, negatively charged lipid vesicles) because – at least in vitro – these counterbalance tau's positive charges and can cause precipitation (Kampers et al., 1996; Goedert et al., 1996). In addition, the tau sequence contains hotspots which have an increased propensity for β-structure; they include the hexapeptide motifs ^{275}VQIINK^{280} and ^{306}VQIVYK^{311} (von Bergen et al., 2000). This preferred conformation is noticeable even in solution and leads to an amyloid-like interaction, resulting in fibrils with cross-β-structure, as judged by X-ray scattering and NMR spectroscopy (Barghorn et al., 2004, Mukrasch et al., 2005).

2. Inhibition of transport by tau

As with any macroscopic traffic system, regulation can be achieved at different levels, for example at the levels of the tracks (microtubules or tau), motors (kinesin or dynein), cargo adaptors (kinesin or dynein light chains or associated proteins), or by influencing their interactions (e.g. by phosphorylation) (Mandelkow et al., 2004, Roy et al., 2005; Terwel et al., 2002). Among the proteins involved, tau stands out as the protein most visibly associated with AD, followed by regulatory proteins such as kinases whose activities are again noticeable by way of tau phosphorylation.

2.1 Activity of motor proteins in presence of tau

When we tried to study the functions of tau in cells we noticed not only the stabilization of microtubules (as expected), but also a general inhibition of intracellular traffic. The general properties of the transport inhibition by tau are best visualized in cells that allow easy transfection or microinjection of traffic components and subsequent time-resolved imaging (e.g. CHO cells, Vero cells etc.). For example, in normal CHO cells, mitochondria are distributed rather homogeneously, corresponding to the cell's ubiquitous need for ATP. This is achieved by the network of microtubules which radiate out from the MTOC throughout the cell and act as tracks for kinesin and dynein motor proteins which can move cargoes, including mitochondria, in both directions along microtubules. As a control, if the microtubules are destroyed (e.g. by nocodazol) or the motors perturbed (e.g. dynein inhibition by dynamitin), the homogeneous distribution breaks down. Remarkably, if tau is elevated in such cells, the mitochondria gradually congregate at the cell center around the MTOC. This is due to a preferential inhibition of their transport towards the cell periphery, resulting in a net flux towards the cell center. Similar observations can be made with other microtubule cargoes such as peroxisomes, intermediate filaments, the ER, exocytotic vesicles or recycling endosomes (measured via VSV-G transport or the rate of transferrin release). In all cases the transport towards the periphery was preferentially retarded (Ebneth et al., 1998).

Quantitative analysis of vesicle and organelle movements revealed the following properties (Trinczek et al., 1999): Tau does not change the speed of a vesicle or organelle while it is in motion along a microtubule, indicating that tau does not affect the mechanics of movement as such. However, tau reduces the attachment of vesicles to the microtubule tracks, the run lengths along microtubules, and the reversal frequencies, particularly in the direction towards the cell periphery. The net effect is that tau imposes a bias favoring centripetal flow towards the cell center. This effect depends strictly on tau's affinity for microtubules, such that tightly binding tau variants (e.g. with 4 repeats) have a bigger impact than more weakly binding variants (e.g. 3-repeat tau, or phosphorylated tau). The concept of tau as a traffic inhibitor was verified directly by electron microscopy and TIRF microscopy of single kinesin movements on tau-decorated microtubules (Santarella et al., 2004; Seitz et al., 2002). The results show that kinesin outcompetes tau in equilibrium binding studies due to its higher affinity and cooperative binding; however, at the lower concentrations of motor proteins in the cytosol, tau can appreciably reduce the attachment rate of motors to microtubules. The effect is greater for kinesin than for dynein, leading to preferential retardation of anterograde traffic.

The above results revealed an unexpected property of tau: In the context of AD, the anomalous behavior of tau is usually thought to result from its dissociation from microtubules (resulting from phosphorylation) which therefore become unstable and break down. By contrast, the experiments discussed here show that even tau bound to microtubules can interfere with the cell's physiological functions by inhibiting the attachment of motor proteins to microtubules and thus inhibiting transport.

2.2 Traffic inhibition by tau in neurons

In cells with a compact shape the effects of this inhibition are moderate because the deficiencies in directed transport can be compensated in part by diffusion; e.g. ATP can diffuse throughout the cell even when the mitochondria are not homogeneously distributed. This aspect becomes much more serious for elongated polarized cells such as neurons. For example, N2a neuroblastoma cells differentiated by retinoic acid develop extended neurites (Fig. 2a, b) (Stamer et al., 2002). The neurites contain mitochondria for the generation of chemical energy, peroxisomes for detoxification of H_2O_2, neurofilaments and microtubules for structural stability and intracellular transport, and transport vesicles carrying supplies for the growth cone. Their distribution was studied in normal cells with endogenous tau only, and in cells where tau was elevated by transfection with human tau. In control cells the mitochondria are distributed throughout the cell body and the neurites by microtubule-based transport (Fig. 2a). However, in cells with elevated tau the mitochondria are nearly absent from the neurites and instead accumulate in the cell body (Fig. 2b). Similar observations hold for neurofilaments which also rely on microtubules for their transport. The interference between motors and tau is diagrammed in Fig. 2c-e.

By contrast, microtubules and tau are present throughout the neurites, showing that their transport is not perturbed. Thus the tracks for axonal transport are present in the neurite, but the transport along them is impaired because tau interferes with the access of motor proteins. This implies that neurites lack ATP and protection against oxidative stress, which would offer an explanation for the reduced growth of the neurites.

Analogous observations can be made with primary retinal ganglion cells transfected with adenovirus encoding human tau (Fig. 2f, g). Normally these neurons have mitochondria throughout the cell body and the axons, but after transfection with tau the organelles disappear from the axon and congregate in the cell body. This means that the elevation of tau has similar effects on intracellular transport in primary neurons and neuron-like cell models. As in non-neuronal cells there is a preferential inhibition of plus-end directed transport by kinesin motors along microtubules so that minus-end directed transport by a dynein-like motor becomes dominant.

2.3 Inhibition of APP trafficking by tau

In the context of AD the transport of APP (amyloid precursor protein) deserves special attention. APP is initially transported by a microtubule-based mechanism to the axon by Golgi-derived vesicles (Amaratunga et al., 1995).

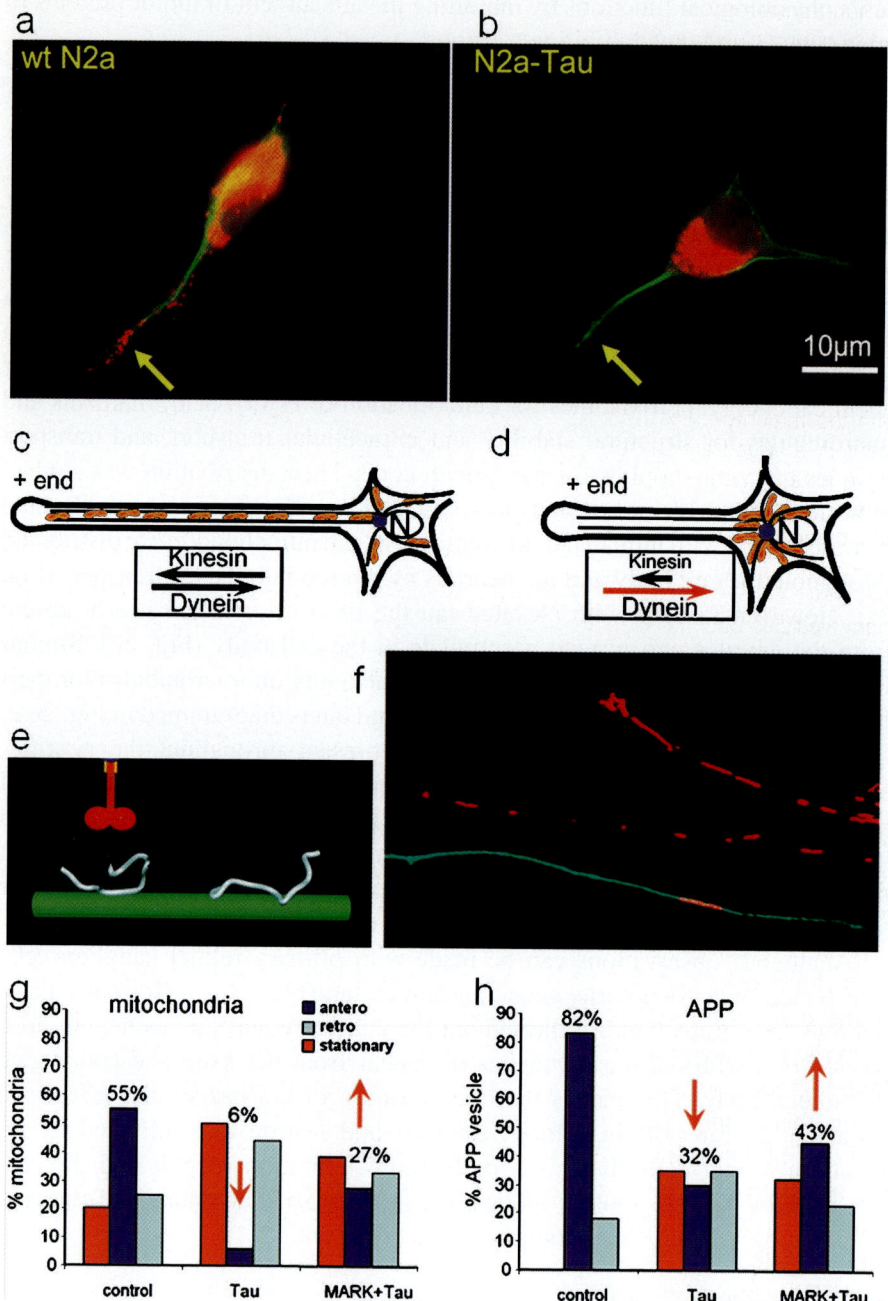

Figure 2. (a, b) Inhibition of microtubule-based transport by tau in N2a neuroblastoma cells. N2a cells, differentiated by retinoic acid, and stained for microtubules (green) and mitochondria (red). (a) control cell with mitochondria dispersed throughout the neurites.

(b) tau-transfected cell where the anterograde flow of mitochondria is inhibited, and mitochondria cluster in the cell body. (c, d) Diagram of neurons illustrating the particle flow along microtubules. (c) Control cell, the anterograde and retrograde movements (by kinesin or dynein, resp.) are regulated by the cell to achieve a balanced particle distribution. (d) With excess tau on the microtubule surface, both types of movement are inhibited, but the effect is greater on anterograde movements, resulting in a net retrograde shift in the distribution. (e) Relationship between microtubules, tau, and motor proteins. Microtubules (green) serve as tracks for motor proteins (red). The attachment of motors can be blocked by tau (blue). Phosphorylation can remove tau (not shown) and thus free the tracks for transport. (f) Movement of mitochondria in retinal ganglion cell axons. A group of RGC axons showing dispersed and actively moving mitochondria (red). One cell has been transfected by tau (blue); its axon contains only few mitochondria which show little remaining movements. (g) Quantification of mitochondrial movements. In the control cells (left), most mitochondria are mobile, with the majority (55%) moving in the anterograde direction. In tau-transfected cells (center), a large fraction (50%) is immobile during the observation period, and only 5% move anterogradely. When tau and MARK2 are cotransfected (right) the inhibition of anterograde transport is relieved (rising to 27% again) because tau is phosphorylated and detaches from the microtubules. (h) Quantification of movements of APP-YFP vesicles in retinal ganglion neurons. Left, control cell without transfected tau shows that most APP-vesicles (~80%) move anterogradely. Center, in the presence of tau many vesicles become immobile, and net transport is now reversed. Right: When tau is phosphorylated by MARK it becomes detached from microtubules, and anterograde flow becomes again dominant. (Adapted from Stamer et al., 2002, and Mandelkow et al., 2004).

By analogy with the experiments described above one would therefore expect that tau would interfere with the anterograde transport of APP. This is indeed the case: Human APP labeled with YFP was expressed in retinal ganglion cells whose axons have a well defined polarity so that transport directions are readily identified. Without tau, APP-vesicles move rapidly in both directions, but the anterograde direction predominates (80%, Fig. 2h). If the cells are co-transfected with tau the APP vesicles become depleted from the axon and moving vesicles show predominantly a retrograde direction (anterograde drops from 80% to 40%). By contrast, tau gradually spreads along the axons. These results emphasize an intriguing link between tau and the inhibition of anterograde APP trafficking. A similar effect of tau on the transport of APP can be demonstrated in cultured hippocampal neurons transfected with APP-YFP and CFP-htau40 using recombinant adenovirus. APP transfection alone yields fluorescent APP-vesicles in the cell body and neurites, while transfection with APP plus tau causes APP to be restricted largely to the cell body.

APP gives rise to the cleavage product $A\beta$ which aggregates into amyloid oligomers and fibers in AD. It has been proposed that APP serves as a cargo adaptor for kinesin, and that overexpression of APP leads to inhibition of transport and increased generation of Aβ since APP vesicles contained all secretases necessary for Aβ cleavage (β-secretase BACE1 and the γ-secretase complex containing presenilin, Kamal & Goldstein, 2002). By this hypothesis, Aβ cleavage could occur in transit. On the other hand, the colocalization has not been verified by other investigators (Lazarov et al., 2005). It will therefore be an interesting question whether the retardation of APP vesicle trafficking by tau has an influence on the generation of Aβ. This issue is currently under investigation, and our preliminary results show that the inhibition does not lead to enhanced Aβ generation, and that in general APP does not colocalize with BACE1 on the same vesicles (Goldsbury et al., 2006).

2.4 Consequences of transport inhibition for neurons

If organelles disappear one would expect deficiencies in local metabolism because diffusion from the cell body could not compensate the loss of local production. This would for example lead to reduced ATP levels due to loss of mitochondria and an increased sensitivity to oxidative damage due to loss of peroxisomes and thus of catalase (Stamer et al., 2002). These predictions can be tested by exposing N2a cells to H_2O_2 and observing the time-course of neurite loss. Indeed, in tau-transfected cells the degradation of neurites is much more rapid than in the non-transfected controls. It is noteworthy that the higher vulnerability is restricted to the neurites but not the cell body where the mitochondria and peroxisomes are still present and functional. A similar result is seen on the level of synapses. One observes a dramatic decay of dendritic spines in 4 week old cultures after transfection with tau (Thies et al., unpublished). Thus it appears that tau does not have a direct negative effect on the biochemical pathways of neurons, but rather that the viability of neurites and synapses is reduced by tau's effect on the transport and redistribution of cell components.

2.5 Tau phosphorylation by MARK and rescue of transport inhibition

As mentioned above, the effects of tau on traffic inhibition require binding to microtubules. One might therefore expect that a release of tau from microtubules might alleviate the traffic inhibition again. This can be tested in retinal ganglion cells which in their normal state show rapid movement of cell organelles or vesicles in axons. The expression of tau causes a redirection of these movements into the retrograde direction towards the cell body and leads to depletion of organelles and vesicles in axons (Fig. 2b, f). However, this inhibition of anterograde flow can be rescued by transfection with MARK, the protein kinase that phosphorylates tau in the KXGS motifs of the repeat domain and thereby detaches it from microtubules (Fig. 2g, h). As a test of this relationship, transfection of cells with a KXGA-mutant of tau (which cannot be phosphorylated by MARK) causes traffic inhibition (because the KXGA mutant binds tightly to microtubules), but the traffic inhibition cannot be relieved because the KXGA mutant of tau cannot be phosphorylated by MARK and therefore tau remains attached to the microtubules. By the same token, transfection with dominant negative versions of MARK cannot relieve the inhibition either. Taken together, tau must be able to fulfill seemingly contradictory requirements, stay on the tracks to keep them intact, but jump off readily when a motor protein passes through (not unlike railroad workers who move off the track when a train approaches).

These relationships reveal puzzling janus-faced aspects of the functions of tau. On one hand, the binding of tau enables traffic by generating stable microtubules. On the other hand, too much tau bound to microtubules can prevent the motor proteins from attaching to microtubules. This spatial and functional paradox of tau comes about because tau protein – unlike the ties of a railroad – lies on top of the tracks rather than underneath and therefore creates a stumbling block on the rails while at the same time tying them together. The solution to the structure-function paradox presumably lies in local regulation and equilibria: Cargoes generally carry more than one motor, and even a single processive motor attached to a microtubule can propel the cargo. Therefore, traffic could remain functional if tau were removed locally, for example by phosphorylation.

3. Correlation with transgenic animal models

How can we link the observations on tau and transport to the brain in transgenic animal models or in Alzheimer's disease? In AD, early changes include loss of synapses and reactions with certain conformation-dependent tau antibodies (Alz50, MC-1, or phosphorylation-dependent antibodies, e.g. 12E8, AT-8, Jicha & Davies, 1999; Augustinack et al., 2002), while neurofibrillary tangles or microtubule breakdown appear at later stages. If the early changes were due to traffic inhibition by tau, one would expect a greater occupancy of tau on the microtubule tracks. This could be achieved by a higher tau concentration and/or a tighter binding of tau to microtubules, which could be achieved by lower phosphorylation at critical sites (e.g. KXGS motifs in repeat domain or Ser214, Illenberger et al., 1998), or an increase in 4-repeat isoforms (as is the case in some forms of FTDP-17, Hutton et al., 2001). There is evidence that tau protein is elevated globally in AD (Khatoon et al., 1992), but more important is the possibility that tau protein is locally elevated in affected neurons or even in neuronal compartments. It is not possible to settle this point with current technology (Galvin & Ginsberg, 2005). We note however that tau-transgenic mouse models have relied on an elevated expression of tau, and moreover on aggregation-prone tau mutants (e.g. P301L), in order to achieve Alzheimer-like changes (in terms of tau aggregation or immunoreactivity). These changes are however late compared with the onset of traffic inhibition due to tau clustering on the microtubule surface (Terwel et al., 2002). We note also that in cell models even a small elevation of tau over endogenous tau ($\sim$2-fold) is sufficient to cause substantial traffic inhibition. A local concentration change (or affinity change) of this magnitude would be hardly detectable in mouse or human brains.

Since tau is primarily an axonal protein, the effects of traffic inhibition can be considered from two angles. A slow-down in axonal traffic would have direct consequences for presynaptic terminals, axonal growth cones, or delivery of synaptic vesicles, including delivery of APP, ion channels, energy sources such as mitochondria. This in turn would impair the communication with the dendrites. Since the survival of the downstream neurons depends on input from the axons, this could lead to the observed dendritic decay (Walsh & Selkoe, 2004). A case in point would be the transport of essential growth factors between hippocampal neurons and basal forebrain cholinergic neurons. Thus, a traffic jam caused by tau could lead to the decay of cholinergic neurons observed in AD.

One persistent observation in AD brain and tau transgenic mice is the early re-distribution of tau from axon-only to ubiquitous in the somatodendritic compartment (Oddo et al., 2004; Santacruz et al., 2005, Andorfer et al., 2003, Lewis et al., 2001; Gotz et al., 2001; Higuchi et al., 2002). The sorting mechanism is not well understood. Fetal tau shows no preferential sorting, similar to other MAPs, and even in the adult stage it appears that the regulation is not very tight and can be overruled by a small elevation of tau (Kanai & Hirokawa 1996). This can be demonstrated by microinjection of tau into cells, transfection of cells, and expression in transgenic mice. Thus, the penetration of tau into the "wrong" compartment can have a direct effect on traffic in dendrites, independently and above the effects on axons (Stamer et al., 2002).

Perhaps the best demonstration of tau's toxicity to axonal traffic is provided by several tau transgenic mouse lines expressing tau under the control of pan-neuronal promoters such as Thy-1 or prion promotors which express tau especially in the spinal cord. Because excess tau is expressed in motor neurons from the fetal stage onwards, these mice develop motor deficits which prevent their use for analyzing cognitive deficits by motion analysis (e.g. water maze test). Histopathologically these mice show clustering of neurofilaments in the cell bodies of motor neurons and axonal swellings, demonstrating the inhibition of transport (Terwel et al., 2002). In later developments of transgenic mice this problem was circumvented by using the CaMK-II promotor which is more restricted to the hippocampus and entorhinal regions.

A number of transgenic animal models of tauopathy have been described by now (see citations above), however, the origin of tau's toxicity to neurons is still not well defined. Expression of normal human tau alone tends to have little effect on measurable changes in tau (such as altered conformation, phosphorylation, neurofibrillary tangle formation), and therefore the changes were exacerbated by choosing tau mutants (known from FTDP-17), enhancing the activity of kinases (cdk5), or combining tau mutations with amyloidogenic mutations in APP or PS1). The emerging data show that tau can be toxic already in a pre-fibrillar form, that the effects of pre-fibrillar tau are reversible while neurofibrillar tangles are not, that the tau effects are downstream of those induced by Aβ (in those cases where both proteins are expressed in mutant form). In general, the appearance of tauopathy (without or with amyloid) is linked to synapse loss and neurodegeneration. The mechanistic interpretation invoked by most authors includes two stages of tau toxicity: Initially, the activation of kinases which "hyperphosphorylate" tau, cause its detachment from microtubules and hence microtubule breakdown, and later, the aggregation of unbound tau leading to neurofibrillary tangles and obstruction of cell interior (the second stage may include an oligomeric intermediate form of tau which can be toxic on its own). With the background on transport inhibition discussed above, we propose an extension to three stages of tau toxicity (Fig. 3):

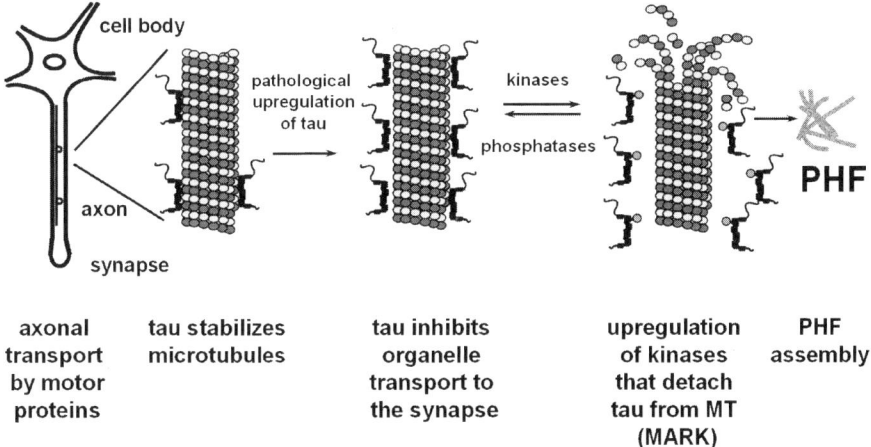

| axonal transport by motor proteins | tau stabilizes microtubules | tau inhibits organelle transport to the synapse | upregulation of kinases that detach tau from MT (MARK) | PHF assembly |

Figure 3. Model of tau functions in neurodegeneration. Left: In a healthy axon, microtubules are stabilized by tau and serve as tracks for axonal transport of vesicles and organelles. Center: Tau bound to microtubules is elevated early during neurodegeneration. Even though microtubules are stable, the excess tau blocks traffic into the axon because it interferes with motor proteins. This makes axons vulnerable and causes the decay of synapses. Right: Phosphorylation of tau occurs through the cell's attempt to clear the inhibitory tau off the microtubule tracks. Microtubules become unstable, and the detached tau is free to aggregate into neurofibrillary tangles.

- A (local) excess of tau bound to microtubules inhibits traffic of vesicles, organelles, cytoskeletal elements and other cargos; this can cause starvation of cell processes, damage to synapses, and redistribution of tau from axon-only to somatodendritic (so that dendrites, too, suffer from transport inhibition). In transgenic models, this is achieved by the overexpression of tau or its variants. Note that this stage of toxicity is based on apparently "healthy" and functional tau. This stage would also explain the toxicity of excess 4-repeat tau observed with some FTDP-17 mutations (Lee et al., 2001; Hutton et al., 2001). The tau-induced traffic damage is best observed in mice expressing tau in motor neurons (Terwel et al., 2002), or in Drosophila models by the degeneration of the neuromuscular junction (Chee et al., 2005).
- The second stage consists of a cellular reaction against excess missorted tau which consists of the activation of kinases. The most potent kinase to detach surplus tau from microtubules is MARK/Par-1 which would explain why the corresponding phosphorylation sites in the repeats of tau occur very early, both in AD and in our mouse models (Augustinack et al., 2002, and unpublished data). The detached tau can in turn be hyperphosphorylated by further kinases (as shown most directly for the Drosophila model by Nishimura et al., 2004). Note that this overshoot kinase activity takes place not only in the axon but also in the somatodendritic compartment, wherever excess tau is located. As a consequence, microtubule tracks become destabilized and break down, and an aggregation-competent pool of unbound tau is generated. Certain conformation-dependent antibodies (Alz50, MC-1) are likely indicators of this stage.
- The third stage leads to the aggregation of tau into oligomers, paired helical filaments, and finally their accretion into neurofibrillary tangles. While the previous stages are reversible, e.g. by silencing the expression of tau (SantaCruz et al., 2005), the third stage gradually becomes irreversible due to the long life-time of the aggregates and their secondary chemical modifications (oxidation, cross-linking etc.).

4.   Summary

The established physiological functions of tau include the stabilization of microtubules and the promotion of neurite outgrowth (Mandelkow & Mandelkow, 1998; Garcia & Cleveland, 2001). More recently it was recognized that tau can regulate the transport of cell components by molecular motors along microtubules (Ebneth et al., 1998). Tau influences the rates of attachment and detachment of motors from microtubules (Trinczek et al., 1999). The result is that movements towards the cell center become predominant. This leads to the gradual retraction of cell components such as the ER, mitochondria or peroxisomes. Starting from these observations one can ask what the effect of tau would be on highly asymmetric cells such as neurons. Tau is the predominant MAP in axons (Binder et al., 1985); if the plus-end directed transport were retarded by tau this would hinder the supply of material into the axons and dendrites, with subsequent damage to synapses. This may be significant in the context of Alzheimer's disease where the affected neurons appear to contain higher levels of tau protein (Khatoon et al., 1992).

We therefore analyzed the distribution of cytoskeletal components, organelles, and Golgi-derived vesicles and their dependence on tau. A consistent interpretation of the combined data is that tau inhibits transport along microtubules, preferentially in the anterograde direction. The consequence is that organelles tend to be excluded from cell processes, and vesicles are strongly reduced. A further consequence is that axonal and dendritic compartments become more vulnerable. The movement of organelles, vesicles and neurofilaments is consistent with the view that these components are transported anterogradely by a kinesin-dependent transport (Hollenbeck & Saxton, 2005). Remarkably, the inhibition of anterograde transport does not apply to microtubules or microtubule-associated proteins such as tau, presumably because their transport takes place by a different type of mechanism, for example by dynein-dependent transport along actin filaments (Baas & Buster, 2004).

The transport inhibition can have profound consequences for the survival of cell processes. Thus, tau-transfected cells become highly sensitive to oxidative stress, consistent with the lack of peroxisomes. Similarly, the exclusion of mitochondria from the cell processes means that ATP is locally depleted. This is tantamount to a loss of mitochondrial function in the neurites.

One can speculate what these observation mean for neurodegenerative disorders such as Alzheimer's disease and other tauopathies which are characterized by locally elevated and aggregated tau protein. One of the earlierst detectable signs is the faulty distribution of tau in the somatodendritic compartment, the loss of synapses and retrograde degeneration ("dying back") of neurons which is accompanied by a decay of intracellular transport (Terry et al., 1991; Coleman & Yao, 2003). This may be accompanied by an elevation of non-fibrillar forms of Aβ (Walsh & Selkoe, 2004). APP is considered to have neurotrophic functions and is transported on kinesin-driven vesicles. It has been proposed that the C-terminal domain of APP interacts directly with a kinesin light chain (Kamal & Goldstein, 2002), although this notion has recently been challenged (Lazarov et al., 2005). Independently of these issues, the local elevation of tau presents obstacles to traffic, such that APP-vesicles tend to disappear from the axons and dendrites and APP becomes trapped in the cell body. This could potentially lead to an elevation of Aβ generated in the trans-Golgi network (Greenfield et al., 1999), although this has not observed experimentally for reasons discussed elsewhere (Goldsbury et al., 2006). By contrast, the transport infrastructure is less dependent on tau than vesicle or organelle traffic since microtubule tracks survive long after mitochondria, APP-vesicles and others have disappeared from the cell body, but eventually microtubules disappear as well when the cell processes degenerate. We can therefore distinguish three types of tau toxicity related to intraneuronal transport: (i) Tau causes a slowdown of anterograde traffic which leads to starvation of neurites and missorting of tau into the somatodendritic compartment. (ii) The cell responds by activating kinases, which results in the detachment of tau from microtubules and their destabilization, (iii) unbound tau aggregates into neurofibrillary tangles which obstruct the cytoplasm.

Acknowledgement

We thank Dr. J. Biernat and Dr. A. Marx for fruitful discussions and help with the design of figures. This work was supported in part by the Deutsche Forschungsgemeinschaft.

References

Andreadis A (2005) Tau gene alternative splicing: expression patterns, regulation and modulation of function in normal brain and neurodegenerative diseases. Biochim Biophys Acta 1739:91–103.

Amaratunga A, Leeman SE, Kosik KS, Fine RE (1995) Inhibition of kinesin synthesis in vivo inhibits the rapid transport of representative proteins for three transport vesicle classes into the axon. J Neurochem 64:2374–2376.

Andorfer C, Kress Y, Espinoza M, de Silva R, Tucker KL, Barde YA, Duff K, Davies P (2003) Hyperphosphorylation and aggregation of tau in mice expressing normal human tau isoforms. J Neurochem 86:582–590.

Augustinack JC, Schneider A, Mandelkow EM, Hyman BT (2002) Specific tau phosphorylation sites correlate with severity of neuronal cytopathology in Alzheimer's disease. Acta Neuropath 103:26–35.

Baas PW, Buster DW (2004) Slow axonal transport and the genesis of neuronal morphology. J Neurobiol 58:3–17.

Baas PW, Karabay A, Qiang L (2005) Microtubules cut and run. Trends Cell Biol 15:518–524.

Barghorn S, Davies P, Mandelkow E (2004) Tau paired helical filaments from Alzheimer's disease brain and assembled in vitro contain beta structure in the core domain. Biochemistry 43:1694–1703.

Bhaskar K, Yen SH, Lee G (2005) Disease-related modifications in tau affect the interaction between Fyn and Tau. J Biol Chem 280:35119–35125.

Biernat J, Gustke N, Drewes G, Mandelkow EM, Mandelkow E (1993) Phosphorylation of Ser262 strongly reduces binding of tau to microtubules: distinction between PHF-like immunoreactivity and microtubule binding. Neuron 11:153–163.

Biernat J, Wu YZ, Timm T, Zheng-Fischhofer Q, Mandelkow E, Meijer L, Mandelkow EM (2002) Protein kinase MARK/PAR-1 is required for neurite outgrowth and establishment of neuronal polarity. Mol Biol Cell 13:4013–4028.

Binder LI, Frankfurter A, Rebhun LI (1985) The distribution of tau in the mammalian central nervous system. J Cell Biol 101:1371–1378.

Binder LI, Guillozet-Bongaarts AL, Garcia-Sierra F, Berry RW (2005) Tau, tangles, and Alzheimer's disease. Biochim Biophys Acta 1739:216–223.

Braak H, Braak E (1991) Neuropathological stageing of Alzheimer-related changes. Acta Neuropathol 82:239–259.

Brown AJ, Hutchings C, Burke JF, Mayne LV (1999) Application of a rapid method (targeted display) for the identification of differentially expressed mRNAs following NGF-induced neuronal differentiation in PC12 cells. Mol Cell Neurosci:119–130.

Buee L, Bussiere T, Buee-Scherrer V, Delacourte A, Hof PR (2000) Tau protein isoforms, phosphorylation and role in neurodegenerative disorders. Brain Res Brain Res Rev 33:95–130.

Cassimeris L, Spittle C (2001) Regulation of microtubule-associated proteins. Int Rev Cytol 210:163–226.

Chee FC, Mudher A, Cuttle MF, Newman TA, MacKay D, Lovestone S, Shepherd D (2005) Over-expression of tau results in defective synaptic transmission in Drosophila neuromuscular junctions. Neurobiol Dis 20:918–928.

Coleman PD, Yao PJ (2003) Synaptic slaughter in Alzheimer's disease. Neurobiol Aging 24:1023–1027.

Crowther RA, Goedert M (2000) Abnormal tau-containing filaments in neurodegenerative diseases. J Struct Biol 130:271–279.

Drewes G, Ebneth A, Preuss U, Mandelkow EM, Mandelkow E (1997) MARK, a novel family of protein kinases that phosphorylate microtubule-associated proteins and trigger microtubule disruption. Cell 89:297–308.

D'Souza I, Schellenberg GD (2005) Regulation of tau isoform expression and dementia. Biochim Biophys Acta 1739:104–115.

Ebneth A, R. Godemann K, Stamer S, Illenberger B, Trinczek, Mandelkow E (1998) Overexpression of tau protein inhibits kinesin-dependent trafficking of vesicles, mitochondria, and endoplasmic reticulum: implications for Alzheimer's disease. J Cell Biol 143:777–794.

Flood DG, Coleman PD (1990) Hippocampal plasticity in normal aging and decreased plasticity in Alzheimer's disease. Prog Brain Res 83:435–443.

Galvin JE, Ginsberg SD (2005) Expression profiling in the aging brain: a perspective. Ageing Res Rev 4:529–547.

Garcia ML, Cleveland DW (2001) Going new places using an old MAP: tau, microtubules and human neurodegenerative disease. Curr Opin Cell Biol 13:41–48.

Goedert M, Jakes R, Spillantini MG, Hasegawa M, Smith MJ, Crowther RA (1996) Assembly of microtubule-associated protein tau into Alzheimer-like filaments induced by sulphated glycosaminoglycans. Nature 383:550–553.

Goldsbury C, Mocanu M, Thies E, Kaether C, Haass C, Biernat E, Mandelkow E-M, Mandelkow E (2006) Inhibition of APP trafficking by tau protein does not increase the generation of amyloid-beta. Traffic 7:873–888.

Götz J, Chen F, van Dorpe J, Nitsch RM (2001) Formation of neurofibrillary tangles in P301l tau transgenic mice induced by Abeta 42 fibrils. Science 293:1491–1495.

Greenfield JP, Tsai J, Gouras GK, Hai B, Thinakaran G, Checler F, Sisodia SS, Greengard P, Xu H (1999) Endoplasmic reticulum and trans-Golgi network generate distinct populations of Alzheimer beta-amyloid peptides. Proc Natl Acad Sci USA 96:742–747.

Higuchi M, Ishihara T, Zhang B, Hong M, Andreadis A, Trojanowski J, Lee VM (2002) Transgenic mouse model of tauopathies with glial pathology and nervous system degeneration. Neuron 35:433–446.

Hirokawa N, Takemura R (2005) Molecular motors and mechanisms of directional transport in neurons. Nat Rev Neurosci 6:201–214.

Hollenbeck PJ, Saxton WM (2005) The axonal transport of mitochondria. J Cell Sci 118:5411–5419.

Hutton M, Lewis J, Dickson D, Yen SH, McGowan E (2001) Analysis of tauopathies with transgenic mice. Trends Mol Med 7:467–470.

Illenberger S, Zheng-Fischhofer Q, Preuss U, Stamer K, Baumann K, Trinczek B, Biernat J, Godemann R, Mandelkow EM, Mandelkow E (1998) The endogenous and cell cycle-dependent phosphorylation of tau protein in living cells: implications for Alzheimer's disease. Mol Biol Cell 9:1495–1512.

Jicha GA, Berenfeld B, Davies P (1999) Sequence requirements for formation of conformational variants of tau similar to those found in Alzheimer's disease. J Neurosci Res 55:713–723.

Kamal A, Goldstein LS (2002) Principles of cargo attachment to cytoplasmic motor proteins. Curr Opin Cell Biol 14:63–68.

Kampers T, Friedhoff P, Biernat J, Mandelkow EM, Mandelkow E (1996) RNA stimulates aggregation of microtubule-associated protein tau into Alzheimer-like paired helical filaments. FEBS Lett 399:344–349.

Kanai Y, Hirokawa N (1995) Sorting mechanisms of tau and MAP2 in neurons: Suppressed axonal transit of MAP2 and locally regulated microtubule binding. Neuron 14:421–432.

Khatoon S, Grundke-Iqbal I, Iqbal K (1992) Brain levels of microtubule-associated protein tau are elevated in Alzheimer's disease: a radioimmuno-slot-blot assay for nanograms of the protein. J Neurochem 59:750–753.

Kopke E, Tung YC, Shaikh S, Alonso AC, Iqbal K, Grundke-Iqbal I (1993) Microtubule-associated protein tau. Abnormal phosphorylation of a non-paired helical filament pool in Alzheimer disease. J Biol Chem 268:24374–24384.

Lazarov O, Morfini GA, Lee EB, Farah MH, Szodorai A, DeBoer SR, Koliatsos VE, Kins S, Lee VM, Wong PC, Price DL, Brady ST, Sisodia SS (2005) Axonal transport, amyloid precursor protein, kinesin-1, and the processing apparatus: revisited. J Neurosci 25:2386–2395.

Lee G, Cowan N, Kirschner M (1988) The primary structure and heterogeneity of tau protein from mouse brain. Science 239:285–288.

Lee VM, Goedert M, Trojanowski JQ (2001) Neurodegenerative tauopathies. Annu Rev Neurosci 24:1121–1159.

Lewis J, Dickson DW, Lin WL, Chisholm L, Corral A, Jones G, Yen SH, Sahara N, Skipper L, Yager D, Eckman C, Hardy J, Hutton M, McGowan E (2001) Enhanced neurofibrillary degeneration in transgenic mice expressing mutant tau and APP. Science 293:1487–1491.

Mandelkow E-M, Mandelkow E (1998) Tau in Alzheimer's disease. Trends in Cell Biol 8:425–427.

Mandelkow E-M, Thies E, Trinczek B, Biernat B, Mandelkow E (2004) MARK/PAR1 kinase is a regulator of microtubule-dependent transport in axons. J Cell Biol 167:99–110.

Morishima-Kawashima M, Hasegawa M, Takio K, Suzuki M, Yoshida H, Titani K, Ihara Y (1995) Proline-directed and non-proline-directed phosphorylation of PHF-tau. J Biol Chem 270:823–829.

Mukrasch MD, Biernat J, von Bergen M, Griesinger C, Mandelkow E, Zweckstetter M (2005) Sites of tau important for aggregation populate {beta}-structure and bind to microtubules and polyanions. J Biol Chem 280:24978–24886.

Nishimura I, Yang Y, Lu B (2004) PAR-1 kinase plays an initiator role in a temporally ordered phosphorylation process that confers tau toxicity in Drosophila. Cell 116:671–682.

Novak M, Kabat J, Wischik CM (1993) Molecular characterization of the minimal protease resistant tau unit of the Alzheimer's disease paired helical filament. EMBO J 12:365–370.

Oddo S, Billings L, Kesslak JP, Cribbs DH, LaFerla FM (2004) Abeta immunotherapy leads to clearance of early, but not late, hyperphosphorylated tau aggregates via the proteasome. Neuron 43:321–332.

Raff M, Whitmore A, Finn J (2002) Axonal self-destruction and neurodegeneration. Science 296:868–871.

Roy S, Zhang B, Lee VM, Trojanowski JQ (2005) Axonal transport defects: a common theme in neurodegenerative diseases. Acta Neuropathol (Berl) 109:5–13.

Santacruz K, Lewis J, Spires T, Paulson J, Kotilinek L, Ingelsson M, Guimaraes A, DeTure M, Ramsden M, McGowan E, Forster C, Yue M, Orne J, Janus C, Mariash A, Kuskowski M, Hyman B, Hutton M, Ashe KH (2005) Tau suppression in a neurodegenerative mouse model improves memory function. Science 309:476–481.

Santarella RA, Skiniotis G, Goldie KN, Tittmann P, Gross H, Mandelkow EM, Mandelkow E, Hoenger A (2004) Surface-decoration of microtubules by human tau. J Mol Biol 339:539–553.

Schweers O, Schonbrunn-Hanebeck E, Marx A, Mandelkow E (1994) Structural studies of tau protein and Alzheimer paired helical filaments show no evidence for beta-structure. J Biol Chem 269:24290–24297.

Seitz A, Kojima H, Oiwa K, Mandelkow EM, Song YH, Mandelkow E. 2002. Single-molecule investigation of the interference between kinesin, tau and MAP2c. EMBO J 21:4896–4905.

Selkoe D (2002) Alzheimer's disease is a synaptic failure. Science 298:789–791.

Simons M, Ikonen E, Tienari PJ, Cid-Arregui A, Monning U, Beyreuther K, Dotti CG (1995) Intracellular routing of human amyloid protein precursor: axonal delivery followed by transport to the dendrites. J Neurosci Res 41:121–128.

Stamer K, Vogel R, Thies E, Mandelkow E, Mandelkow EM (2002) Tau blocks traffic of organelles, neurofilaments, and APP vesicles in neurons and enhances oxidative stress. J Cell Biol 156:1051–1063.

Stoothoff WH, Johnson GV (2005) Tau phosphorylation: physiological and pathological consequences. Biochim Biophys Acta 1739:280–297.

Tanzi RE, Bertram L (2005) Twenty years of the Alzheimer's disease amyloid hypothesis: a genetic perspective. Cell 120:545–555.

Terry RD, Masliah E, Salmon DP, Butters N, DeTeresa R, Hill R, Hansen LA, Katzman R (1991) Physical basis of cognitive alterations in Alzheimer's disease: synapse loss is the major correlate of cognitive impairment. Ann Neurol 30:572–580.

Terwel D, Dewachter I, Van Leuven F (2002) Axonal transport, tau protein, and neurodegeneration in Alzheimer's disease. Neuromolecular Med 2:151–165.

Timm T, Li XY, Biernat J, Jiao J, Mandelkow E, Vandekerckhove J, Mandelkow E-M (2003) MARKK, a Ste-20-like kinase, activates the polarity-inducing kinase MARK/PAR-1. EMBO J 22:5090–5101.

Trinczek B, Ebneth A, E Mandelkow M, Mandelkow E (1999) Tau regulates the attachment/detachment but not the speed of motors in microtubule-dependent transport of single vesicles and organelles. J Cell Sci 112:2355–2367.

von Bergen M, Friedhoff P, Biernat J, Heberle J, Mandelkow EM, Mandelkow E (2000) Assembly of tau protein into Alzheimer paired helical filaments depends on a local sequence motif 306-VQIVYK-311 forming beta structure. Proc Natl Acad Sci USA 97:5129–5134.

Walsh DM, Selkoe DJ (2004) Deciphering the molecular basis of memory failure in Alzheimer's disease. Neuron 44:181–193.

Wille H, Drewes G, Biernat J, Mandelkow EM, Mandelkow E (1992) Alzheimer-like paired helical filaments and antiparallel dimers formed from microtubule-associated protein tau in vitro. J Cell Biol 118:573–584.

Wittmann CW, Wszolek MF, Shulman JM, P Salvaterra M, Lewis J, Hutton M, Feany MB (2001) Tauopathy in Drosophila: neurodegeneration without neurofibrillary tangles. Science 293:711–714.

Yan Y, Brown A (2005) Neurofilament polymer transport in axons. J Neurosci 25:7014–7021.

Chapter 15

Growth Factors in Alzheimer's Disease*

A. H. Nagahara[1] and M. H. Tuszynski[1,2]

[1]*Departments of Neurosciences-0626*
University of California
San Diego La Jolla, California 92093, USA

[2]*Veterans Affairs Medical Center*
San Diego, California 92161, USA

1. Introduction

Growth factors are proteins that are naturally expressed in the nervous system during development and throughout adult life. First discovered more than 50 years ago, nervous system growth factors were originally found to support the survival and patterning of neural projections during development. When first applied therapeutically to the adult brain approximately 20 years ago, it was discovered that growth factors also extensively prevent the death of specific populations of neurons in animal models of neurodegenerative disease, including Alzheimer's disease. This led to considerable enthusiasm regarding the potential of growth factors to treat neurodegenerative disorders early in the course of disease, reducing neuronal degeneration and stimulating neuronal function to slow disease progression. However, the implementation of clinical testing of growth factor therapy for neurological disease has been constrained by the dual need to achieve adequate concentrations of these proteins in specific brain regions containing degenerating neurons, while preventing spread of growth factors to non-targeted regions to avoid adverse effects. This chapter will review the biology of growth factors, their potential to treat disease, means of practical delivery, and the results of the first human trial of nerve growth factor (NGF) gene delivery for Alzheimer's disease.

---

* **Correspondence to:** Email: mtuszynski@ucsd.edu

2.   Growth factors

The term "neurotrophic factor" literally refers to a molecule that attracts neurons, but in conventional use refers to proteins with the properties of supporting neuronal survival, migration or neuritic outgrowth. Some neurotrophic factors also support the mature phenotypic state of cells, upregulating a number of intracellular signaling pathways (Kaplan and Miller, 2000) that modulate neuronal function throughout life. Some neurotrophic factors have further direct effects on synaptic function, influencing neurotransmitter release and synaptic efficacy (Aicardi et al., 2004).

To date, more than 50 molecules with neurotrophic factor properties have been identified in the brain (for review see Tuszynski, 1999). Most of these growth factors can be grouped into distinct families, including: 1) the "classic neurotrophins," consisting of nerve growth factor (NGF), brain-derived neurotrophic factor (BDNF), neurotrophin-3 (NT-3), and neurotrophin-4/5 (NT-4/5); 2) the Transforming Growth Factor (TGF)-β family, including glial cell line-derived neurotrophic factor (GDNF), neurturin, artemin, persephin and the bone morphogenic proteins (BMPs); 3) the cytokine growth factor family, including ciliary neurotrophic factor (CNTF), leukemia inhibitory factor (LIF), and cardiotropin-1; 4) the epidermal growth factor (EGF) family, consisting of EGF, transforming growth factor-alpha (TGF-α), the neuregulins, and NTAK (neural and thymus-derived activator for ErbB kinases); 5) the "insulin-like" growth factors, including insulin-like growth factor I (IGF-I) and IGF-II; and 6) the fibroblast growth factor (FGF) family, of which at least 24 different proteins have been identified, including acidic FGF (FGF 1) and basic FGF (FGF 2).

The biological effects of growth factors differ greatly between families, and between members of the same family. The spectrum of survival-enhancing effects and functional-stimulating effects of a particular growth factor determine its potential relevance to a specific disease. For example, the growth factors GDNF and neurturin robustly promote the survival and stimulate the function of nigral neurons, hence they are both therapeutic candidates for the treatment of Parkinson's disease (Gash et al., 1996; Kordower et al., 2000). Three other growth factors are of potential interest in the treatment of AD: NGF, to address the cholinergic component of neuronal degeneration, and BDNF or NT-4/5, to address cortical and hippocampal neuronal degeneration.

3.   NGF and Alzheimer's Disease

NGF, a member of the neurotrophin family, was the first growth factor identified and subsequently emerged as an intriguing therapeutic candidate for AD (Levi-Montalcini, 1987). NGF was discovered serendipitously over

50 years ago by Rita Levi-Montalcini and Viktor Hamburger in the course of examining properties of a sarcoma cell line in vitro (Levi-Montalcini and Hamburger, 1951). A role for NGF as an essential survival factor for peripheral sensory and sympathetic neurons during development of the nervous system was soon established (Levi-Montalcini and Hamburger, 1953; Levi-Montalcini et al., 1954). Yet a subsequent finding regarding NGF neurobiology revolutionized the perception that the adult CNS was an inflexible, non-plastic structure: in 1986, Franz Hefti (Hefti, 1986) and others (Kromer et al., 1981; Williams et al., 1986) reported that infusion of NGF into the adult rat brain completely prevented the death of basal forebrain cholinergic neurons after injury. The ability of neurotrophins to prevent neuronal degeneration in the adult brain ushered in the modern era of effort directed at enhancing CNS plasticity, preventing neuronal loss, and promoting CNS regeneration with growth factors.

Shortly following reports appeared that NGF prevented lesion-induced degeneration of basal forebrain cholinergic neurons in adult rats, Fischer and colleagues reported that *spontaneous* atrophy of basal forebrain cholinergic neurons in *aged* rats was also be prevented by NGF infusion (Fischer et al., 1987). Further, NGF significantly ameliorated memory deficits in aged rats (Fischer et al., 1987; Markowska et al., 1994) or adult lesioned rats (Tuszynski and Gage, 1995). When tested in a mouse model of Down's syndrome, trisomy 16, in which mice have an extra copy of the amyloid precursor protein (APP) gene, Mobley and colleagues reported that NGF also reversed cholinergic atrophy and improved cognitive deficits (Holtzman et al., 1993; Cooper et al., 2001). Addressing yet another mechanism of neuronal damage, NGF was reported to prevent the degeneration of neurons following excitotoxic lesions to the basal forebrain cholinergic Nucleus Basalis of Meynert (Dekker et al., 1992) or cortical lesions (Liberini et al., 1993). Experiments in primates further demonstrated that NGF prevented degeneration of basal forebrain cholinergic neurons in monkeys (Fig. 1) (Tuszynski et al., 1990; Koliatsos et al., 1991; Tuszynski et al., 1991; Emerich et al., 1994; Kordower et al., 1994b; Tuszynski et al., 1996).

In 1983, human NGF was cloned (Ullrich et al., 1983), bringing forth the prospect of molecular therapy using the human molecule. Thus, NGF prevented the death and stimulated the functional state of cholinergic neurons following a variety of mechanisms of neuronal damage, including injury, excitotoxicity, aging, and amyloid overproduction. NGF showed efficacy at the cellular level in both rodents and primates, and improved cognition in a variety of models in rodents.

Collectively, these findings established a theoretical and practical framework in which clinical trials for the treatment of Alzheimer's disease could

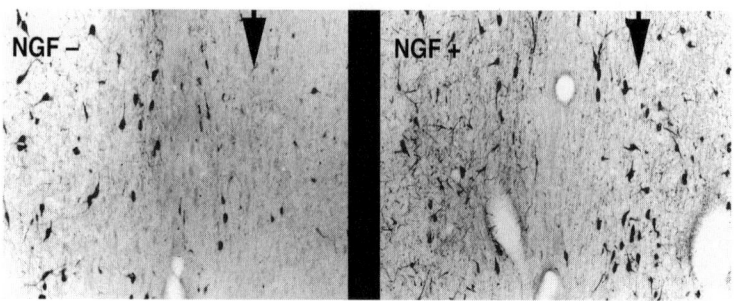

Figure 1. NGF prevents cholinergic neuronal death. (**A**) Fornix transections on the right side of the brain (arrow) cause cholinergic neurons to degenerate and die one month after the lesion. Neurons labeled choline acetyltransferase (ChAT). (**B**) NGF treatment prevents the death of the majority of cholinergic neurons after fornix lesions in rodents and primates (shown).

move forward. A clinical program of NGF in AD would test two key hypotheses: 1) that NGF could prevent cholinergic neuronal degeneration in AD, and 2) that addressing only the cholinergic component of cell decline in AD would meaningfully impact cognitive function. Although a number of neuronal populations undergo degeneration in Alzheimer's disease, cholinergic neuronal loss is particularly severe (Perry et al., 1978; Candy et al., 1983; Perry et al., 1985; Mufson et al., 2003a). Cholinergic neurons of the basal forebrain provide virtually all of the cholinergic innervation to the hippocampus and neocortex, and cholinergic systems play a major role in modulating neural activity in these target regions (Sofroniew et al., 1990; Howe and Mobley, 2001; Conner et al., 2003, 2005). Cholinergic systems have been shown to modulate memory and attention in rodents, primates and humans (Voytko et al., 1994; Wenk, 1997; Ridley et al., 2005). Indeed, of all the neuronal systems that degenerate in AD, the loss of cholinergic neurons correlates best with reduction in synapse number and cognitive impairment (Perry et al., 1985; Masliah et al., 1991). Highlighting the importance of cholinergic systems, the only drugs approved for the treatment of early and mid-stage AD are the class of cholinesterase inhibitors. These drugs, while only modestly elevating acetylcholine levels in the brain, have had detectable benefits on cognition in several independent clinical trials (Davis et al., 1992; Grundman and Thal, 2000). Unlike the cholinesterase inhibitors, however, NGF might both stimulate the cholinergic function of cells and prevent cell death in AD.

NGF is normally made throughout life in the neocortex and hippocampus (Korsching et al., 1985; Whittemore et al., 1986), and is taken up by specific receptors on cholinergic axon terminals (Dawbarn et al., 1988; Kiss et al., 1988; Holtzman et al., 1992) (see below). NGF, through receptor activation and retrograde transport to the cholinergic neuronal soma (Howe and Mobley,

2004), modulates the functional state of cholinergic neurons in the intact animal (Sofroniew et al., 1993; Conner et al., 2003, 2005). In AD, although production of NGF is not diminished in the hippocampus or cortex (Crutcher et al., 1993; Scott et al., 1995; Fahnestock et al., 1996; Narisawa-Saito et al., 1996; Hellweg et al., 1998; Hock et al., 2000; Mufson et al., 2003b), an apparent decrement of retrograde NGF transport leads to reduced levels within the cholinergic basal forebrain (Mufson et al., 1995; Scott et al., 1995). Impairment of NGF retrograde transport may arise from the cytoskeletal dysfunction that is a hallmark of AD pathology (Dai et al., 2002; Mandelkow et al., 2003; Stokin et al., 2005). The relative deficiency of NGF in basal forebrain neurons may contribute to cholinergic decline in AD (Mufson et al., 1995; Sofroniew et al., 2001; Tuszynski, 2002; Salehi et al., 2004), and therapeutic NGF delivery adjacent to cholinergic cell bodies could bypass the transport defect (Emerich et al., 1994; Hu et al., 1997; Smith et al., 1999; Conner et al., 2001). Whereas perturbations in levels of the high-affinity NGF receptor trkA and the lower affinity neutorophin receptor p75 occur in AD (Higgins and Mufson, 1989; Mufson et al., 1989; Mahadeo et al., 1994; Salehi et al., 1996; Boissiere et al., 1997; Mufson et al., 1997; Salehi et al., 2000; Savaskan et al., 2000; Chu et al., 2001; Mufson et al., 2002; Counts et al., 2004), sufficient quantities of these receptors remain expressed to retain NGF responsiveness.

Some reports suggest that the primary form of secreted NGF in the adult brain is a pro-form of the molecule that retains amino acid sequences normally involved in growth factor secretion from the cell (Lee et al., 2001; Fahnestock et al., 2004). The pro-form of NGF binds with higher avidity to the p75 receptor and, in this state, has been implicated in promoting cellular apoptosis (Harrington et al., 2004). If correct, these findings would suggest that therapeutically administered NGF should be of the "mature" configuration, lacking pro-NGF.

The preceding body of knowledge established a rationale for clinical trials of NGF to address the cholinergic component of cell loss in AD. However, a simple means of delivering NGF to the brain remained a challenge. It is a relatively large and polar molecule, thus it does not cross the blood brain barrier and requires central administration. Most studies examining effects of NGF in animal models infused the trophic factor intracerebroventricularly, thereby directly exposing diverse CNS regions to the growth factor. When delivered in this manner, NGF successfully rescues degenerating cholinergic neurons, but causes weight loss (Williams, 1991), sympathetic axon sprouting around the cerebral vasculature (Isaacson et al., 1990), and migration and proliferation of Schwann cells into a thick layer surrounding the medulla and spinal cord; this layer is densely penetrated by nociceptive sensory and sympathetic axons (Emmett et al., 1996; Winkler et al., 1997). These adverse affects originate from cells of the nervous system that express NGF receptors, suggesting that broad, non-targeted delivery of NGF is problematic. Indeed, three AD patients in

Sweden received mouse NGF intracerebroventricular infusions, and developed pain and weight loss after several months (Eriksdotter Jonhagen et al., 1998). Thus, the intracerebroventricular infusion of NGF for the treatment of AD is an unsatisfactory method of delivery. Instead, to be adequately assessed in human clinical trials, NGF delivery must meet the dual requirements of central administration in sufficient doses to achieve efficacy while restricting delivery to targeted brain regions.

One means of accurately targeting NGF delivery to the CNS is *gene delivery*, whereby either: a) autologous cultured cells are genetically modified to produce NGF and are then grafted into the region of the cholinergic basal forebrain cholinergic (*ex vivo* gene therapy), where the cells active as biological pumps for local delivery of the growth factor, or b) cells already within the basal forebrain region are genetically modified to produce elevated levels of NGF (*in vivo* gene therapy). Using *ex vivo* gene therapy, lesion-induced degeneration of basal forebrain cholinergic neurons has been prevented in both rodents and primates (Rosenberg et al., 1988; Emerich et al., 1994; Kordower et al., 1994b; Tuszynski et al., 1996), without the development of the adverse effects that follow intracerebroventricular NGF protein infusions. Further, age-related atrophy of basal forebrain cholinergic neurons and behavioral impairment is reversed in rodents (Chen and Gage, 1995). Subsequent studies in aged monkeys using autologous fibroblasts as vehicles of *ex vivo* NGF gene delivery showed complete reversal of age-related neuronal atrophy, restoration of cortical levels of cholinergic innervation, and maintenance of NGF gene expression for at least a year (Fig. 2) (Smith et al., 1999; Conner et al., 2001).

In subsequent dose-escalation and toxicity studies in primates, no adverse effects of *ex vivo* NGF gene delivery were found. Using autologous fibroblasts as vehicles of NGF gene delivery into the cholinergic basal forebrain, monkeys exhibited no weight loss, pain or Schwann cell migration into the CNS over periods of observation lasting up to one year. Doses exceeding the anticipated human dose by 10-fold showed no toxicity in the primate. Gene expression persisted for over one year in the primate and rodent brain, the longest time periods tested. Whereas gene expression over this period declined by approximately 90%, levels of NGF production by ELISA remained 5-fold greater than physiological NGF concentrations in the intact brain, and persistent cellular effects of NGF were detectable. Thus, preclinical studies in primates indicated that NGF could be accurately delivered to the basal forebrain, exert biological efficacy in two models (aging and cell degeneration), sustain gene expression for at least one year, and show no toxicity at doses exceeding those proposed for human trials.

The gene delivery vector used in these studies was derived from the murine Moloney leukemia virus (Rosenberg et al., 1988). The vector was non-virulent and incapable of replication, and as noted above, sustained gene expression

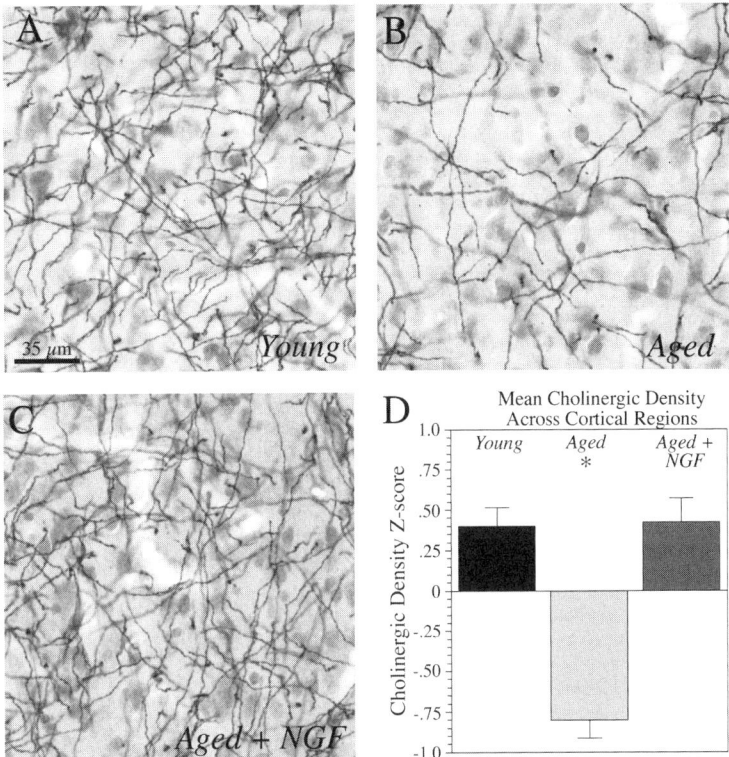

Figure 2. NGF enhances the density of cortical cholinergic terminals. (A) The normal architecture and density of cholinergic neurons in the non-aged, adult rhesus monkey is shown in the temporal cortex using acetylcholine esterase staining. (B) Aged rhesus monkeys show a spontaneous reduction in cholinergic axon density. (C) NGF gene delivery restores AChE labeling to levels observed in non-aged subjects, indicating that NGF ameliorates age-related cholinergic alterations in the cortex. (D) Quantification of the observed effects. From Conner et al., PNAS, 2001.

in the rodent and primate brain for the longest time periods examined. While the vector could not be regulated or turned off in the event that adverse effects developed, no adverse effects of non-regulated expression from this vector were found in extensive and long-term studies in both rodents and primates. While a regulatable vector system is highly desirable, practical and non-immunogenic regulatable systems for human use are not currently available.

Given the apparent safety and absence of toxicity of the non-regulated vector systems used in the preclinical studies summarized above, a rationale could be formulated for proceeding to human testing of NGF ex vivo gene delivery in humans with AD, consisting of the following: 1) Cholinergic cell degeneration is a consistent and significant feature of pathological neuronal degeneration in AD. 2) Cholinergic neurons exhibit robust sensitivity to NGF in pre-clinical studies, with extensive evidence that NGF prevents the death and stimulates the function of cholinergic neurons regardless of mechanism of cell death. 3) There is a pressing need for better AD therapies, and growth factor delivery offers the unique potential to prevent cell loss, rather than merely compensating for cell degeneration after it has occurred. While it is not proven that targeting the cholinergic component of cell loss in AD will have a meaningful impact on overall cognitive function and quality of life over an extended time period in the disease, a cogent rationale for predicting such an outcome exists and is readily testable in a human clinical trial, based on the importance of cholinergic systems in critically regulating the function of diverse cortical systems. Given the availability of a safe and effective means of delivering NGF for prolonged time periods in animal models, including primates, we proceeded with a human clinical trial of ex vivo NGF gene delivery in AD.

3.1   A Phase 1 human trial of ex vivo NGF gene therapy for AD

The results of this clinical trial were recently published (Tuszynski et al., 2005). Eight subjects with early stage "probable AD" by NIH research criteria were recruited into this open label Phase 1 trial to assess the safety and feasibility of NGF gene delivery in AD. Early stage patients were recruited for two reasons: first, because these individuals frequently retain insight and the ability to understand the potential risks of an experimental procedure; this was a particularly important consideration for obtaining informed consent in the first human clinical trial of growth factor gene delivery. Second, one of the hypothetical mechanisms of NGF benefit in AD, prevention of cell death, is most likely to be of benefit in early and mid-stages of the disease when a majority of cholinergic neurons remain viable. The average age of enrolled subjects was 67 years, with a range of 54–76 years.

Subjects underwent skin biopsies to obtain primary autologous fibroblasts. These cells were cultivated and expanded in vitro, and were genetically modified using the same MLV vectors described above to express and secrete human NGF (Fig. 3).

Quantities of NGF produced by cells in vitro prior to implantation ranged from 50–75 ng NGF/10^6 cells/day, similar to amounts observed in preclinical animal studies. Nearly all of the NGF secreted from transduced cells was the full-length protein implicated in promoting cell survival; pro-NGF was virtually undetectable (Tuszynski et al., 2005). After testing for sterility and purity, cells were harvested and frozen until subsequent injection into patients. Three dose cohorts were enrolled: Subjects 1 and 2 received a total of 1.25×10^6 autologous fibroblasts into the non-dominant, right nucleus basalis; subjects 3– 6 received a total of 2.5×10^6 cells, injected into both left and right basal nuclei;

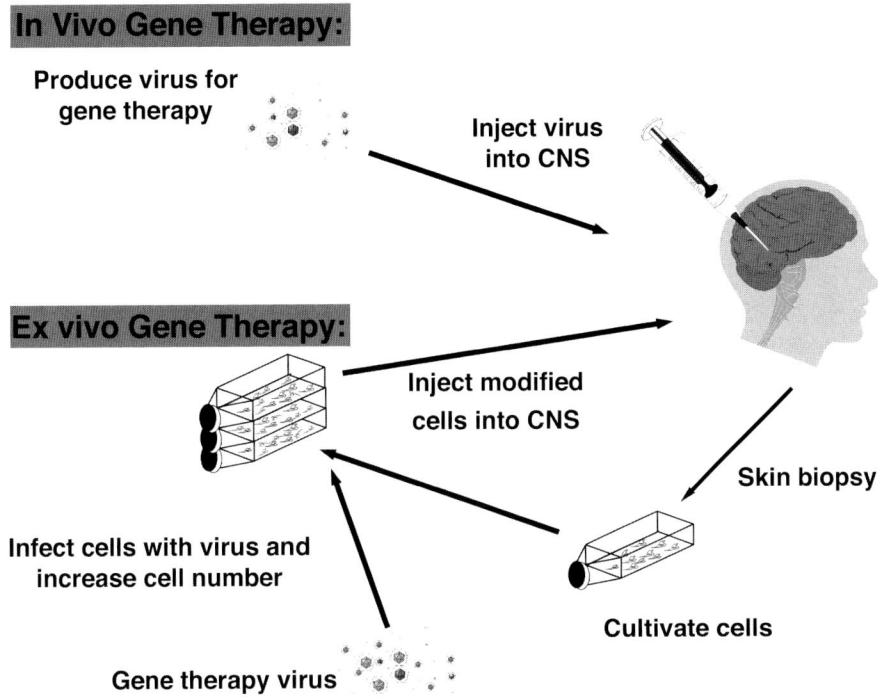

Figure 3. Schematic of gene transfer. *In vivo* gene therapy utilizes injection of viral vectors carrying therapeutic genes into the brain. *In vivo* gene delivery vectors, such as AAV and lentivirus, are capable of genetically modifying non-dividing cells of the adult CNS. In contrast, *ex vivo* gene therapy transduces donor host cells that are capable of dividing *in vitro*; donor cells are obtained from biopsies. These cells are characterized *in vitro* for production of the desired gene product, and then introduced into specific brain regions where they act as biological minipumps for the local delivery of growth factors.

and subjects 7 and 8 received injection of 5×10^6 cells also injected both left and right basal nucleus.

The mean Mini-Mental Status Exam (MMSE) score at the time of subject enrollment into the study was 27. Subjects underwent cell injection into the brain at staggered intervals ranging from 3 to 7 months between dose cohorts, to allow sufficient time for safety assessment before dosing of the next group. This prolonged period of pre-treatment observation allowed repeated assessment of the MMSE over a mean period of 11 months before injection of NGF-secreting cells into the brain. Baseline MRI scans and, in some cases, PET scans were also obtained.

Cells were injected into the nucleus basalis region using standard stereotaxic targeting techniques. On study initiation, patients underwent cell injections into the nucleus basalis in an awake but lightly sedated state to assist patient monitoring. However, two subjects moved abruptly while the injection needle was in the brain, causing subcortical hemorrhage. Neither hemorrhage required drainage. One subject with a hemorrhage gradually improved and was discharged with a moderate worsening of his baseline language deficit. The second subject was initially stuporous but was gradually improving. However, 5 weeks post-injection he sustained a pulmonary embolus leading to myocardial infarction and sudden death. Subsequently, all subjects underwent general anesthesia or (in a recent trial) deep sedation to avoid movement, and all subsequent cell injections have been completed safely. Patients were discharged 1–2 days post-treatment.

With mean follow-up of two years, there were no long-term adverse events of NGF gene delivery in AD. No weight loss, pain or MRI abnormalities occurred as a result of the treatment. Serial PET scans were obtained in four bilaterally treated subjects, and demonstrated significant increases in cortical glucose uptake when measured 6–8 months after the first scan, reversing a pattern of decline normally observed in AD over time (Fig. 4) (Potkin et al., 2001; Alexander et al., 2002).

Autopsy was obtained in the patient who died five weeks post-treatment. The patient had multiple cortical plaques (Braak stage 4) and neurofibrillary tangles distributed throughout the cortex, as well as cortical Lewy bodies. Robust NGF gene expression by in situ hybridization was evident in the region of autlogous cell implantation. Immunocytochemistry revealed a robust trophic response of host cholinergic axons to the growth factor source (see Fig. 4). This finding provided the first direct evidence in the human brain that degenerating neurons retain sensitivity to growth factors, a prerequisite if growth factors are to be of value in treating human neurodegenerative disorders.

Cognitive testing was performed in all subjects. Conclusions should not be drawn from a small cohort of subjects lacking blinded assessments and comparison to a placebo control group. With these caveats, testing on the Mini-Mental Status Examination (MMSE) and Alzheimer's Disease Assessment Scale – Cognitive subcomponent (ADAS-Cog) scales suggested possible reductions in rate of decline after receiving NGF. Examining annual rates of decline over a two-year period, the mean rate of decline on the ADAS-Cog was reduced by approximately one-third in the period of 6–18 months and 12–24 months compared to the period of 1–12 months after receiving NGF (Fig. 5).

Similar effects were observed when examining *median* ADAS-Cog decline scores. Measurement of the MMSE also showed possible reductions in rate of cognitive decline beginning approximately 6-months post-treatment (Fig. 5). The period of 6–18 months post-treatment is of particular interest because, based on aged non-human primate studies (Smith et al., 1999; Conner et al., 2001), it is projected that 6 months would be required to remodel cortical

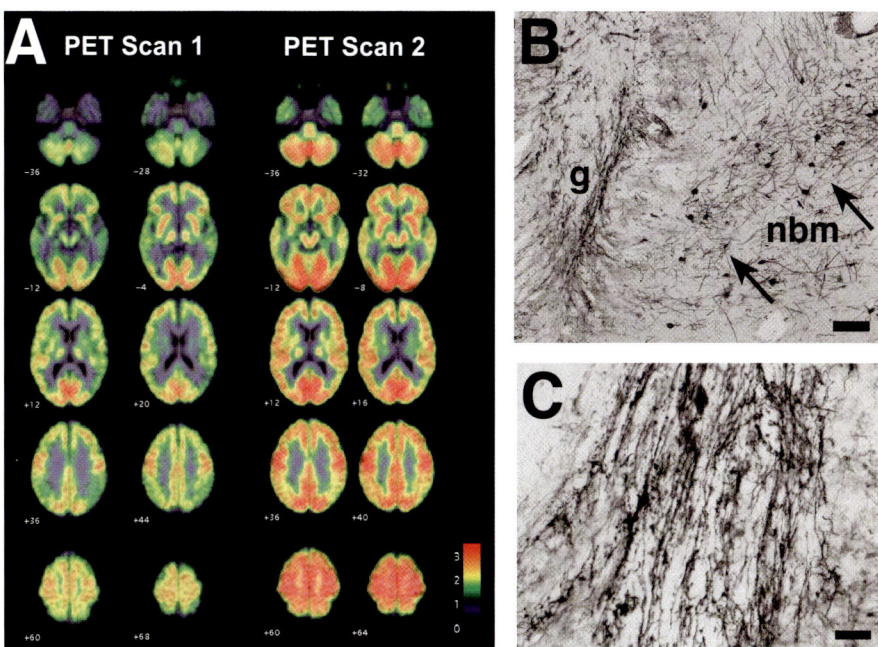

Figure 4. PET Scan and trophic response. (**A**) FDG PET scans in four subjects treated with NGF, overlaid on standardized MRI templates. Representative axial sections, with 6–8 months between first and second scan, showing interval increases in brain metabolism in diverse cortical regions. Flame scale indicates FDG use /100g tissue/min; red color indicates more FDG use than blue. (**B**) Immunocytochemistry for cholinergic neurons (p75) shows graft implant on left (**g**) and adjacent neurons of **nbm** (arrows). (**C**) Higher magnification shows extensive penetration of cholinergic axons into graft, a "trophic" response to NGF in the human brain. Scale bar 82 μm, **B**; 11 μm, **C**. From Tuszynski et al., Nat Med, 2005.

cholinergic terminals after NGF delivery. The strengthening of cortical cholinergic terminals would presumably serve as the mechanism underlying potential effects on cognitive function and PET. Further, it has been documented that NGF transgene expression persists for at least 18 months in the primate brain. In this human trial, in the epoch of 6–18 months post-treatment, 5 of 6 subjects demonstrated stabilization or improvement in cognitive function. Overall, over the two-year period of post-treatment observation, rate of cognitive decline was reduced by approximately 50%.

In summary, this human trial of ex vivo NGF gene delivery demonstrated four points: 1) NGF can be delivered safely to the brain over an extended time period using gene delivery, but subjects must undergo cell injections under general anesthesia or deep sedation. 2) Degenerating cholinergic neurons of the human brain can exhibit trophic responses to growth factors. 3) Broad regions

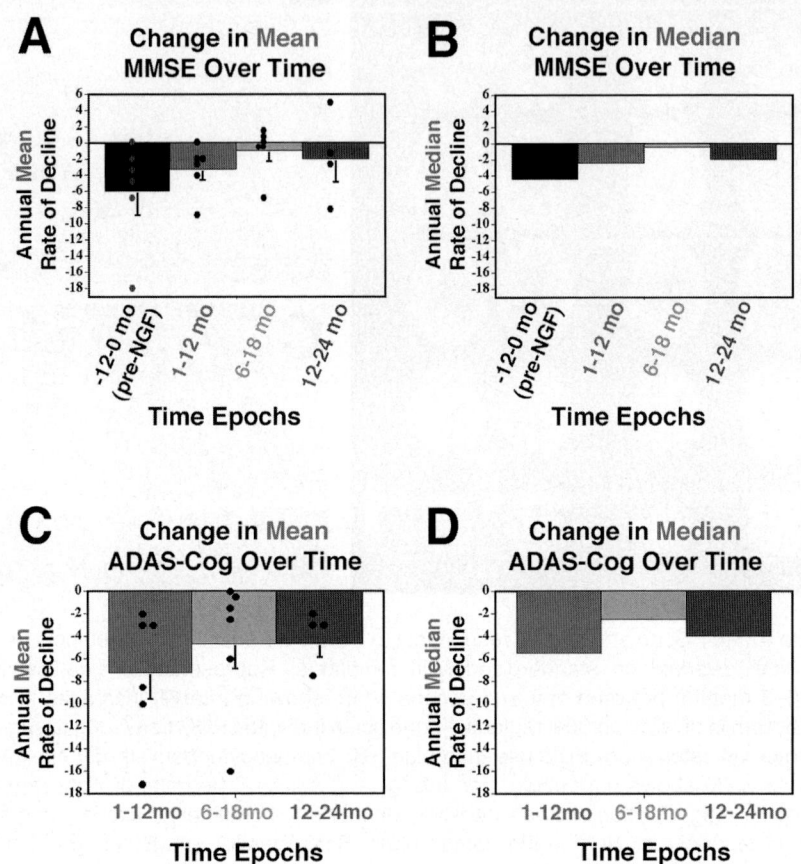

Figure 5. Cognitive outcomes. (**A**) Mean annualized change in MMSE score in year prior to treatment, and in time epochs of 1–12 mo, 6–18 mo, and 12–24 mo post-treatment. Individual subject data points shown in circles. Of note, during one-year period beginning 6 mo after treatment, when sufficient time passed for NGF to enhance cholinergic projections, 2 of 6 patients show improved MMSE scores, one patient has no decline, and two patients decline only one point. Overall decline was reduced 51% compared to pre-operative rate for the mean 22 mo post-treatment period. (**B**) Given wide range in MMSE scores, median data also are shown and parallel observations of mean scores. (**C**) Mean annualized changes in ADAS-Cog over time epochs of 1–12 mo, 6–18 mo, and 12–24 mo post-treatment. Individual subject data points shown in circles. As with MMSE, rate of decline slows after sufficient time has passed (6 mo) for NGF to enhance cholinergic systems. Rate of decline is reduced by 36% at 6–18 mo, compared to 1–12 mo. (**D**) Given wide range in ADAS-Cog, median data also are shown. Rate of decline is slowed by 55% at 6–18 mo compared to 1–12 mo. From Tuszynski et al., Nat Med, 2005.

of the cortex demonstrate enhanced cortical glucose metabolism following NGF gene transfer. 4) There is sufficient indication of a possible effect on cognition to justify future, larger, controlled and blinded clinical trials of NGF gene delivery in AD.

3.2 AAV-NGF gene delivery for AD

The preceding clinical program utilized *ex vivo* NGF gene transfer. While useful for proof-of-principle regarding potential benefits of trophic factor therapy for neurodegenerative disease, *ex vivo* techniques suffer from the disadvantage that they are generally cumbersome, labor-intensive and expensive. A simpler approach for potentially broad application to human disease is *in vivo* gene delivery, wherein vectors capable of modifying non-dividing cells of the brain are injected intracerebrally. As noted above, several *in vivo* gene delivery vectors hypothetically meet the requirements for targeted and regionally restricted delivery of growth factors to the brain. One of these vectors, adeno-associated virus (AAV), has now been employed in six clinical trials in neurological disorders, including a trial of AAV-NGF gene delivery in AD.

In preclinical studies, AAV-NGF gene delivery and lentiviral-NGF gene delivery prevent cholinergic neuronal death and enhance cognitive function to degrees similar to those reported using NGF protein infusions and *ex vivo* NGF gene transfer (Klein et al., 2000; Blesch et al., 2005). Extensive preclinical studies also demonstrated that non-regulated AAV-NGF gene delivery was safe in the CNS over extended time periods exceeding one year. Based on these data, a phase I clinical trial has been performed at Rush University, sponsored by the biotechnology industry, in which AAV-NGF has been injected into the brain bilaterally in six subjects with early-to-moderate AD. Findings of the study will be reported in the near future. Of note, unlike *ex vivo* gene delivery approaches in which transgene expression slowly declines over time, anecdotal reports suggest that transduction of neurons with AAV leads to no evident decrement in gene expression for several years in the rodent and primate CNS. Thus, *in vivo* gene delivery may achieve a sustained and constant supply of growth factor over time.

3.3 Other means of NGF delivery to the CNS

The preceding sections have dealt with gene delivery as a means of accurately targeting and restricting growth factor delivery to the brain. Other methods of delivering neurotrophins to the CNS have also been reported in the literature, include binding active components of neurotrophic factors to carrier molecules that naturally cross the blood brain barrier (e.g., transferrin-linked NGF transport; (Friden et al., 1993; Kordower et al., 1994a)), generating small peptide analogs of the growth factors that cross the blood brain barrier to activate neurotrophin receptors (Massa et al., 2003), or direct intraparenchymal infusions of growth factors (Gill et al., 2003; Slevin et al., 2005). The benefit of simple peripheral administration of a growth factor is obvious: the risk and expense of cranial injection is avoided, and doses can be adjusted and terminated as needed. However, all of the peripherally administered approaches to growth factor delivery, if effectively simulating the brain, will suffer from the drawback that they are *non-targeted*, with the risk of causing adverse events from stimulation of non-targeted systems that will ultimately limit their usefulness. Direct intraparenchymal infusions of growth factors are another means of accurately targeting growth factors in sufficient doses to elicit biological efficacy, while limiting spread to avoid non-targeted brain regions (Gill et al., 2003; Slevin et al., 2005). These approaches have the added benefit that they can be discontinued in the event of toxicity. However, the practicality of intraparenchymal infusion is reduced by the difficulty of spreading the growth factor evenly throughout a target region from either a single point-source tip of a catheter, or from a catheter containing multiple pores arranged linearly. Implanted hardware is also vulnerable to malfunction and infection (Gill et al., 2003). Further efforts to develop this technology in Parkinson's disease face an uncertain future.

4. Other targets of growth factor therapy in AD: Brain-derived neurotrophic factor (BDNF)

Recent evidence indicates that another growth factor, BDNF, can potently protect *cortical* circuits that degenerate in AD (Nagahara et al., 2005). Specifically, BDNF protein infusions or gene delivery prevent the death of entorhinal cortical neurons after axotomy, or following exposure to Aβ *in vitro*. In aged rats, BDNF infusions to the entorhinal cortex reverse age-related decrements in hippocampal-dependent forms of learning and memory, including spatial memory in the Morris water maze and contextual fear conditioning (Nagahara et al., 2005). Finally, in transgenic mice overexpressing the Swedish and Indiana amyloid mutations (Mucke et al., 2000; Palop et al., 2003), BDNF gene delivery after the onset of neurodegeneration significantly increases synapse number, stimulates phosphorylation of the transcriptional activator CREB, and improves performance on two cognitive tasks sensitive to the integrity of the hippocampal formation, spatial memory in the Morris water maze and contextual fear conditioning (Nagahara et al., 2005).

BDNF is normally produced in the entorhinal cortex throughout life, and is anterogradely transported into the hippocampus where it influences synaptic plasticity and neuronal function (Altar et al., 1997; Conner et al., 1998).

Importantly from a practical perspective, BDNF administered therapeutically in the models described above is actively transported into the hippocampus, and restores synapse density in the hippocampal dentate gyrus in amyloid mutant mice. Thus, delivery to the practical entorhinal target can provide a "gateway" for improving the functional state of the hippocampus more broadly. This provides an intriguing potential future target in AD. Neuronal degeneration occurs early in the entorhinal and cortex in AD (Gomez-Isla et al., 1996; Kordower et al., 2001), likely causing alterations in short-term memory that are characteristic of AD and its precursor state, mild cognitive impairment. Levels of BDNF and its receptor trkB decline in the entorhinal cortex and neocortex in AD (Narisawa-Saito et al., 1996; Connor et al., 1997; Allen et al., 1999; Ferrer et al., 1999; Hock et al., 2000; Michalski and Fahnestock, 2003). The entorhinal cortex could readily be targeted by vector injections in the human brain, providing a potential mechanism of neuroprotection early in the disease to prevent subsequent neurodegeneration.

5.   Conclusions

Growth factors have the potential to potently influence neuronal survival and function. Their broad efficacy against a variety of pathogenic insults likely arises because growth factors influence final common mechanisms that mediate cell atrophy and death, including apoptosis-related mechanisms (bcl-2, bax, and the caspases) as well as genes important for cell function (ERK/MAP kinase, CREB, others). As such, the growth factors offer the potential to treat neurodegenerative disorders either alone or as combination therapies with other anti-AD therapies, including anti-amyloid approaches. The transition to clinical trials using methods that generate high yet locally restricted levels of growth factors should, over the next few years, indicate whether they will ultimately have a useful role in the treatment of AD.

References

Aicardi G, Argilli E, Cappello S, Santi S, Riccio M, Thoenen H, Canossa M (2004) Induction of long-term potentiation and depression is reflected by corresponding changes in secretion of endogenous brain-derived neurotrophic factor. Proc Natl Acad Sci USA 101:15788–15792.

Alexander GE, Chen K, Pietrini P, Rapoport SI, Reiman EM (2002) Longitudinal PET Evaluation of Cerebral Metabolic Decline in Dementia: A Potential Outcome Measure in Alzheimer's Disease Treatment Studies. Am J Psychiatry 159:738–745.

Allen SJ, Wilcock GK, Dawbarn D (1999) Profound and selective loss of catalytic TrkB immunoreactivity in Alzheimer's disease. Biochem Biophys Res Commun 264:648–651.

Altar CA, Cai N, Bliven T, Juhasz M, Conner JM, Acheson AL, Lindsay RM, Wiegand SJ (1997) Anterograde transport of brain-derived neurotrophic factor and its role in the brain. Nature 389:856–860.

Blesch A, Pfeiffer A, Conne JM, Britton W, Verma I, Tuszynski MH (2005) Regulated lentiviral NGF genetransfer controls rescue of medial septal cholinergic neurons. Molec Ther 11:916–925.

Boissiere F, Hunot S, Faucheux B, Hersh LB, Agid Y, Hirsch EC (1997) Trk neurotrophin receptors in cholinergic neurons of patients with Alzheimer's disease. Dement Geriatr Cogn Disord 8:1–8.

Candy JM, Perry RH, Perry EK, Irving D, Blessed G, Fairbairn AF, Tomlinson BE (1983) Pathological changes in the nucleus basalis of Meynert in Alzheimer's and Parkinson's diseases. J Neurosci 54:277–289.

Chen KS, Gage FH (1995) Somatic gene transfer of NGF to the aged brain: Behavioral and morphological amelioration. J Neurosci 15:2819–2825.

Chu Y, Cochran EJ, Bennett DA, Mufson EJ, Kordower JH (2001) Down-regulation of trkA mRNA within nucleus basalis neurons in individuals with mild cognitive impairment and Alzheimer's disease. J Comp Neurol 437:296–307.

Conner JM, Lauterborn JC, Gall CM (1998) Anterograde transport of neurotrophin proteins in the CNS – a reassessment of the neurotrophic hypothesis. Rev Neurosci 9:91–103.

Conner JM, Chiba AA, Tuszynski MH (2005) The basal forebrain cholinergic system is essential for cortical plasticity and functional recovery following brain injury. Neuron 46:173–179.

Conner JM, Culberson A, Packowski C, Chiba A, Tuszynski MH (2003) Lesions of the basal forebrain cholinergic system impair task acquisition and abolish cortical plasticity associated with motor skill learning. Neuron 38:819–829.

Conner JM, Darracq MA, Roberts J, Tuszynski MH (2001) Non-tropic actions of neurotrophins: Subcortical NGF gene delivery reverses age-related degeneration of primate cortical cholinergic innervation. Proc Nat Acad Sci 98:1941–1946.

Connor B, Young D, Yan Q, Faull RL, Synek B, Dragunow M (1997) Brain-derived neurotrophic factor is reduced in Alzheimer's disease. Brain Res Mol Brain Res 49:71–81.

Cooper JD, Salehi A, Delcroix JD, Howe CL, Belichenko PV, Chua-Couzens J, Kilbridge JF, Carlson EJ, Epstein CJ, Mobley WC (2001) Failed retrograde transport of NGF in a mouse model of Down's syndrome: Reversal of cholinergic neurodegenerative phenotypes following NGF infusion. Proc Natl Acad Sci 98:10439–10444.

Counts SE, Nadeem M, Wuu J, Ginsberg SD, Saragovi HU, Mufson EJ (2004) Reduction of cortical TrkA but not p75(NTR) protein in early-stage Alzheimer's disease. Ann Neurol 56:520–531.

Crutcher KA, Scott SA, Liang S, Everson WV, Weingartner J (1993) Detection of NGF-like activity in human brain tissue: increased levels in Alzheimer's disease. J Neurosci 13:2540–2550.

Dai J, Buijs RM, Kamphorst W, Swaab DF (2002) Impaired axonal transport of cortical neurons in Alzheimer's disease is associated with neuropathological changes. Brain Res 948:138–144.

Davis KL, Thal LJ, Gamzu ER, Davis CS, Woolson RF, Gracon SI, Drachman DA, Schneider LS, Whitehouse PJ, Hoover TM (1992) A double-blind, placebo-controlled multicenter study of tacrine for Alzheimer's disease. The Tacrine Collaborative Study Group. New Engl J Med 327:1253–1259.

Dawbarn D, Allen SJ, Semenenko FM (1988) Coexistence of choline acetyltransferase and nerve growth factor receptors in the rat basal forebrain. Neurosci Lett 94:138–144.

Dekker AJ, Gage FH, Thal LJ (1992) Delayed treatment with nerve growth factor improves acquisition of a spatial task in rats with lesions of the nucleus basalis magnocellularis: evaluation of the involvement of different neurotransmitter systems. Neuroscience 48:111–119.

Emerich DW, Winn S, Harper J, Hammang JP, Baetge EE, Kordower JH (1994) Implants of polymer-encapsulated human NGF-secreting cells in the nonhuman primate: Rescue and sprouting of degenerating cholinergic basal forebrain neurons. J Comp Neurol 349:148–164.

Emmett CJ, Stewart GR, Johnson RM, Aswani SP, Chan RL, Jakeman LB (1996) Distribution of radioiodinated recombinant human nerve growth factor in primate brain following intracerebroventricular infusion. Exp Neurol 140:151–160.

Eriksdotter Jonhagen M, Nordberg A, Amberla K, Backman L, Ebendal T, Meyerson B, Olson L, Seiger, Shigeta M, Theodorsson E, Viitanen M, Winblad B, Wahlund LO (1998) Intracerebroventricular infusion of nerve growth factor in three patients with Alzheimer's disease. Dement Geriatr Cogn Disord 9:246–257.

Fahnestock M, Yu G, Coughlin MD (2004) ProNGF: a neurotrophic or an apoptotic molecule? Prog Brain Res 146:101–110.

Fahnestock M, Scott SA, Jette N, Weingartner JA, Crutcher KA (1996) Nerve growth factor mRNA and protein levels measured in the same tissue from normal and Alzheimer's disease parietal cortex. Brain Res Mol Brain Res 42:175–178.

Ferrer I, Marin C, Rey MJ, Ribalta T, Goutan E, Blanco R, Tolosa E, Marti E (1999) BDNF and full-length and truncated TrkB expression in Alzheimer disease. Implications in therapeutic strategies. J Neuropathol Exp Neurol 58:729–739.

Fischer W, Wictorin K, Bjorklund A, Williams LR, Varon S, Gage FH (1987) Amelioration of cholinergic neuron atrophy and spatial memory impairment in aged rats by nerve growth factor. Nature 329:65–68.

Friden PM, Walus LR, Watson P, Doctrow SR, Kozarich JW, Backman C, Hoffer B, Bloom F, Granholm AC (1993) Blood-brain barrier penetration and in vivo activity of an NGF conjugate. Science 259:373–377.

Gash DM, Zhang Z, Ovadia A, Cass WA, Yi A, Simmerman L, Russell D, Martin D, Lapchak PA, Collins F, Hoffer BJ, Gerhardt GA (1996) Functional recovery in parkinsonian monkeys treated with GDNF. Nature 380:252–255.

Gill SS, Patel NK, Hotton GR, O'Sullivan K, McCarter R, Bunnage M, Brooks DJ, Svendsen CN, Heywood P (2003) Direct brain infusion of glial cell line-derived neurotrophic factor in Parkinson disease. Nat Med 9:589–595.

Gomez-Isla T, Price JL, McKeel DW, Morris JC, Growdon JH, Hyman BT (1996) Profound loss of layer II entorhinal cortex neurons occurs in very mild Alzheimer's disease. J Neurosci 16:4491–4500.

Grundman M, Thal LJ (2000) Treatment of Alzheimer's disease: rationale and strategies. Neurol Clin 18:807–828.

Harrington AW, Leiner B, Blechschmitt C, Arevalo JC, Lee R, Morl K, Meyer M, Hempstead BL, Yoon SO, Giehl KM (2004) Secreted proNGF is a pathophysiological death-inducing ligand after adult CNS injury. Proc Natl Acad Sci USA 101:6226–6230.

Hefti F (1986) Nerve growth factor (NGF) promotes survival of septal cholinergic neurons after fimbrial transection. J Neurosci 6:2155–2162.

Hellweg R, Gericke CA, Jendroska K, Hartung HD, Cervos-Navarro J (1998) NGF content in the cerebral cortex of non-demented patients with amyloid-plaques and in symptomatic Alzheimer's disease. Int J Dev Neurosci 16:787–794.

Higgins GA, Mufson EJ (1989) NGF receptor gene expression is decreased in the nucleus basalis in Alzheimer's disease. Exp Neurol 106:222–236.

Hock C, Heese K, Hulette C, Rosenberg C, Otten U (2000) Region-specific neurotrophin imbalances in Alzheimer disease: decreased levels of brain-derived neurotrophic factor and increased levels of nerve growth factor in hippocampus and cortical areas. Arch Neurol 57:846–851.

Holtzman DM, Li Y, Chen K, Gage FH, Epstein CJ, Mobley WC (1993) Nerve growth factor reverses neuronal atrophy in a Down syndrome model of age-related neurodegeneration. Neurology 43:2668–2673.

Holtzman DM, Li Y, Parada LF, Kinsman S, Chen CK, Valletta JS, Zhou J, Long JB, Mobley WC (1992) p140trk mRNA marks NGF-responsive forebrain neurons: evidence that trk gene expression is induced by NGF. Neuron 9:465–478.

Howe CL, Mobley WC (2004) Signaling endosome hypothesis: A cellular mechanism for long distance communication. J Neurobiol 58:207–216.

Howe CL, Mobley WC (2001) Nerve growth factor effects on cholinergic modulation of hippocampal and cortical plasticity. In: Neurobiology of the Neurotrophins (Mocchetti I, ed). Johnson City, TN: F.P. Graham Publishing.

Hu L, Cote SL, Cuello AC (1997) Differential modulation of the cholinergic phenotype of the nucleus basalis magnocellularis neurons by applying NGF at the cell body or cortical terminal fields. Exp Neurol 143:162–171.

Isaacson LG, Saffran BN, Crutcher KA (1990) Intracerebral NGF infusion induces hyperinnervation of cerebral blood vessels. Neurobiology of Aging 11:51–55.

Kaplan DR, Miller FD (2000) Neurotrophin signal transduction in the nervous system. Curr Opin Neurobiol 10:381–391.

Kiss J, McGovern J, Patel AJ (1988) Immunohistochemical localization of cells containing nerve growth factor receptors in the different regions of the adult rat forebrain. Neurosci Lett 27:731–748.

Klein RL, Hirko AC, Meyers CA, Grimes JR, Muzyczka N, Meyer EM (2000) NGF gene transfer to intrinsic basal forebrain neurons increases cholinergic cell size and protects from age-related, spatial memory deficits in middle–aged rats. Brain Res 875:144–151.

Koliatsos VE, Clatterbuck RE, Nauta HJ, Knusel B, Burton LE, Hefti FF, Mobley WC, Price DL (1991) Human nerve growth factor prevents degeneration of basal forebrain cholinergic neurons in primates. Ann Neurol 30:831–840.

Kordower JH, Charles V, Bayer R, Bartus RT, Putney S, Walus LR, Friden PM (1994a) Intravenous administration of a transferrin receptor antibody-nerve growth factor conjugate prevents the degeneration of cholinergic striatal neurons in a model of Huntington's disease. Proc Nat Acad Sci 91:9077–9080.

Kordower JH, Chu Y, Stebbins GT, DeKosky ST, Cochran EJ, Bennett D, Mufson EJ (2001) Loss and atrophy of layer II entorhinal cortex neurons in elderly people with mild cognitive impairment. Ann Neurol 49:202–213.

Kordower JH, Winn SR, Liu Y-T, Mufson EJ, Sladek JR, Hammang JP, Baetge EE, Emerich DF (1994b) The aged monkey basal forebrain: Rescue and sprouting of axotomized basal forebrain neurons after grafts of encapsulated cells secreting human nerve growth factor. Proc Nat Acad Sci 91:10898–10902.

Kordower JH, Emborg ME, Bloch J, Ma SY, Chu Y, Leventhal L, McBride J, Chen EY, Palfi S, Roitberg BZ, Brown WD, Holden JE, Pyzalski R, Taylor MD, Carvey P, Ling Z, Trono D, Hantraye P, Déglon N, Aebischer P (2000) Neurodegeneration prevented by lentiviral vector delivery of GDNF in primate models of Parkinson's disease. Science 290:767–773.

Korsching S, Auburger G, Heumann R, Scott J, Thoenen H (1985) Levels of nerve growth factor and its mRNA in the central nervous system of the rat correlate with cholinergic innervation. EMBO J 4:1389–1393.

Kromer LF, Bjorklund A, Stenevi U (1981) Regeneration of the septohippocampal pathways in adult rats is promoted by utilizing embryonic hippocampal implants was bridges. Brain Research 210:173–200.

Lee R, Kermani P, Teng KK, Hempstead BL (2001) Regulation of cell survival by secreted proneurotrophins. Science 294:1945–1948.

Levi-Montalcini R (1987) The nerve growth factor 35 years later. Science 237:1154–1162.

Levi-Montalcini R, Hamburger V (1951) Selective growth stimulating effects of mouse sarcoma on the sensory and sympathetic nervous system of the chick embryo. J Exp Zool 116:321–362.

Levi-Montalcini R, Hamburger Y (1953) A diffusible agent of mouse sarcoma, producing hyperplasia of sympathetic ganglia and hyperneurotization of viscera in the chick embryo. J Exp Zool 123:233–288.

Levi-Montalcini R, Meyer H, Hamburger V (1954) In vitro experiments on the effects of mouse sarcoma 180 and 37 on the spinal and sympathetic ganglia of the chick embryo. Cancer Research 14:49–57.

Liberini P, Pioro EP, Maysinger D, Ervin FR, Cuello AC (1993) Long-term protective effects of human recombinant nerve growth factor and monosialoganglioside GM1 treatment on primate nucleus basalis cholinergic neurons after neocortical infarction. Neuroscience 53:625–637.

Mahadeo D, Kaplan L, Chao MV, Hempstead BL (1994) High affinity nerve growth factor binding displays a faster rate of association than p140trk binding. Implications for multisubunit polypedtide receptors. J Biol Chem 269:6884-6891.

Mandelkow EM, Stamer K, Vogel R, Thies E, Mandelkow E (2003) Clogging of axons by tau, inhibition of axonal traffic and starvation of synapses. Neurobiol Aging 24:1079–1085.

Markowska AL, Koliatsos VE, Breckler SJ, Price DL, Olton DS (1994) Human nerve growth factor improves spatial memory in aged but not in young rats. J Neurosci 14:4815–4824.

Masliah E, Terry RD, Alford M, DeTeresa R, Hansen LA (1991) Cortical and subcortical patterns of synaptophysinlike immunoreactivity in Alzheimer's disease. Am J Pathol 138:235–246.

Massa SM, Xie Y, Longo FM (2003) Alzheimer's therapeutics: neurotrophin domain small molecule mimetics. J Mol Neurosci 20:323–326.

Michalski B, Fahnestock M (2003) Pro-brain-derived neurotrophic factor is decreased in parietal cortex in Alzheimer's disease. Brain Res Mol Brain Res 111:148–154.

Mucke L, Masliah E, Yu GQ, Mallory M, Rockenstein EM, Tatsuno G, Hu K, Kholodenko D, Johnson-Wood K, McConlogue L (2000) High-level neuronal expression of abeta 1-42 in wild-type human amyloid protein precursor transgenic mice: synaptotoxicity without plaque formation. J Neurosci 20:4050–4058.

Mufson EJ, Bothwell M, Kordower JH (1989) Loss of nerve growth factor receptor-containing neurons in Alzheimer's disease: A quantitative analysis across subregions of the basal forebrain. Exp Neurol 105:221–232.

Mufson EJ, Conner JM, Kordower JH (1995) Nerve growth factor in Alzheimer's disease: defective retrograde transport to nucleus basalis. Neuroreport 6:1063–1066.

Mufson EJ, Ginsberg SD, Ikonomovic MD, DeKosky ST (2003a) Human cholinergic basal forebrain: chemoanatomy and neurologic dysfunction. J Chem Neuroanat 26:233–242.

Mufson EJ, Lavine N, Jaffar S, Kordower JH, Quirion R, Saragovi HU (1997) Reduction in p140-TrkA receptor protein within the nucleus basalis and cortex in Alzheimer's disease. Exp Neurol 146:91–103.

Mufson EJ, Ikonomovic MD, Styren SD, Counts SE, Wuu J, Leurgans S, Bennett DA, Cochran EJ, DeKosky ST (2003b) Preservation of brain nerve growth factor in mild cognitive impairment and Alzheimer disease. Arch Neurol 60:1143–1148.

Mufson EJ, Ma SY, Dills J, Cochran EJ, Leurgans S, Wuu J, Bennett DA, Jaffar S, Gilmor ML, Levey AI, Kordower JH (2002) Loss of basal forebrain P75(NTR) immunoreactivity in subjects with mild cognitive impairment and Alzheimer's disease. J Comp Neurol 443:136–153.

Nagahara AH, Schroeder BE, Wang L, Torres R, Blesch A, Rockenstein E, Masliah E, Koo E, Tuszynski MH (2005) BDNF gene delivery into entorhinal cortex reverses behavioral deficits in APP transgenic mice. Society for Neuroscience Abstract 30:206–209.

Narisawa-Saito M, Wakabayashi K, Tsuji S, Takahashi H, Nawa H (1996) Regional specificity of alterations in NGF, BDNF and NT-3 levels in Alzheimer's disease. Neuroreport 7:2925–2928.

Palop JJ, Jones B, Kekonius L, Chin J, Yu GQ, Raber J, Masliah E, Mucke L (2003) Neuronal depletion of calcium-dependent proteins in the dentate gyrus is tightly linked to Alzheimer's disease-related cognitive deficits. Proc Natl Acad Sci USA 100:9572–9577.

Perry EK, Tomlinson BE, Blessed G, Bergmann K, Gibson PH, Perry RH (1978) Correlation of cholinergic abnormalities with senile plaques and mental test scores in senile dementia. British Medical Journal 2:1457–1459.

Perry EK, Curtis M, Dick DJ, Candy JM, Atack JR, Bloxham CA, Blessed G, Fairbairn A, Tomlinson BE, Perry RH (1985) Cholinergic correlates of cognitive impairment in Parkinson's disease: comparison with Alzheimer's disease. J Neurol Neurosurg Psychiatry 48:413–421.

Potkin SG, Anand R, Alva G, Fallon JH, Keator D, Carreon D, Messina J, Wu JC, Hartman R, Fleming K (2001) Brain metabolic and clinical effects of rivastigmine in Alzheimer's disease. Int J Neuropsychopharm 4:223–230.

Ridley RM, Baker HF, Leow-Dyke A, Cummings RM (2005) Further analysis of the effects of immunotoxic lesions of the basal nucleus of Meynert reveals substantial impairment on visual discrimination learning in monkeys. Brain Res Bull 65:433–442.

Rosenberg MB, Friedmann T, Robertson RC, Tuszynski M, Wolff JA, Breakefield XO, Gage FH (1988) Grafting genetically modified cells to the damaged brain: restorative effects of NGF expression. Science 242:1575–1578.

Salehi A, Delcroix JD, Swaab DF (2004) Alzheimer's disease and NGF signaling. J Neural Transm 111:323–345.

Salehi A, Verhaagen J, Dijkhuizen PA, Swaab DF (1996) Co-localization of high-affinity neurotrophin receptors in nucleus basalis of Meynert neurons and their differential reduction in Alzheimer's disease. Neuroscience 75:373–387.

Salehi A, Ocampo M, Verhaagen J, Swaab DF (2000) P75 neurotrophin receptor in the nucleus basalis of meynert in relation to age, sex, and Alzheimer's disease. Exp Neurol 161:245–258.

Savaskan E, Muller-Spahn F, Olivieri G, Bruttel S, Otten U, Rosenberg C, Hulette C, Hock C (2000) Alterations in trk A, trk B and trk C receptor immunoreactivities in parietal cortex and cerebellum in Alzheimer's disease. Eur Neurol 44:172–180.

Scott SA, Mufson EJ, Weingartner JA, Skau KA, Crutcher KA (1995) Nerve growth factor in Alzheimer's disease: Increased levels throughout the brain coupled with declines in nucleus basalis. J Neurosci 15:6213–6221.

Slevin JT, Gerhardt GA, Smith CD, Gash DM, Kryscio R, Young B (2005) Improvement of bilateral motor functions in patients with Parkinson disease through the unilateral intraputaminal infusion of glial cell line-derived neurotrophic factor. J Neurosurg 102:216–222.

Smith DE, Roberts J, Gage FH, Tuszynski MH (1999) Age-associated neuronal atrophy occurs in the primate brain and is reversible by growth factor gene therapy. Proc Nat Acad Sci 96:10893–10898.

Sofroniew MV, Howe CL, Mobley WC (2001) Nerve growth factor signaling, neuroprotection, and neural repair. Annu Rev Neurosci 24:1217–1281.

Sofroniew MV, Galletly NP, Isacson O, Svendsen CN (1990) Survival of adult basal forebrain cholinergic neurons after loss of target neurons. Science 247:338–342.

Sofroniew MV, Cooper JD, Svendsen CN, Crossman P, Ip NY, Lindsay RM, Zafra F, Lindholm D (1993) Atrophy but not death of adult septal cholinergic neurons after ablation of target capacity to produce mRNAs for NGF, BDNF, and NT3. J Neurosci 13:5263–5276.

Stokin GB, Lillo C, Falzone TL, Brusch RG, Rockenstein E, Mount SL, Raman R, Davies P, Masliah E, Williams DS, Goldstein LS (2005) Axonopathy and transport deficits early in the pathogenesis of Alzheimer's disease. Science 307:1282–1288.

Tuszynski MH (1999) Neurotrophic factors. In: CNS Regeneration: Basic Science and Clinical Advances. (Tuszynski MHaK, J.H, ed), pp 109–158. San Diego: Academic Press.

Tuszynski MH (2002) Gene therapy for neurodegenerative disorders. Lancet Neurol 2002:51–57.

Tuszynski MH, Gage FH (1995) Bridging grafts and transient NGF infusions promote long-term CNS neuronal rescue and partial functional recovery. Proc Nat Acad Sci 92:4621–4625.

Tuszynski MH, U H-S, Gage FH (1991) Recombinant human nerve growth factor infusions prevent cholinergic neuronal degeneration in the adult primate brain. Ann Neurol 30:625–636.

Tuszynski MH, U HS, Amaral DG, Gage FH (1990) Nerve growth factor infusion in primate brain reduces lesion-induced cholinergic neuronal degeneration. J Neurosci 10:3604–3614.

Tuszynski MH, Roberts J, Senut MC, U H-S, Gage FH (1996) Gene therapy in the adult primate brain: intraparenchymal grafts of cells genetically modified to produce nerve growth factor prevent cholinergic neuronal degeneration. Gene Therapy 3:305–314.

Tuszynski MH, Thal L, Pay M, Salmon DP, U HS, Bakay R, Patel P, Blesch A, Vahlsing HL, Ho G, Tong G, Potkin SG, Fallon J, Hansen L, Mufson EJ, Kordower JH, Gall C, Conner J (2005) A phase 1 clinical trial of nerve growth factor gene therapy for Alzheimer disease. Nat Med 11:551–555.

Ullrich A, Gray A, Berman C, Dull TJ (1983) Human beta-nerve growth factor gene sequence highly homologous to that of mouse. Nature 303:821–825.

Voytko ML, Olton DS, Richardson RT, Gorman LK, Tobin JR, Price DL (1994) Basal forebrain lesions in monkeys disrupt attention but not learning and memory. J Neurosci 14:167–186.

Wenk GL (1997) The nucleus basalis magnocellularis cholinergic system: one hundred years of progress. Neurobiol Learn Mem 67:85–95.

Whittemore SR, Ebendal T, Larkfors L, Olson L, Seiger A, Stromberg I, Persson H (1986) Developmental and regional expression of B nerve growth factor messenger RNA and protein in the rat central nervous system. Proc Natl Acad Sci USA 83:817–821.

Williams LR (1991) Hypophagia is induced by intracerebroventricular administration of nerve growth factor. Exp Neurol 113:31–37.

Williams LR, Varon S, Peterson GM, Wictorin K, Fisher W, Bjorklund A, Gage FH (1986) Continuous infusion of nerve growth factor prevents basal forebrain neuronal death after fimbria-fornix transection. Proc Natl Acad Sci USA 83:9231–9235.

Winkler J, Ramirez GA, Kuhn HG, Peterson DA, Day-Lollini PA, Stewart GR, Tuszynski MH, Gage FH, Thal LJ (1997) Reversible Schwann cell hyperplasia and sprouting of sensory and sympathetic neurites after intraventricular administration of nerve growth factor. Ann Neurol 41:82–93.

Index

Note: Citations derived from figures are indicated by an *f*; citations from tables are indicated by a *t*.

Printed in the United States of America.